THE AMPHIBIAN EAR

THE AMPHIBIAN
EAR

By Ernest Glen Wever

Princeton University Press

Copyright © 1985 by Princeton University Press

Published by Princeton University Press, 41 William Street,
Princeton, New Jersey 08540
In the United Kingdom: Princeton University Press, Guildford, Surrey

Library of Congress Cataloging in Publication Data will be found
on the last printed page of this book

ISBN 0-691-08365-7

This book has been composed in Linotron Times Roman

Clothbound editions of Princeton University Press books are printed on acid-free paper, and
binding materials are chosen for strength and durability.

Printed in the United States of America by Princeton University Press
Princeton, New Jersey

Frontispiece: A specimen of the grass frog, *Rana pipiens*. Drawing by Anne Cox.

CONTENTS

PREFACE

This study of the ear in the amphibians has developed out of a general consideration of the nature of the ear and hearing in vertebrates and the problem of the origin of this sense. From this consideration it has become clear that over the long period of vertebrate evolution the ear has taken two distinct courses of development out of the equilibratory organs of fishes, one leading to the reptilian type of ear, which we share with the other mammals and also the birds, and another type presented by the amphibians that differs markedly both in its structural design and in its manner of operation in response to sounds. An examination of this alternative design and its many distinctive characteristics is the central objective of this book.

Unfortunately the fossil record does not present a clear indication of the course of evolution of the ears of vertebrates, and this course must be inferred from indirect evidence, mainly the structural and functional characteristics of the ears of animals now living. The study of the ears of the various forms of amphibians presently available, as pursued in the following pages, provides our best evidence from which the course of early evolutionary developments in this group of vertebrates can be determined.

The treatment begins with a discussion of the general characteristics of the amphibia, which occur in three distinct forms, the frogs, salamanders, and caecilians, with an inquiry into the problem of their origins from primitive fishes and their variations from one another. For each of the three groups this discussion includes a consideration of the functions of hearing both as providing warnings of impending danger and as significant signals facilitating the many forms of social interactions. In this discussion the available species are examined in detail as regards the structure of their ears and the performances of these sense organs in response to sound stimuli. This treatment falls into three parts in which the frogs, urodeles, and caecilians are examined in turn.

For each of these three groups the discussion begins with a detailed presentation of the ear structures in a representative specimen, with these structures examined in a series of sections along the anteroposterior axis, and then considered further in sections cut in other planes that bring out additional details. Other species are then examined so far as available specimens will permit, with special consideration of departures from the typical form.

The performances of the ears of these various species are assessed in terms of the inner ear potentials generated by the auditory hair cells in response to sounds and recorded with an electrode inserted into the perilymph space adjacent to the endolymph cavity in which the hair cells are contained.

In this treatment, to simplify the representations of structure, all the figures are for the right ear.

References are given at the end of the text, cited by author, title, and place and date of publication.

This presentation brings out the manner of performance of a type of ear whose design and action in response to sounds differs in many respects from the ear possessed by ourselves and the other higher vertebrates, and thus presents an alternative solution to the problem of hearing in contrast to the one with which we are most familiar.

The experimental studies on which this treatment is based have extended over many years, with participation by numerous friends and colleagues, whose contributions are gratefully acknowledged. Of special significance is the work of Jerry Palin and Joseph M. Pylka, who designed, constructed, and maintained the electronic equipment employed in the measurements of inner ear potentials. I am especially indebted to Wilmer C. Ames for many forms of assistance, including equipment assembly and the care of experimental animals, and to Anne Cox for the highly competent preparation of specimens for histological study.

Participating in a number of the observations were Carl Gans, Yehudah L. Werner, Jack A. Vernon, and William F. Strother, who contributed both to the planning and execution of the experiments and to the interpretation of the results.

The experimental animals were mainly obtained from commercial sources, but with the addition of a number of species to extend the coverage of the study. In this relation I am much indebted to Joseph M. Pylka, who supplied a number of specimens from the Atlantic seaboard area and made a special collecting trip to the Panama Canal Zone where he was ably assisted by M. H. Clark attached to a branch of the Smithsonian Institution in this zone. A number of laboratory-raised specimens came from the Amphibian Facility of the University of Michigan, and others were supplied by Martha Paton.

I am much indebted to James G. McCormick and Gloria Gamble for extensive bibliographic assistance. In the preparation of the manuscript I have had the benefit of the highly effective secretarial pool of the Department of Psychology of Princeton University under Arlene Kronewitter, made available through the kindness of the Chairman of the Department, John M. Darley.

Finally, I express my special gratitude to Carl Gans and Yehudah L. Werner, who have provided continued advice and counsel, and have critically examined many parts of the manuscript. I am also greatly indebted to Shyam M. Khanna for much encouragement and advice, with particular regard to problems of the physical action of sounds on the ear.

This research has been made possible by grants, extending over many years, from the National Institutes of Health (NINCDS), Public Health Service.

PART I. INTRODUCTION

1. NATURE AND ORIGIN

OF THE AMPHIBIA

The name Amphibia literally signifies "both lives" and refers to the pro-
clivity of many of these animals for alternating between aquatic and terrestrial habitats. This name was first used broadly by Linnaeus in reference to
a wide variety of animals, including such species as seals, turtles, and croc-
odiles as well as frogs and salamanders, that live sometimes in the water and
sometimes on land. Soon thereafter the term was applied more selectively to
a particular class of vertebrates, a class including the frogs, salamanders,
and caecilians among modern types, together with a number of ancient animals known only as fossils that are commonly considered to be distant rel-
atives of these.

The living amphibians—the anurans, urodeles, and caecilians—form three
somewhat disparate orders that constitute a relatively small group among the
vertebrates, a group dwarfed in numbers by the varied forms of fishes and
exceeded also by the three classes of higher vertebrates, the reptiles, birds,
and mammals. The amphibians are located after the fishes and before the
other classes in the suborder Vertebrata as an indication of their phylogenetic rank: these animals are considered to have evolved out of certain of the
fishes, and some of their earlier forms are regarded as having given rise,
through the long and complex process of evolution, to all the higher animals. It is estimated that there are about 2100 species of amphibians now
living, out of a total of something like 40,000 vertebrates of all kinds.

AMPHIBIAN CHARACTERISTICS

On the basis of general body form and habits the three groups of modern
amphibians would not seem to belong together in one class, but closer study
reveals many characters in common. Parsons and Williams (1963) in a study
primarily of tooth structure and then of numerous other features concluded
that these three groups form a natural monophyletic assembly, to which they
applied the collective name of Lissamphibia, a name first used for this group
by Gadow in 1901.

A pedicellate tooth structure is the most distinctive characteristic of this
assembly. This tooth, as shown in Fig. 3-9 below, consists of two portions,

a basal pedicel and a peripheral crown connected to the base part by a hinge. These two elements consist of dentine covered by a dense enamel layer, and the hinge consists of uncalcified dentine or fibrous tissue to provide flexibility. It is supposed that such a tooth structure may give an increased facility in seizing prey; the prey object readily enters the mouth but tends to catch if pulled away. Somewhat similar teeth occur in teleost fishes, but in these the hinge is between the tooth and the jaw rather than within the tooth itself. The true pedicellate tooth is not found in any other living vertebrates, but has recently been discovered in certain fossil forms belonging to the labyrinthodonts, in two closely related species from the Lower Permian. These are *Doleserpeton annectens* and *Tersomius texensis*, and their discovery has led to new speculations about the origin of the Lissamphibia (Bolt, 1969, 1977).

Other features now to be described are not exclusive to the amphibians, but taken as a whole serve to characterize this group.

1. Ectothermy. The amphibians are "cold-blooded"; their body temperature is close to that of their immediate environment. However, this is true of the other lower vertebrates, the fishes and reptiles, as well.

2. Skin Character. The skin is soft; it is smooth in some species and rough in others; its surface lacks the layer of keratin and lipids found in reptiles, and also lacks special coverings such as hair and scales. In the caecilians, however, minute scales may be found deep in the grooves between the body segments, though a dissection is necessary to reveal them. These scale remnants are indicative of a descent from ancient fishes.

3. Skin Glands. The skin is kept moist by numerous mucous glands distributed over its surface. The secretion produced by the glands of many species is at least mildly toxic, and seems to serve in some degree to repel predators. In many species there are also more specialized skin glands, such as the paratoid glands in the shoulder areas of frogs, that produce highly toxic secretions.

4. Body Form. There are marked variations in body form in the three groups. This form in salamanders is elongate, well streamlined, with legs of moderate size and a well-developed tail; Romer (1966) regarded this form as the original, basic one for the class, and considered the types in frogs and caecilians as highly specialized. There are two pairs of legs in all salamanders except the sirens, which have only the anterior pair. The frogs are specialized for leaping or hopping; they have compact bodies, short forelegs, long, well-developed hindlegs, and they have no tails. The caecilians are elongated, wormlike in appearance, without legs, and have a compact skull that in most species is utilized for burrowing.

5. The Egg. The amphibian egg lacks a shell, but is covered by several layers of gelatinous membrane. For the most part the eggs are laid in water or in moist places, or a variety of means are employed to prevent them from drying, and often also to protect them from predators. In some species the

female curls around the cluster of eggs and both guards them and keeps them moist; in other species the eggs are carried in vesicles in the skin of the back; in one genus the eggs are retained in the male's vocal pouch, and in another genus in his stomach. In a few species the eggs remain in the oviducts of the female, are fertilized internally, and the young hatch out as small individuals closely approaching the adult form.

6. Metamorphosis. Typically the larval and adult forms are greatly dissimilar; and at the end of a larval stage there is a rapid metamorphosis in which the new adult form emerges, with profound changes of both structure and function. This metamorphosis is most complete in frogs, in which legs appear and the tail dwindles away, the gills are lost and usually replaced by lungs, and the notochord is largely replaced by a jointed vertebral column, or else is incorporated into such a column. There is developed a three-chambered heart, with pulmonary and systemic divisions of the circulatory system in an arrangement that permits only a slight mixing of spent and oxygenated bloodstreams.

7. Respiratory Patterns. Respiration in amphibians takes a variety of forms, and usually varies greatly between larval and adult stages. Gaseous interchange through the skin is a general feature; the skin is moist and well supplied with capillaries, and cutaneous respiration is present in some degree in all species. This form of respiration alone can sustain life under certain conditions, as during hibernation when the body temperature is low and activity is minimal.

A more efficient form of respiration is buccopharyngeal, in which throat movements facilitate an interchange of gases between dense capillary beds in the lining of the mouth and pharynx and either the air or the water in contact with these surfaces. Aquatic larvae when in need of added oxygen will speed up their throat movements, or sometimes will rise to the surface and take in a bubble of air that is then held near the buccopharyngeal surface. Terrestrial animals such as frogs use integumentary respiration much of the time but supplement it by involvement of the lungs as described below.

Gills are present in all larval amphibians and are retained in those salamanders that fail to metamorphose, such as *Necturus maculosus*. These gills are often obvious as branched plumes arising from the branchial arches.

At the terrestrial stage most of the amphibians are equipped with lungs, which arise from branchial pouches. In some species, however, the lungs are reduced or even absent: in caecilians the left lung is usually rudimentary, and in many salamanders the lungs are small, or as in all species of the plethodontid family are entirely lacking. In both urodeles and frogs buccopharyngeal respiration is regularly combined with the pulmonary type.

Respiratory Patterns in the Frog. — The respiratory process has been extensively studied in frogs, where it is surprisingly complex (De Jongh and

Gans, 1969; Gans, 1974). Rhythmic throat pulsations constitute a sustaining phase, in which the mouth is closed and the nostrils are held open, the glottis is closed also, and moderate cyclic variations of pressure are produced within the buccal cavity. Thus an airflow in and out of the nostrils aids olfaction and refreshes the oxygen contents of the buccopharyngeal region. The lungs are not involved in this type of air flow.

The Pulmonary Cycle. — After the sustaining phase, at rather irregular intervals (after something like 50 throat pulsations in one frog species studied), a new type of activity intervenes, known as the pulmonary or ventilatory cycle. In this pattern there are five successive steps: (1) with the nostrils remaining open, the floor of the mouth is depressed more than usual and the posterior part of the buccal cavity is widened, causing an increased inflow of outside air. (2) Then the glottis opens, permitting a rapid escape of the spent air from the lungs and out the nostrils. (3) Next the nostrils are closed, and the floor of the mouth is vigorously elevated, forcing the air out of the buccal cavity into the lungs. (4) Near the peak pressure the glottis is closed and the nostrils are opened, so that the air in the lungs is maintained at a high pressure. (5) Finally, further rhythmic pulsations of the floor of the mouth are resumed.

The Inflation Cycle. — The pulmonary cycles are arranged in inflation cycles. In these each pulmonary cycle produces a stepwise increase in the volume of air contained in the lungs, mainly by limiting step 2 (as described above) and emphasizing step 3. After a peak value of lung pressure is attained, the glottis and nostrils are opened, and the air flows out of both lungs and buccal cavity.

The respiratory pattern becomes still more complex in amphibians during vocalization.

THE ORIGIN OF THE AMPHIBIA

It is generally agreed that sometime in the Devonian period, about 350 million years ago, the Amphibia originated from the lobe-finned fishes, the crossopterygians, which inhabited freshwater streams and pools and had developed sturdy fins that could be used like legs, increasing their agility as bottom feeders and also enabling them to survive when their pools dried up. These appendages eventually were converted into true legs, and so the first land vertebrates were produced.

The fossil record presents two major groups of these early amphibians, the labyrinthodonts and the lepospondyls, and these gave rise to several other distinct lines from some of which existing groups are generally considered to have had their origins. The labyrinthodonts produced two general lines, the temnospondyls and the anthracosaurs, distinguished by the structure of the vertebrae; in the temnospondyls the vertebral centrum was derived mainly as an enlargement of the intercentrum, whereas in the anthracosaurs the centrum developed from another element, the pleurocentrum.

The temnospondyls divided into two groups, the stereospondyls along one line and the rhachitomes and neorhachitomes along another; but all these appear finally to have become extinct without producing any modern representatives.

Among the anthracosaurs there are two groups, the embolomeres, which appeared toward the end of the Mississippian period (about 310 million years ago) and continued well into the Permian (about 200 million years ago), and the seymouriamorphs, which appeared somewhat later toward the middle of the Pennsylvanian and continued into the beginning of the Triassic (from about 280 to about 180 million years ago).

The embolomeres were aquatic fish predators throughout the Pennsylvanian but disappeared in the early Permian. The seymouriamorphs display a mixture of amphibian and reptilian characteristics, with vertebrae like those of primitive reptiles; Romer (1966) pointed to this group as demonstrating that there is no clear-cut distinction between amphibians and reptiles in their skeletal characteristics.

The lepospondyls are distinguished by a vertebral centrum that forms a single spool-like cylinder around the notochord much like the structure in the modern amphibians. There are three divisions of these: the aistopods, microsaurs, and nectrideans. In the aistopods the legs are lost and the body is long and snakelike, with as many as two hundred vertebrae. These forms reached a peak in the Pennsylvanian, and rapidly disappeared thereafter.

The microsaurs attained their highest level of development in the Pennsylvanian, and nearly disappeared in the Permian period about 200 million years ago. They possessed a number of specialized characters, including a single large bone in the temporal region of the skull, only three digits in the hand, an incompletely roofed skull, and a single occipital condyle.

These fossil forms provide a background for consideration of the specific ancestry of existing amphibians.

THEORIES OF AMPHIBIAN ANCESTRY

With three existing orders of amphibians to be accounted for, a theory of ancestry can take five different forms: (1) these three orders can be considered as having completely distinct origins—a polyphyletic theory; (2) all can be regarded as having a common origin—a monophyletic theory; or (3) there can be three types of diphyletic theory according to what two orders are linked and what one is taken as separate: (a) anurans and caudates paired and apodans considered as separate, (b) caudates and apodans paired and anurans left separate, or (c) anurans and apodans paired and caudates left separate. All these five possibilities have had proponents except the last.

The Polyphyletic Theory. — Only Herre (1935) has favored the theory that the three orders have had distinct origins, and he presented no specific evidence to support this position.

Diphyletic Theories. — One form of diphyletic theory was vigorously defended by Jarvik (1942, 1955), in which he proposed that the anurans, along with the labyrinthodonts and the amniotes in general, were derived from one group of crossopterygians, the Osteolepiformes, whereas the urodeles and apodans came from another crossopterygian group, the Porolepiformes. This view was accepted by Romer (1945) and also by Holmgren (1949) and has been generally popular.

Another form of diphyletic theory, in which urodeles and anurans are closely related while the apodans are considered as of separate ancestry, was favored by Eaton (1959). He indicated numerous points of similarity between anurans and urodeles that seemed to favor a common derivation, including the broad form of the skull with its palatal openings, the presence of an operculum and the opercular muscle, and similarities of the vertebrae in which these forms resemble the temnospondyls.

The Monophyletic Theory. — The evidence linking all three of the living orders of amphibians in a single ancestry is treated in detail by Parsons and Williams (1963), with a discussion of 18 features in which resemblances are seen in addition to the presence of pedicellate teeth. These features include several peculiarities of skull and palatal structure: the skull is broad, with large orbital openings; the palate is of the advanced type, with wide separation of the pterygoids; an operculum is present in the oval window; the occipital condyles are paired; and the mentomeckelian bones are present in the mandible.

In the anurans an otic notch is present and closely resembles the one in temnospondyls. In the apodans there is a notch that is at least vaguely similar to that primitive type, but such a notch is absent in urodeles.

To these skeletal features may be added some further ones: the presence of the amphibian papilla in all three orders of modern amphibia, the presence of a peculiar element in the retina known as the green rod (not known elsewhere), fat bodies in association with the gonads, special structure of the skin glands, and the extensive utilization of cutaneous respiration.

ANCESTRY OF THE LISSAMPHIBIA

Parsons and Williams further considered the ancestry of the Lissamphibia in relation to four groups of fossil forms: the temnospondyls, anthracosaurs, nectrideans, and microsaurs. From the evidence they found themselves unable to propose even a tentative hypothesis concerning amphibian ancestry and concluded that a solution of this problem will have to await the accumulation of further evidence.

Others have been somewhat more venturesome concerning the problem of amphibian ancestry. As just mentioned, Eaton (1959) expressed the opinion that the anurans probably were derived from the temnospondyls, and he considered that the urodeles show sufficient points of resemblance to these to have had a similar origin. The apodans, however, he believed to have developed separately, being derived from the lepospondyls. Estes (1965) was generally favorable to an acceptance of the temnospondyls as the most probable ancestor of all the modern amphibians.

Schmalhausen in his monograph on the origin of terrestrial vertebrates (1959, 1968) took a much more positive position on this problem. He traced the development of early vertebrates from the crossopterygian fishes through the ichthyostegids, which lived in the Carboniferous period (about 300 million years ago), to certain of the labyrinthodonts, which represent the earliest terrestrial forms, and from these by somewhat different routes of development to the three existing orders of amphibians.

The rhachitomes were a group of labyrinthodonts that are pictured as living on the banks of streams and lakes, feeding largely on worms and arthropods, and taking to the water on occasion to escape from reptilian predators. These rhachitomes, it is supposed, developed keen sensory capacities and superior leaping ability and became the anurans. *Protobatrachus,* a fossil form discovered in deposits from the Lower Triassic (laid down about 200 million years ago) is clearly frog-like, and other fossils of similar structure are known from the Permian period.

The urodeles, in Schmalhausen's view, were derived from the microsaurs, a group of lepospondyls occurring in the Pennsylvanian period, and these developed largely through a process of neoteny: through the persistence in adult life of numerous larval characteristics. This process involved a retention of gills along with cutaneous respiration and produced a form that was well adapted to an existence in cold mountainous regions, where a great many of these amphibians still live.

The apodans, according to Schmalhausen, were derived from those terrestrial microsaurs that had adopted burrowing habits, which led eventually during the Lower Permian to an elongation of the body, the loss of limbs and limb girdles, and the reduction of the tail to a mere remnant. The one group of apodans *(Typhlonectes)* that is now aquatic is considered to have become so secondarily.

These suggestions of Eaton, Estes, and Schmalhausen are among the more positive considerations of the problem of amphibian ancestry in which a continuity of descent is predicated from the ancient fossil amphibians to the present living forms. It must be emphasized, however, that the evidence connecting these ancient forms with the ones now living is exceedingly tenuous and uncertain; the fossil record presents an enormous gap of the order of 250 million years between the presumed ancestors—the crossopterygian fishes of the Devonian period—and the more recent forms that are undoubted amphibians. This long period needs to be filled in before continuity can be accepted with any degree of confidence. Many fossils have been collected that belong to this critical period, but few have been thoroughly studied, and still others need to be sought in the strata representing this period. Many persons now share the opinion expressed by Parsons and Williams that the evidence is still insufficient for a decision on the question of amphibian ancestry.

FUNCTION OF HEARING IN AMPHIBIA

There are two basic functions served by the sense of hearing in the lower animals: the reception of signals that give warning of the presence of predators and other dangers, and the use of sounds in social relations, which at the primitive level consists mainly of the use of mating calls for the assembly and interaction of the two sexes.

Our knowledge about the protective functions of hearing in amphibia is severely limited. It is frequently observed that frogs on the bank of a pool or stream when approached will either become immobile, by which they often escape notice, or they leap into the water, thus reaching a region of greater safety. It is not usually clear, however, whether these animals are responding to the visual or the auditory stimuli produced by the intruder.

Hearing ability as it relates to both the escape from danger and mating activities is of greater biological significance, and no doubt these relationships contributed greatly to the early development of hearing in the evolution of this group. The earliest indication of this development comes from fossil forms that possess an otic notch, an opening on each side of the skull in the posterior region between squamosal and tabular bones that is supposed to have been the site of attachment of a tympanic membrane. Such a notch is found in the labyrinthodonts dating from the Carboniferous period 350 million years ago. It is further supposed that these early amphibians are the predecessors of the later ones, but this supposition is open to question.

A discussion of the biological uses of hearing in amphibians is practically limited to the anurans, for only in these is vocal activity obvious in the breeding season. The caecilians have not been studied in this respect, as their usual concealment in subterranian haunts and their apparent inability to produce sounds has not encouraged this sort of exploration. The salamanders have not been examined to any great extent either, though some species are known to produce sounds. It is entirely possible that vocal activity plays a role in their breeding assemblies much as it does in anurans, but this matter still awaits investigation.

Many have described the vocal activity of male frogs in the mating season, and it has been established for some species at least that females respond to these sounds by close approach, and males often respond by croaking, but these responses occur only if two conditions are satisfied: the sounds must have a character that ordinarily is specific to the particular species, and the listener must be in a receptive physiological state. Many earlier experiments produced results that seemed confusing or contradictory because these two conditions had not been met.

The nature of this pattern of response has been worked out mainly by the use of recorded and synthetic sounds. The experiments of Capranica (1965, 1966) on the bullfrog *Rana catesbeiana* used synthetic stimuli, and the vocal responses of males were observed. Males are most suitable for these observations because they respond with a croak of their own, a behavior that is readily observed and quantifiable, whereas the female's response is more subtly behavioral, involving approach in some instances but in others consisting only of a degree of submissiveness that is not easily observed.

In Capranica's experiments, after a study of the character of a number of natural mating calls a large series of synthetic calls was prepared, varying in such features as the number of croaks in a call, the duration of each croak, the intervals between, frequency composition, and the temporal relations of these components. These synthetic stimuli were then presented to a group of 18 males maintained in a terrarium (and accompanied by a small number of females) to ascertain the relative effectiveness of each synthetic call in evoking a response by at least one male in the group.

From these observations it was established that an effective call for this species must contain acoustic energy in two distinct frequency regions, one around 200 Hz and the other around 1400 Hz, with a minimum of energy in the intervening region around 500–600 Hz. A further characteristic, which was observed in the natural calls and maintained in the synthetic ones, is a periodicity of the spectral envelope of 100 per second, produced in the artificial stimuli by the use of a pulse train of 100 per second applied to the resonant circuits used in generating the sound.

Further experiments carried out by a number of investigators on several frog species have clarified these relations further and have disclosed the roles of the two auditory papillae in the mating process (see Capranica, 1976). In the bullfrog, *Rana catesbeiana*, three types of auditory sensorineural elements have been identified and studied by recording the nerve action in response to sounds. One type of neural element is most sensitive to the low frequencies—those below 500 Hz—and is inhibited by other tones of higher frequencies and intensities; a second type is responsive to tones between 500 and 900 Hz and is not inhibitable by other tones; and a third type is most sensitive to tones around 1000–1700 Hz and also is not inhibited by other tones. The first and second types of element serve to innervate the amphibian papilla, and the third type supplies the basilar papilla.

It is remarkable that through these sensorineural actions it is made possible for a female to identify the mating call of a male of her own species with a high degree of accuracy, and with a minimum demand on higher neural processes.

CLASSIFICATION OF THE AMPHIBIA

In vertebrate classification in general an extensive use has been made of the structure of the vertebrae, because this variable feature of the bony skeleton is well preserved in fossils, and it has been possible in terms of this feature to work out a series of stages leading from the crossopterygian fishes (from which the amphibians are believed to have been derived) through the extinct forms to the reptiles and higher vertebrates. The ancient amphibians fall into place in this series, but unfortunately the existing ones fail to do so. In the amphibians now living the vertebral segments in the early embryonic stage develop in a simple manner from mesenchyme without exhibiting the successive transformations of cartilaginous elements that can be inferred from the examination of fossils (Romer, 1962, pp. 161–168). Consequently it is not possible to rely on vertebral characteristics to differentiate these modern groups, and it becomes necessary to look for other diagnostic attributes.

In early attempts to develop classification systems for the different amphibian groups a small number of features were used, often only one or two, such as the structure of the pectoral girdle, form of the pupil, or the presence or absence of the tongue, but the later and more satisfactory arrangements have employed a multiplicity of characters. A chosen system of classification will be presented for each of the three existing amphibian orders.

CLASSIFICATION OF THE ANURANS

Of the many proposals for the systematic arrangement of the frogs, none has been as influential, or has endured for so long a time, as that of Noble, presented in final form in his textbook, *The Biology of the Amphibia*, in 1931. This scheme was widely accepted and dominated its field for nearly three decades. Only recently has there been a renewed interest in this problem, stimulated by the accumulation of further evidence on structural relationships, new approaches in the study of larval types, and the investigation of biochemical and chromosome characteristics.

Dowling and Duellman's classification system, to be adopted here, represents an attempt to assemble the several anuran families in an arrangement that reflects their sharing of as many as 24 characters.

The order Anura is given two suborders, the Archaeobatrachia and the Neobatrachia, which are Reig's names for the ancient (and more primitive) and the recent (and more advanced) species. On the borderline between these two groups is the superfamily Pelobatoidea, the members of which have qualities of an intermediate nature. The arrangement, taken from Dowling and Duellman's report but modified to make the intermediate position of the Pelobatoidea more obvious, is presented in Table 1-I.

Table 1-I

Classification of the Anurans (after Dowling and Duellman, 1978)

	Class Amphibia (Linnaeus, 1758)
	Subclass Lissamphibia (Haeckel, 1866)
	Superorder Salientia (Laurenti, 1768)
	Order Anura (Giebel, 1847)
The Archaic Frogs	Suborder Archaeobatrachia (Reig, 1958)
	Superfamily Discoglossoidea (Günther, 1858)
	Family Leiopelmatidae (Mivart, 1869)
	Family Discoglossidae (Günther, 1858)
	Superfamily Pipoidea (Bonapart, 1831)
	Family Paleobatrachidae (Cope, 1865) + *
	Family Pipidae (Bonapart, 1831)
	Family Rhinophrynidae (Günther, 1858)
The Intermediate Frogs	Superfamily Pelobatoidea (Stannius, 1856)
	Family Pelobatidae (Stannius, 1856)
	Family Pelodytidae (Cope, 1866)*
The Advanced Frogs	Suborder Neobatrachia (Reig, 1958)
	Superfamily Bufonoidea (Gmelin, 1815)
	Family Myobatrachidae (Schlegel, 1850)*
	Family Leptodactylidae (Berg, 1838)
	Family Bufonidae (Gmelin, 1815)
	Family Brachycephalidae (Günther, 1858)
	Family Rhinodermatidae (Günther, 1858)
	Family Dendrobatidae (Cope, 1865)
	Family Pseudidae (Fitzinger, 1843)*
	Family Hylidae (Hallowell, 1857)
	Family Centrolenidae (Taylor, 1951)
	Superfamily Microhyloidea (Parker, 1934)
	Family Microhylidae (Parker, 1934)
	Superfamily Ranoidea (Linnaeus, 1758)
	Family Sooglossidae (Griffiths, 1963)*
	Family Ranidae (Linnaeus, 1758)
	Family Hyperoliidae (Laurenti, 1951)
	Family Rhacophoridae (Parker, 1934)

* Indicates those families for which no specimens were available for the present study.
+ Known only as fossils.

CLASSIFICATION OF THE CAUDATES

Like the classification for the anurans, Noble's classification system for the salamanders, presented in 1931, was widely accepted and strongly influenced further systematic developments. There has been general agreement in recognizing the eight families in this order, but some differences over their groupings. Noble's system placed these families in five suborders, whereas Dowling and Duellman in a later scheme, which will be presented here, reduced the suborders to four by including the Proteida in the suborder Salamandroidea and at the same time shifted the Plethodontidae out of this suborder and bracketed them with the Ambystomidae (along with three other families now extinct). Under consideration here is whether the plethodontids and salamandrids are sufficiently similar to be located together in the same group as Noble had them, or whether the plethodontids are more suitably linked with the ambystomatids as the arrangement of Dowling and Duellman provides. The location of the Proteidae is also debatable, and an isolated position as favored by Noble raises fewer questions than the one in which they are bracketed with the Salamandridae and Amphiumidae. In general, the small number of caudate families keeps the problem of systematic arrangement a relatively simple one, and either of these two systems is workable. The Dowling and Duellman (1978) system is shown in Table 1-II.

Table 1-II

Classification of the Caudates (after Dowling and Duellman, 1978)

Class Amphibia (Linnaeus, 1758)
 Subclass Lissamphibia (Haeckel, 1866)
  Order Caudata (Oppel, 1811)

 Suborder Cryptobranchoidea (Dunn, 1922)
  Family Hynobiidae (Cope, 1859)
  Family Cryptobranchidae (Cope, 1889)
 Suborder Sirenoidea = Meantes (Linnaeus, 1776)
  Family Sirenidae (Gray, 1825)
 Suborder Salamandroidea (Sarasin, 1890)
  Family Salamandridae (Gray, 1825)
  Family Proteidae (Tschudi, 1839)
  Family Amphiumidae (Cope, 1866)
 Suborder Ambystomatoidea (Noble, 1931)
  Family Ambystomatidae (Hallowell, 1858)
  Family Plethodontidae (Gray, 1850)
  (3 extinct families omitted)

Classification of the Gymnophionans

The Gymnophiona or caecilians are limbless, elongated animals generally of burrowing habit, though some are aquatic, found in tropical and subtropical areas in Asia, Africa, and South and Central America. They are distinguished by the presence of a pair of tentacles, the absence of legs, a short tail or almost none at all, and a division of the body longitudinally into a series of segments indicated by deep skin folds. Scales are usually present, but are concealed deep in the segmental folds. Eyes are usually present also, but often are reduced or concealed; their degree of functioning is uncertain. The tentacles are located between eye and nostril, and evidently are sense organs serving a form of tactile function. It has been suggested that this organ may be useful for such purposes as the finding and recognition of mates, or perhaps of food substances, or even the sensing of temperature or moisture; its true function is thus unknown.

Early the caecilians were considered a single family. The name has varied; Daudin (1803) referred to them as "Les caeciles"; Gray (1825) used the name Caeciliadae; Boulenger (1882) called them Coeciliidae and recognized 11 genera and 32 species; and Cope (1882) first employed the present name of Caeciliidae.

Taylor in 1968 split away two groups from the general category of Caeciliidae to obtain three families: the Ichthyophiidae with 4 genera in Asia and South America, the Typhlonectidae with 4 genera, all of aquatic habit and confined to South America, and the remaining forms, still called Caeciliidae, including about 20 genera widely distributed in Asia, Africa, and

Table 1-III

Classification of the Gymnophionans (after Taylor, 1968, 1969, and Nussbaum, 1977, 1979)

Order Gymnophiona (Caecilians)
  Family Ichthyophiidae (Taylor, 1968)
    Subfamily Ichthyophinae (Taylor, 1968)
    Subfamily Uraeotyphlinae (Nussbaum, 1979)*
  Family Typhlonectidae (Taylor, 1968)
  Family Caeciliidae (Cope, 1889)
    Subfamily Caeciliinae (Taylor, 1969)
    Subfamily Dermophinae (Taylor, 1969)
  Family Scolecomorphidae (Taylor, 1969)*
  Family Rhinatrematidae (Nussbaum, 1977)*

---

*Indicates families or subfamilies for which no specimens were available for the present study.

South America. Then Taylor (1969) separated a further group from the Caeciliidae to form a fourth family, the Scolecomorphidae, consisting of 6 African species of the genus *Scolecomorphus*.

Later, in 1977, Nussbaum removed the genera *Epicrionops* with 8 species and *Rhinatrema* with a single species from the Ichthyophiidae and constructed for these a new family of Rhinatrematidae. *Epicrionops* occurs in regions of South America, in Ecuador, Peru, Colombia, and Venezuela; and *Rhinatrema* is found in Cayenne and Surinam. In 1979 Nussbaum split off the subfamily Uraeotyphlinae to contain the species *Uraeotyphlus*.

The caecilians as a group have not been extensively studied, and no doubt there will be further classificatory revisions. The present arrangement is shown in Table 1-III.

2. EXPERIMENTAL METHODS

Three types of procedures are used in the study of the ear and its performance in response to sounds; these are anatomical methods for the examination of the receptive structures, behavioral observations of the animal's acoustic discriminations and reactions, and electrophysiological procedures in the analysis of the ear's internal processes. These procedures have been developed over many years in the course of studies on the ears of higher vertebrates, and with minor modifications are found applicable to the amphibians. An earlier study of the ear of reptiles (Wever, 1978b) dealt with these procedures at considerable length and should be consulted for many of the finer details.

ANATOMICAL PROCEDURES

The anatomical study of the ear is carried out in two stages; the first deals with the gross morphology, in which the general conformation of the structures is observed, and the second deals with the fine histology, in which additional details are worked out. Consideration of these anatomical features then leads to an identification of the mechanisms by which sound vibrations are received and a delineation of the pathways through which the vibrations are conducted inward to the final receptive organs of the auditory system.

Gross Anatomy

The first step toward an understandning of the ear structures in a given amphibian species is best taken by a careful dissection under an operating microscope provided with multiple objectives giving magnifications in three or four steps from about 6 to 30 times, first using preserved specimens in which the tissues, being somewhat hardened by the preservative, will withstand a degree of rough handling, and then continuing the study in specimens freshly killed, preferably by an overdose of anesthetic, so that the blood vessels remain filled and are easily recognized. Thereafter the finer details of structure may be filled in by the use of serial sections passing through the ear region, usually cut in conventional planes, perhaps beginning with the transverse plane going from anterior to posterior and then using the frontal plane going from dorsal to ventral.

Sometimes it is of value to use the sagittal plane in a third manner of sectioning, going from the left side to the midline and then continuing to the right side. Initially this plane is not as useful as the others because most of the ear structures are cut obliquely and are difficult to follow, but in the later stage of the study of a given species these sections often bring out features that are not easily perceived in the other views.

When ample material was available, at least one specimen of a given species was sectioned in each of these three planes, but when there were only two specimens the sectioning in the sagittal plane was omitted. However, when but a single specimen could be obtained, as was the case for a few rare species, a strategem was used in which one half of the head was sectioned sagittally and the other half frontally. This was done by sectioning from one side (say the left) to the midline, removing the remaining piece of tissue from the mounting block, giving it a fresh coating of celloidin to cover the cut surface, and then orienting it on another block for sectioning in the frontal plane. By this procedure the structures were presented in two planes, and most of the relations could be worked out.

In the study of certain structures like the muscles that have oblique courses through the head a special plane of sectioning was often useful; in the present study an oblique plane was employed in the investigation of the middle ear muscles.

FINE HISTOLOGY

The more delicate details of ear structure are such as to tax to the limit our technical resources of tissue preparation. The first and most critical requirement is the process of fixation, in which the cellular elements are killed and their contained enzymes are made inactive so that these are prevented from autolysing and thus altering the tissues. These destructive processes begin as soon as the tissues are deprived of oxygen and proceed with great rapidity. To prevent them it is necessary to introduce a fixative through the blood stream so as to reach the head tissues promptly, and indeed that the fixative be itself the killing agent so as to leave no interval between killing and fixation processes.

The ear is a highly complex structure, combining several tissues of different characters, including bone, cartilage, muscle, and tendon and other connective tissues, as well as assemblies of cellular elements of several types. No one histological treatment is altogether satisfactory for all of these, and one must be chosen that is a good compromise, preserving all the tissues at least reasonably well, with the most favorable rendering of the auditory hair cells in their relations to the tectorial bodies. Shrinkage and swelling of the tissues must be kept to a minimum to avoid a serious disruption of their complex relationships. Therefore all changes of solutions must be made in small steps, to keep the internal forces small. The processing thus cannot be hurried; the whole procedure requires several months.

The Perfusion Procedure. — The tissues of the head are fixed by perfusion through the circulatory system by use of a cannula introduced into the truncus arteriosus of the heart, which involves opening the chest and the heart's exposure on its ventral side. The cannula is inserted into the ventricle through a fine slit in its ventral wall and then pushed anteriorly into the truncus, where it is secured by a ligature passed around this structure as shown in Fig. 2-1. Then at once the sinus venosus is exposed to view by carefully lifting the posterior end of the ventricle, and an opening made in its wall to permit the escape of blood and perfusion fluids as the injection proceeds. Pulsating pressure is used as this form of pressure helps to prevent the formation of clumps and blockages in the finer vessels and capillaries.

First amphibian saline (Ringer's) solution is used to clear the heart and blood vessels, and this is immediately followed by the fixative solution. To the Ringer's solution is added a small quantity (0.3 ml per liter) of amyl nitrite as a vasodilator, to aid in keeping the vessels and capillaries open.

The clearing of the heart and vessels with the Ringer's solution is brief,

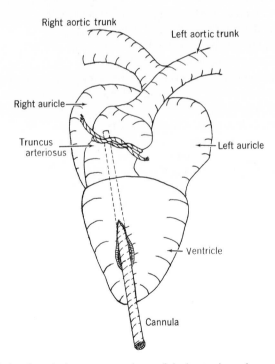

Right aortic trunk

Left aortic trunk

Right auricle

Truncus arteriosus

Left auricle

Ventricle

Cannula

Fig. 2-1. Perfusion through the truncus arteriosus of the heart, shown in a specimen of *Rana catesbeiana*. Scale 3.5X.

taking about half a minute after the flow is well started and continuing only until the escaping fluid begins to lose its color. The intention is to remove a sufficient amount of blood to prevent coagulation in the vessels, but not to take so long that anoxia in the tissues will reach an appreciable level. Then by use of a three-way stopcock a transfer is made from saline to fixing fluid, and the perfusion is continued.

This perfusion with the fixative must be thorough; it requires about half an hour in the larger animals in which the flow is relatively rapid, up to an hour in the smaller ones in which the vessels are narrow and a finer cannula must be used. This long period of perfusion is essential because the solution is only brought in the near vicinity of the innermost tissues of the ear; to reach the hair cells themselves this fluid must spread farther by diffusion, for in all ears the final receptive structures are somewhat isolated from the circulatory system with its noisy pulsations. An extended perfusion period serves to build up a high concentration of fixative in the tissues surrounding the final receptive structures so that an adequate diffusion into these structures can occur.

The Choice of Fixative. — Hundreds of fixative solutions have been proposed. Most of these are mixtures of toxic fluids like formaldehyde solutions along with compounds of heavy metals like mercuric bichloride that kill the tissues and render their contained enzymes inactive.

A proper fixative must have certain physical characteristics of diffusibility as well as the capability of preserving the tissues: it must penetrate the mass of tissue so that all parts undergo much the same physical and organic alterations in response to the reagents and there are no appreciable dislocations of form, and yet the penetration must be rapid enough to denature the enzymes present before they begin to modify the tissues. Many reagents and combinations were tried over a number of years in the study of mammalian and reptilian ears, and several were found satisfactory, with the one known as Maximow's solution the final choice. Extensive further tests were made on amphibian specimens over a period of three years, using this fixative and several others, and again Maximow's solution was found to be the most suitable.

It is worth mentioning that of all the fixative solutions investigated the worst was osmic tetroxide, one that was used extensively by the early anatomists, notably by Retzius, and has had a resurgence of application in recent years, especially in ultramicroscopy. The reason for the poor performance of this fixative became readily apparent: it acts so rapidly that the surface layer of tissue, or those regions surrounding the capillaries, are made impervious so that deep penetration is prevented, and the bulk of the specimen remains essentially unfixed. Moreover, this surface action has a disruptive effect so that the physical relations of superficial and deep structures are altered. Such a reagent may have limited use as a purely surface fixative, when only the outer layer of the material is of concern, but is altogether unsuitable for the complex tissues constituting the ear.

At the end of the perfusion procedure a block of tissue containing the ear was dissected out, suitably trimmed, and placed in a jar containing about 50 times its volume of fixative solution for a period of 7–10 days, and then kept for a similar period in 10% formol solution, always with daily changes of solution. Thereafter the specimen can be held in 10% formol solution indefinitely, until needed for further processing.

Further Processing. — The further treatment involves three steps: decalcification, dehydration, and embedding in celloidin.

Decalcification was carried out in 0.5% nitric acid in 10% formol solution with daily changes for a period of 10–20 days. This acid is chosen because all its salts are soluble.

Dehydration then followed, with 13 steps of increasing concentration of ethyl alcohol, beginning with 10% and increasing daily by 10% until 90% was reached, then 95% for two days, and finally 100% for two days. This final 100% alcohol must be truly absolute, for even a trace of water in the tissues impairs the infiltration of celloidin solutions.

Celloidin embedding began with transfer of the tissue block to the celloidin solvent. This required two steps: first placing the block in a mixture of 1 part anhydrous ether to 3 parts of anhydrous alcohol for 1 day, then a transfer to a 50:50 mixture for 1 day; thereafter the block was exposed for 3 weeks each in celloidin solutions of 4%, 8%, 12%, and 16%, all with the container sealed. Then the lid of the container was slightly displaced to permit slow evaporation so as to obtain a firm consistency at the end of about 4 weeks.

The specimen then was cut out in a block of celloidin, hardened in chloroform vapor for 4 days or more, and finally placed in 80% alcohol for 2 weeks or longer with three changes at first to remove the last traces of chloroform. The specimen was stored in 80% alcohol until needed for sectioning.

For sectioning the block of tissue was mounted on a fiber base, oriented according to the plane of sectioning to be used, and the sections cut with a sliding microtome at thicknesses of 20 μm. Every section through the ear region was saved.

Staining was carried out in small groups, with adjacent sections kept well separated in different marked dishes to make easier their later identification and arrangement in a continuous series. This arrangement was made with the aid of a sorting machine, an optical device that presented enlarged views of six sections at a time.

When serial sections became available for a particular species these were examined together with the study of dissected specimens to work out the anatomical relations in final detail.

BEHAVIORAL INDICATIONS OF AUDITORY SENSITIVITY

Among the higher animals the best determinations of auditory capability are made by a training method in which the animal is induced to make some specific response to a sound either to obtain a food reward or to escape punishment such as an electric shock. Many mammals and several species of birds have been studied in this manner and have yielded meaningful and consistent sensitivity functions. These animals after suitable training respond in some observable way to a range of frequencies that is characteristic of their species and present a particular form of sensitivity function within this range.

The lower vertebrates, however, have yielded to this kind of investigation only in a spotty and generally uncertain manner. Good results have been obtained on several species of fishes, and consistent determinations have been made in turtles of the species *Pseudemys scripta* (Patterson, 1966). The amphibians, on the contrary, have not provided much useful information through this method of study.

General behavioral procedures, such as the observation of startle responses, either as movements or, more often, as the cessation of an ongoing activity at the production of a sound, have given only fragmentary indications of auditory ability. Among frogs the principal behavioral results of value have come from the observation of natural responses, such as the perception of sounds that alert the animal to impending danger and the hearing of their own vocalizations that assist in the mating process. Of interest in this relation are the early observations of Yerkes (1904, 1905).

After extensive observation of the behavior of some North American frogs under natural conditions, Yerkes was convinced that frogs are able to hear one another's croakings. Both males and females react to the croaking of a male of their own species, often by approach, and the males may respond by croaking as well. These animals react also to a cessation of vocalization when others in the same area suddenly become silent in response to an intruder. Frogs on the bank of a pool respond further to the splashing sound made by a frog that has jumped into the water, and then are likely also to take to the safety of the pool.

Courtis (1907) studied toads of an unidentified species (possibly *Bufo americanus*; see Bogert, 1960) and found, much as Yerkes had done with frogs, that both males and females respond to the call of a male, first by orientation to the direction of the sound and then by approach and bodily contact. If this contact is made by one male with another, the second male struggles and often produces a particular sound known as a release call, whereupon the contact is broken. But if the contact is with a female it usually develops into clasping, and mating often follows.

Though Yerkes found it difficult to obtain direct behavioral evidences of hearing in frogs, he was able to design an experiment on frogs of the species *Rana clamitans* to prove sound reception through a modification of responses to other stimuli. A specimen of this species, blindfolded, was supported on a stand with the hindlegs hanging free, and observations were made of the sudden response of leg retraction when the back of the head was struck with a rubber hammer. Then the effects on this reaction were noted when acoustic stimuli (either the sound of a bell or a tap on a metal plate) were produced simultaneously with the blow on the head or preceded it by various intervals.

It was found that the auditory stimulus when simultaneous with the blow on the head or preceding it by times up to 0.35 sec produced a reinforcement of the leg reaction, whereas still longer anticipation times (the sounds occuring 0.4 to 0.9 sec earlier) produced a reduction of this response. Thus convincing evidence of hearing in these animals was obtained.

Yerkes evidently failed to observe a conditioned response—a leg retraction in consequence of the sound alone—in this experimental situation.

CONDITIONED RESPONSE EXPERIMENTS

The failure of Yerkes to obtain stable conditioned responses to sounds in frogs was experienced also by several other investigators in later times, including Blankenagel (1931), Bajanderov and Pegel (1932–33), and Diebschlag (1935), as reported by Kleerekoper and Sibabin (1959).

Respiratory Conditioning Studies. — The efforts of Kleerekoper and Sibabin in this relation are perhaps representative. They worked with two frog species, *Rana pipiens* and *Rana clamitans*, and sought by numerous pairings of tones and electrical shocks to obtain changes in respiratory movements in response to the tonal stimulus alone. Only occasional positive reactions were observed, which were either jumping responses or definite alterations in breathing movements. These reactions were most obvious in two out of their 18 specimens of *Rana clamitans* in which skin lesions had developed as a result of frequent contacts with the wire grid on the floor of the training cage, and these reactions largely ceased after the lesions had healed.

These results do not appear to have been suitable for the delineation of a tonal threshold function for these animals. Although such a function was presented and has often been referred to, a lack of calibration of the sound transducer leaves the results indeterminate.

Conditioning in Salamanders. — A conditioned feeding reaction was used by Ferhat-Akat (1938) to test the hearing of two salamander species. The most extensive studies were made on larvae of the axolotl (*Ambystoma mexicanum*) of about 6 cm body length. After blinding to exclude visual cues, these animals were trained by presenting together a tone and a morsel of meat. Usually 4–15 trials sufficed to establish a reaction to the tone alone, which consisted of raising the head, quivering, and making snapping movements. The tone was then varied to determine frequency limits. The upper limit in these animals was found to be around 194 to 244 Hz when the labyrinths were intact but was reduced by about an octave, to the region of 97 to 122 Hz, when the labyrinths were destroyed and the responses evidently were mediated only by the skin.

Discrimination tests were made also by giving food at the sounding of one tone and punishing by striking the body a light blow with a glass rod at the sounding of another tone. By this procedure it was found that axolotl specimens are able to discriminate a musical fourth, or in some instances a musical fifth.

Similar experiments were carried out on larvae of the spotted salamander, *Salamandra salamandra*, in which the upper limit of frequency reception was found to be about 1035 Hz. Also these animals were shown to be capable of discriminating a major third.

HEARING IN FROGS AS SHOWN BY MATING CALLS

Although it has not been possible thus far to study the auditory capabilities of frogs in detail by the use of conditioning methods, some general information has been obtained by an examination of their responses to mating calls. Capranica and his associates (see Capranica, 1965, 1966, 1976; Capranica and Moffat, 1975) developed a procedure for this study in the bullfrog, *Rana catesbeiana*. These calls were recorded and then presented to both males and females both in their natural form and modified by the use of filters. Such calls were also produced synthetically in electronic circuits. The synthetic calls were most useful because they could be varied in numerous ways to determine the essential features involved in their elicitation of the usual responses.

This call ordinarily consists of a series of croaks, varying in number from 3 to 15, spaced perhaps a second apart. Each croak is a burst of highly complex wave form containing sound energy distributed broadly over the frequency spectrum up to about 3500 Hz, mainly in two general regions, one around 200–300 Hz and the other spread more widely around a maximum at 1400–1500 Hz. A significant feature is the absence of appreciable energy in an intermediate zone around 500–700 Hz; if the energy in this zone is considerable, and especially if it exceeds that in the lower frequency region, the call is ineffective in eliciting responses from either males or females.

There is evidence to show that both auditory papillae are involved in the reception of the mating call in *Rana catesbeiana*, the amphibian papilla serving in response to low and middle frequencies, and the basilar papilla acting in response to the high-frequency components.

ELECTROPHYSIOLOGICAL METHODS

The most extensive information concerning the hearing capabilities of the amphibians comes from electrophysiological observations. Several types of potentials are produced in the ear as a result of sound stimulation or are modified by such stimulation, and many investigators have sought to utilize these potentials in a study of the ear's performances.

GALVANIC SKIN RESPONSES

Early consideration was given to changes in the skin's resistance, or variations in its self-generated potentials, that may occur in response to a great variety of stimuli. Most effective, it would seem, are those stimuli that have biological significance to the animal, and especially those that constitute a threat to its well-being. Sudden sounds, and particularly loud and repeated ones, seem most marked in effectiveness in producing these changes in the skin's electrical state.

Kohlrausch and Schilf (1922) observed that this response was readily elicited in frogs by loud whistles, hand claps, raps upon the floor, and by the croaking of other frogs. Further observations of this kind were reported by Buytendijk and Eerelman in 1930.

Strother (1962) reported results obtained by this method on specimens of the bullfrog *Rana catesbeiana*, with significant differences between males and females: in general the sensitivity function exhibited by this type of response was shifted downward along the frequency scale for males in comparison with females. This shift no doubt is to be ascribed in part to the larger eardrum size in the males of this species, which displaces their sensitivity peak toward the lower frequencies.

More specific to the ear itself, and produced as an essential phase of the ear's response to sounds, are two further types of electrical potentials, the inner ear potentials produced in their hair cells through the action of sounds, and the auditory nerve potentials that transmit the inner ear activities to the brain.

Both of these potentials have had extensive use in the study of hearing and give particular promise in the investigation of this sense in reptiles and amphibians in which other methods (such as behavioral and training procedures) are of limited application.

AUDITORY NERVE POTENTIALS

The activities of the nerve fibers supplying the auditory sense organs of amphibians were first recorded in a general way by Adrian, Craik, and Sturdy (1938) by placing a large electrode in contact with the eighth nerve. A general recording of this kind shows the presence of a response to sounds, but does not permit an analysis of specific processes. Such an analysis must be based upon numerous recordings from individual nerve fibers, which requires a microelectrode that can be applied to a nerve bundle so as to make contact with a single fiber or a small number of fibers, depending on the size relations. The application is then repeated throughout the nerve bundle until an adequate sampling of the contained fibers is obtained.

Microelectrode studies were carried out on amphibian ears in 1960 by Axelrod and also by Glekin and Erdman. In both of these studies multiple types of auditory nerve fibers were encountered, differing in such features as the presence of spontaneous activity, the tonal range of excitability, and the rapidity of adaptation to a sustained stimulus.

Further studies dealing with several species of frogs have given results showing the following neural characteristics (as summarized by Capranica, 1976, and briefly referred to above, p. 24):

1. In any given species there are many nerve fibers that are spontaneously active, some very strikingly and others only moderately so.

2. Every nerve fiber has a characteristic frequency curve of excitability, and within its range it increases the rate of firing as the tonal stimulus is raised in intensity, at first increasing rapidly and then more slowly, until finally a limit is reached.

3. Feng, Narins, and Capranica (1975), working on bullfrogs (*Rana catesbeiana*,) identified three types of auditory nerve fibers, two of which belong to the amphibian papilla and one to the basilar papilla. One of the amphibian papilla types represents the low tones, those in the region of 250 Hz, and the responses of these fibers are inhibitable by simultaneously presented tones higher in frequency than the primary stimuli. The other type of nerve fiber serving the amphibian papilla represents tones of middle frequency and is not inhibitable by other tones presented simultaneously. The third type of auditory nerve fiber serves the basilar papilla, and these fibers represent the high frequencies, those around 1400 Hz, and are not inhibitable by other simultaneously presented tones.

As already indicated, because a microelectrode applied to the eighth nerve only records from one fiber or a small number of fibers, a conception of the nerve's activity as a whole must be built up by repeated applications of the electrode, varying its location within the nerve bundle so as to obtain a wide sampling of its elements. These observations then can be plotted together to obtain a composite picture. This is a laborious procedure and has been carried out in only a few amphibian specimens.

INNER EAR POTENTIAL MEASUREMENTS

A more practical procedure that provides a comprehensive picture of the performance of the ear as a whole is to introduce an electrode into the papillar cavity so as to make contact with the endolymph that bathes the interior of the receptor organ (including the entire assembly of hair cells). An even better procedure (because it avoids the risk of injury to the deep-lying tissues by opening this papillar cavity) is to locate the electrode in the perilymph space that surrounds this inner receptor area. The hair cells radiate their electrical effects directly into the endolymph that contains them, and then this electrical field spreads broadly into the surrounding perilymph space where it may easily and safely be sampled. This procedure was employed in the great majority of the experiments on amphibian ears carried out for the present study.

RECORDING METHODS

The acoustic tests were carried out in a small building especially designed for these experiments. The building was remote from traffic and other noise sources and was heavily sound-proofed. The animal was enclosed in a booth within this building, well isolated both acoustically and electrically from the surroundings.

Aerial Stimulation. — The sound source was a wide-range dynamic loudspeaker (Western Electric Type 555, no longer manufactured) to which was connected a plastic tube of 8 mm inside diameter that led through heavy masonry walls to the chamber containing the experimental animal. In those frogs with a tympanic membrane the end of the tube was sealed to the skin surface around this membrane on the right side of the head, and in all others (a few frogs and all salamanders and caecilians) it was in a corresponding location at the side of the head over the ear region. At the end of this sound tube, so placed as to be barely free of contact with the eardrum or head surface was the end of a probe tube that led to a condenser microphone for monitoring the acoustic input. The arrangement is shown in Fig. 2-2 taken from an earlier report.

Vibratory Stimulation. — A mechanical vibrator of our own design consisted of a Rochelle salt crystal made up of a series of laminal plates cemented together and enclosed in a metal case, with a stiff needle tip extending out of the case to transmit the vibrations to the desired surface, which was either the middle of the tympanic membrane when such a membrane was present or the most sensitive point on the lateral surface of the head located after exploring this area.

Electrode Placement. — The region of the lateral semicircular canal was exposed, the course of this canal was identified by a ridge that appears on the surface of the prootic bone, and a small tapered hole was drilled so as to enter the perilymph space (with care to avoid so deep a penetration as to intrude into the endolymphatic cavity beyond). The active electrode wire, whose tip was tapered to fit the drilled hole, was then inserted and anchored so as to prevent any but an almost imperceptible loss of inner ear fluid. Then a second electrode was placed on inactive tissue a little distance away (but

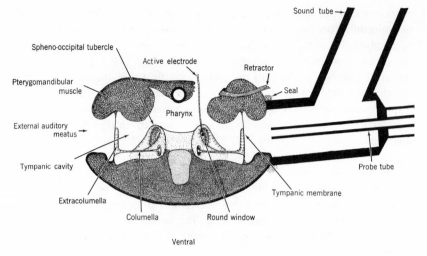

Fig. 2-2. The arrangement of sound and probe tubes. From Wever, *The Reptile Ear*, 1978, p. 45.

not in the perilymph space itself), and a third electrode, which was grounded, was placed in inactive tissue a short distance from the two electrodes just mentioned.

These three electrodes led to a push-pull input circuit of an amplifier designed to minimize stray potentials arising both from the animal and from other sources in the vicinity.

That the insertion of the electrodes into the perilymph space was not detrimental to the ear was shown by repeating the whole procedure: after a set of measurements had been completed the active electrode was removed, its hole was plugged with wax, and then on the following day, or even several days later, the electrode was reinserted and the series of observations was repeated; almost always these second measurements were in full agreement with the first ones. (A failure in such agreement could usually be traced to an activity of the middle ear muscles, which could be suppressed by producing a deeper level of anesthesia.)

This recording method permitted the routine measurement of inner ear potentials produced by sounds down to 0.1 μv, and occasionally, in a well-anesthetized and thus a particularly quiet animal, down to 0.03 μv. These low levels of recording are essential in animals like the amphibians in which the population of hair cells is small.

Routinely the tones were presented at a number of levels, from one that gave a barely perceptible reading to others up to the beginning of overloading. By this means intensity curves were obtained at various frequencies within the range of the species under test, thus defining the ear's effective scope.

The complete experimental arrangement for these studies is diagrammed in Fig. 2-3.

THE INDICATORS OF HEARING ABILITY

The recording of inner ear potentials in response to sounds as just described constitutes the principal means of assessing the functioning of the ears of the species here considered. The alternative procedure, mentioned earlier, in which measurements are made in response to sounds at a further stage along the sensory stream, in the cochlear nerve fibers, can be regarded as an even better means of evaluating the ear's performance, as this method takes into account the action of the auditory nervous system as well as that of the sensory elements. This procedure, however, presents formidable difficulties when a wide range of species is to be considered. Such a neural sensitivity test requires the exposure of the nerve branches supplying both auditory papillae (when two are present as in most species) and the placement of the recording electrode on a good many fibers of each branch in turn until an adequate sampling has been obtained. Each microelectrode placement must be carefully checked to make certain that the contact with the nerve fiber is suitable, and the testing must include a good many tonal stimuli to make sure that the performance of the fiber has been adequately covered. This is a lengthy and exacting task during which the animal must be maintained in good responsive condition. Therefore such experiments have been carried out on only a small number of amphibian specimens, representing relatively few species. Some of these observations are reproduced for comparison with the cochlear

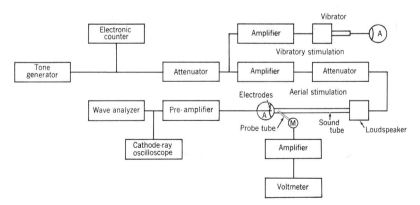

Fig. 2-3. Stimulating and measuring circuits. From Wever, *The Reptile Ear*, 1978, p. 45.

potential data here described, and it is highly desirable that additional experiments of this kind be carried out. Meanwhile the cochlear potential observations will be chiefly relied upon as our best indicators of hearing ability over a wide range of amphibian species.

THE VALIDATION PROBLEM

It is necessary to consider the question of the degree of assurance that may be accorded to the inner ear response functions as indicative of the ear's sensitivity in the amphibians. This problem was confronted earlier in an examination of the reptilian ear (Wever, 1978b), in which evidence was brought forward showing that in general there is a close relationship between inner ear potential functions and other means (especially behavioral methods) of assessing auditory capability. This evidence as cited earlier extended over 16 species that had been examined by both methods; these were mostly mammals but included a few examples among birds, reptiles, and fishes. The degree of relationship found for the two types of sensitivity functions among the various species is impressive: the correlation coefficients range from 0.40 to 0.89 and average 0.68.

There is abundant evidence that these inner ear potentials arise in the activity of the auditory hair cells. There are a number of hereditary conditions in which these hair cells are reduced or absent, and the inner ear potentials are correspondingly deficient. These cells are lacking in certain animal strains, such as albinotic cats and a line of Dalmatian dogs, and these animals are congenitally deaf; in these no inner ear potentials can be found. The relationship is particularly striking in a strain of mice known as the Shakers in which the hair cells are present at birth and these infants are evidently able to hear (and their ears produce a good level of potentials in response to sounds), but degeneration begins at an age of about three weeks (about the time of weaning), when the hair cells begin to be lost and the potentials in response to sounds progressively fade away. Then also the behavioral responses to sounds, which are present in the young animals soon after birth, diminish and finally are lost as the degeneration proceeds (Wever, 1965).

Further evidence on the dependence of the potentials that may be recorded to sounds on the presence and integrity of the hair cells comes from experiments in which these cells are damaged and destroyed by overstimulation with sounds. Alexander and Githler (1951, 1952), working on guinea pigs, obtained a correlation of 0.91 between hair-cell loss from such stimulation and the reduction in cochlear potentials.

Unfortunately, comparable validation data are not available for the amphibians. Worthy of mention, however, are some limited observations by Strother (1959, 1962) on the species *Rana catesbeiana* in which both electrodermal and inner ear potential methods were used, as indicated in Fig.

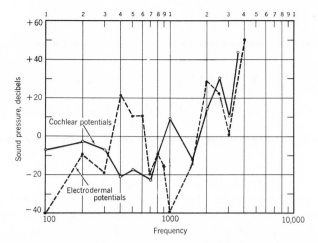

Fig. 2-4. Sensitivity functions for the bullfrog *Rana catesbeiana* as determined by the observation of inner ear potentials (solid line) and of electrodermal potentials (broken line); data from Strother, 1959, 1962. Shown is the sound pressure in decibels relative to 1 dyne per sq cm required for a response of 0.3 μv.

2-4. Plotted in the same graph here are curves for one frog (a female) showing for various tones along the frequency scale the threshold levels for electrodermal responses (the electrical potentials arising in the skin as a result of acoustic stimulation), and for another frog (also a female) of this same species showing an inner ear potential function (the sound intensities required for a response of 0.3 μv). These two functions are at variance in the lower frequency range but are closely similar in the upper range: there are large departures at 100, 400, and 1000 Hz, and then the two curves follow much the same course in their rapid rise in the high frequencies. The correlation coefficient obtained for these two functions is 0.33, a positive but not very impressive relationship. The great variability of the electrodermal responses in the lower frequency range is unexplained, and further study of this problem is needed. Meanwhile it may be argued that a relationship that holds so widely over the phylogenetic scale as to include mammals, birds, reptiles, and fishes ought to apply to the amphibians as well, and it is expected that further investigation will support this expectation. Hence observations of sensitivity in terms of the inner ear potentials will be reported for the amphibian species thus far examined as the best available indications of the functioning of their ears, observations that are useful for comparative purposes even though their absolute significance may remain in some doubt.

3. GENERAL ANATOMY OF
THE AMPHIBIAN EAR

A general treatment of the anatomy of the amphibian ear presents difficulties because the three orders now living, the anurans, caudates, and caecilians, exhibit many different specializations combined with what appear to be various degrees of reduction and degeneration from an earlier, more general form. Parsons and Williams (1963) in their defense of the lissamphibian hypothesis, which holds that the three existing orders of Amphibia had a common origin, tried to formulate what the primordial lissamphibian ought to be like. They listed 19 features that they considered as most probably possessed by this ancestral form, among which they emphasized the nature of the teeth, the skull with its peculiar palatal structure, and the forms of the middle ear ossicles as these are seen in the anurans. Also of great importance in this relation are the form and manner of operation of the auditory papillae.

A consideration of the anatomy of the amphibian ear will begin with an examination of this structure in anurans, and more specifically in ranids, which are the most advanced members of this order. This choice reflects the widespread opinion that members of this group, in spite of a high degree of specialization, are more representative of the original line of tetrapods from which the amphibians arose, and thus better reflect the basic features of ear structure, than either the urodeles or the caecilians. Members of these last two orders appear to have been subject to degenerative changes in the auditory system, as well as in other bodily features, probably as a result of drastic alterations of environment.

For the urodeles this environmental change seems to have been primarily a return to aquatic life after the transition to a terrestrial existence by their early progenitors, with secondary adaptations such as the loss of the tympanic membrane and often of the columella as well. An aquatic animal can perceive sounds in the water without the use of a mechanical transformer such as the middle ear provides, for the impedance of the inner ear structures is itself well matched to that of the water medium, and the transformation provided by a tympanic membrane and ossicular mechanism of the usual form would reduce sensitivity rather than augment it.

Likewise, an animal living underground as many of the caecilians do, with the head often in contact with soil or mud, would receive no benefit from a middle ear transformer. Those few caecilians that have become aquatic have probably done so secondarily, developing from fossorial ancestors. Furthermore, for an animal living in water or underground the presence of an external ear opening would most likely be a handicap, presenting a route for the entrance of waste, soil, and parasites that might lead to infection in deep-lying tissues.

GENERAL VIEW OF THE ANURAN EAR REGION

The following description of anuran ear structures will be based upon the Ranidae as the most highly developed of the anuran families. Three species were examined in detail: the common northern species *Rana pipiens*,[1] the closely similar Florida species *Rana utricularia sphenocephala*, and the larger species *Rana catesbeiana*. Specimens of these three species were used according to their availability at the times the various dissections were made, and the ear structures are so much alike in these animals that close distinctions among them are unnecessary at the present level of treatment.

First in this discussion a general orientation will be provided in a series of skull figures, along with representations of isolated parts that present special difficulties of recognition in the complete skull. These figures also provide reference points and anchorages for elements of the auditory system. The ear's deep location in the head and its complex relations to other structures make this orientation necessary for an understanding of the entrance of sounds and their transmission inward to the final receptive structures.

The Anuran Skull

Figure 3-1 presents a dorsal view of the skull of the bullfrog *Rana catesbeiana*. This skull is noticeably wide, almost as broad as it is long, and is marked by the large eye cavities. The bony frame forms a parabola with the two sides consisting of the maxillaries extended posteriorly by the quadratojugals; the structure is braced by a large inverted "T" formed by the frontoparietals in the middle and the prootics extending laterally in the posterior region. Below in this figure the large tympanic membranes have been retained in their close attachment to the tympanic annulus on either side. Underlying this annulus, and visible only at its anterior and posterior ends, is the squamosal.

---

[1] There appear to be at least four species of frogs that are usually referred to as *Rana pipiens* but that are genetically distinct. These species are not separable by visual examination and in the hands of animal dealers are often hopelessly intermingled. The specimens used in this study came from two different localities, from Vermont (and here are referred to as *Rana pipiens*) or from Florida (referred to as *Rana utricularia sphenocephala*).

No significant hearing differences have been found among these species, but further investigation is needed.

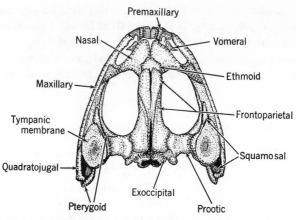

Fig. 3-1. The skull of the bullfrog, *Rana catesbeiana*, in a dorsal view. Natural size.

A ventral view of this skull, seen in Fig. 3-2, shows the continuous rows of teeth along the maxillaries and premaxillaries, ending at the quadratoju- gals. This undersurface of the skull reveals the vomers, each bearing a small patch of slender, pointed teeth. Posterior to this is a pair of palatines that provide crossbracing at the anterior end of the skull and are attached to the T-shaped parasphenoid. Posterior crossbracing is provided by the ptery- goids, which unite medially with the lateral processes of the parasphenoid. At the posterior end are the two occipital condyles, which are extensions of

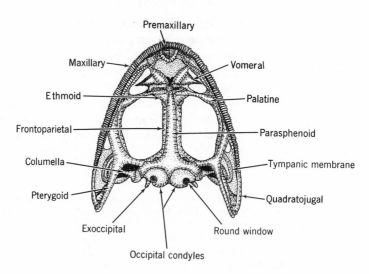

Fig. 3-2. The skull of *Rana catesbeiana* in a ventral view. Natural size.

the occipital bones that form the neck joint surfaces. Openings at the lateral edges of the exoccipitals are the round windows, each covered by a thin membrane. On either side a columella runs to an oval window in the prootic, with its expanded end filling the opening.

The lateral aspect of this skull is shown in Fig. 3-3. Here is provided a broad view of the tympanic membrane, with the ends of the squamosal supporting its anterior and posterior edges. Seen also is the pterygoid that connects the maxillary anteriorly and the quadratojugal posteriorly to parasphenoid and exoccipital.

The squamosal and pterygoid are of special interest because they bear close relations to the ear structures. As these are somewhat difficult to visualize in the intact skull they are shown in isolation in the next two figures.

Figure 3-4 shows the right squamosal as seen from the right side after removal of the tympanic membrane, which frees the dorsal process from its attachment to the parotic crest, and also the detachment of the posterior end of its long, handle-like posterior process from the quadrate. A zygomatic process runs anteroventrally; in the prepared skull as shown here it ends freely, but normally a ligament runs from this process to the upper edge of the maxillary.

Figure 3-5 shows the right pterygoid as seen from a lateroventral, left position (see Fig. 3-2). It has three processes, a dorsal one that is fused to the lateral edges of the exoccipital and prootic, an anterior process that runs along the maxillary and fuses to both this bone and the palatine, and a posterior process that attaches to the articular process of the quadrate.

The head skeleton of the frog contains much cartilage that constitutes what is commonly referred to as the primordial cranium, shown in Fig. 3-6, taken from Gaupp (1896). The posterior portions of these structures are of particular interest. A lateral extension on each side is the tympanic annulus to which the tympanic membrane is attached. Just posterior is the oval window, which normally contains the end of the columella, and below this is the articulatory

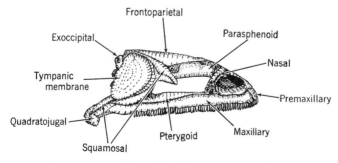

Fig. 3-3. Lateral view of the skull of *Rana catesbeiana*. Natural size.

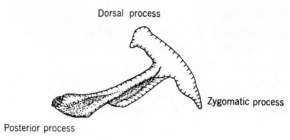

Fig. 3-4. The squamosal bone of *Rana catesbeiana*. Scale 1.5X.

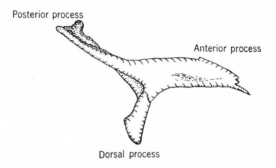

Fig. 3-5. The pterygoid bone of *Rana catesbeiana*. Scale 1.5X.

process of the quadrate by which the skull is suspended. Further details are presented in Figs. 3-7 and 3-8, also taken from Gaupp (1896). Here the lighter areas represent bone and the darker areas represent cartilage; as is apparent, the adult skull remains largely cartilaginous, with the nasal and otic areas almost entirely so.

Tooth Structure. — The presence of what are called pedicellate teeth has already been mentioned as the starting point of the Parsons and Williams hypothesis concerning amphibian origins and relationships. A sketch of such teeth as seen in *Rana catesbeiana* is shown in Fig. 3-9. Represented here is a portion of the upper jaw (the maxillary bone) containing three teeth in place and a socket from which the tooth has been extracted. The fibrous joint between pedicle and crown permits a limited and rather resistant movement of the crown. The root portion of the pedicle is firmly cemented in the socket within the maxillary.

The Palate. — In modern amphibia the palate is of a type to be found in the ancient temnospondyls, a Paleozoic group in which the pterygoid bones are reduced and widely separated. The result is a peculiarly open form of skull structure, as Figs. 3-1 and 3-2 portray.

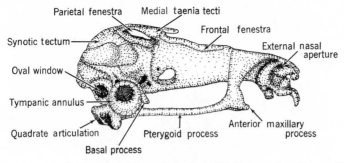

Fig. 3-6. The primordial cranium of *Rana fusca*. Scale 3X. Redrawn from Gaupp, 1896, fig. 17, p. 37.

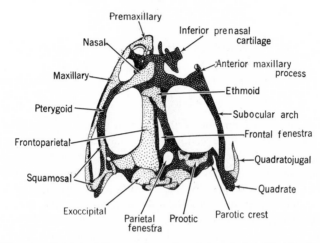

Fig. 3-7. The primordial skull of *Rana esculenta*. Dorsal view. Scale 1.6X. After Gaupp, 1896, fig. 15.

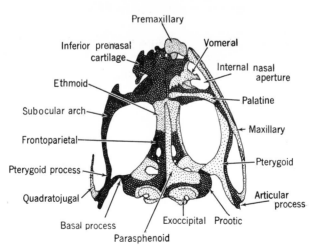

Fig. 3-8. The primordial skull of *Rana esculenta*. Ventral view. Scale 1.6X. After Gaupp, 1896, fig. 16.

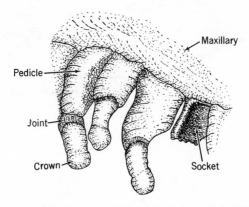

Fig. 3-9. Pedicellate teeth in *Rana catesbeiana*. Scale 20X.

THE ENDOLYMPHATIC LABYRINTH IN ANURANS

Preliminary to an examination of the detailed structure of the frog's laby-rinth, a general view of the endolymphatic labyrinth will be presented for two species of *Rana*. First will be shown two figures redrawn from Birk-mann (1940) in which the labyrinth is represented for the European frog *Rana temporaria*, one viewed laterally (Fig. 3-10) and the other medially (Fig. 3-11). These figures present the forms of the structures after the investing bone and cartilage have been removed and indicate the relations of the acoustic organs, the amphibian and basilar papillae, to the equilibrial system that largely surrounds them. As Fig. 3-10 shows, the two auditory papillae are nearly hidden in their middle position behind the utricle and the superior division of the saccule; these are more clearly seen lying closely adjacent in the me-dial view of Fig. 3-11.

Additional details of labyrinthine structure are shown in a series of five figures representing the green water frog of Europe, *Rana esculenta*, re-drawn from Gaupp who took them, with some slight modifications, from the classical drawings of Retzius (1881).

This structure contains the usual organs of hearing and equilibrium: the amphibian and basilar papillae, the three maculae in utricle, saccule, and lagena, and the three cristae belonging to the anterior, external, and poste-rior semicircular canals.

In all five drawings the right labyrinth is represented, with different ori-entations. Figure 3-12 shows this structure in a lateral view, with dorsal above and anterior to the right. As will be seen, the anterior and posterior semicir-cular canals have their ampullae, which contain the crista organs, at opposite sides of the structure, and between these, united in a broad duct, is the su-perior sinus that runs into the utricular expansion below. The lateral canal

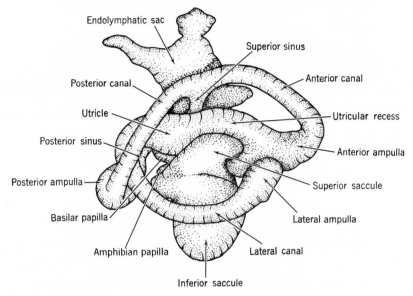

Fig. 3-10. The membranous labyrinth of *Rana temporaria*, in lateral view. Scale 25X. Redrawn from Birkmann, 1940, p. 462.

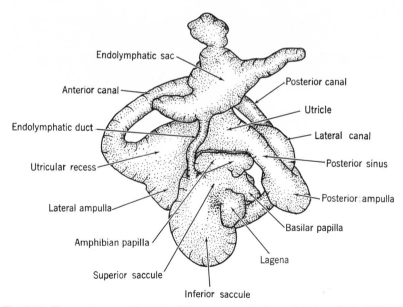

Fig. 3-11. The membranous labyrinth of *Rana temporaria*, in medial view. Scale 25X. Redrawn from Birkmann, 1940, p. 463.

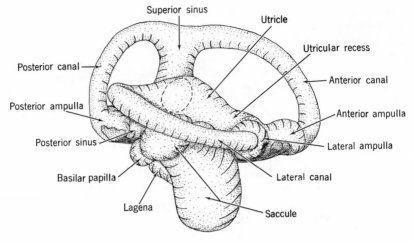

Fig. 3-12. The membranous labyrinth of *Rana esculenta*; lateral view. Redrawn from Gaupp, 1904, p. 684; after Retzius.

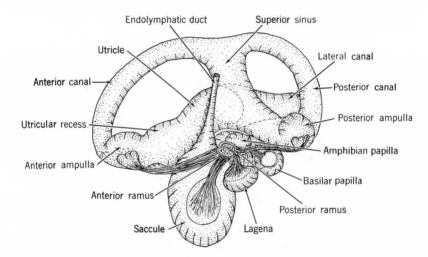

Fig. 3-13. The membranous laybrinth of *Rana esculenta*; medial view. Redrawn from Gaupp, 1904, p. 684; after Retzius.

arises from the utricle and swings around anteriorly with its ampulla closely adjacent to that of the anterior canal.

The saccule is seen to form two prominent expansions beneath the lateral canal, one partly hidden in Fig. 3-12 by the midportion of this canal and another, much larger, that curves ventrally. Also seen here are the sacs containing the basilar papilla and the lagena. The dashed oval in this picture represents the apertura utriculi, an opening within the utricle marked off by a fold of tissue on its inner wall that separates anterior and posterior halves of the inner cavity. The posterior end of the lateral canal runs into the anterior part of this opening, and the superior sinus runs into the posterior part.

Figure 3-13 is a medial view of this structure. In addition to the parts already noted are the amphibian and basilar papillae and the endolymphatic duct running off from the posterior division of the saccule. This duct continues in the cranial cavity and expands into the endolymphatic sac, a tube of highly complex form that runs out into the vertebral canal to form chalk sacs on the spinal ganglia.

Seen here also are the two main divisions of the eighth nerve, an anterior and slightly ventral one with four branches: one to the saccule, another to the utricle that enters its macular endorgan, a third to the anterior ampulla, and a fourth not shown in this figure but partly indicated in Fig. 3-14 extending to the lateral ampulla. The nerve supply to the anterior and lateral ampullae is also shown, somewhat more clearly, in Fig. 3-15.

The second division of the eighth nerve as seen in Fig. 3-13 is posterior and slightly dorsal to the first, and also has four branches. This figure shows all of these: one branch is to the lagena, another to the basilar papilla, a third to the amphibian papilla, and the fourth to the posterior ampulla. According to the later observations of Feng, Narins, and Capranica (1975) there is also in this second division of the nerve a fifth bundle of fibers that runs to the saccule (at least in the bullfrog).

Figure 3-14 is a dorsal view of the labyrinth with the superior sinus in immediate aspect and the anterior and posterior canals diverging laterally from it, embracing the two divisions of the saccule farther ventrally. The posterior sinus, indicated to the lower left, is the most posterior portion of the utricle and borders on the expanded posterior end of the lateral semicircular canal. This portion of the utricle runs outward and expands to form the utricular recess, in which the utricular macula is contained. The utricle communicates with the saccule by way of the utriculosaccular foramen, a slit-like opening along the superior division of the saccule.

Figure 3-15 shows the labyrinth as viewed in the preceding figure, except that the anterior and posterior canals have been largely removed to show the deeper structures more fully. The basilar papilla is now seen extending over the two divisions of the saccule.

Figure 3-16 presents much the same view as Fig. 3-12 except that again the canals are mostly removed to show the relations of saccule and utricle. This figure indicates more precisely the form of the superior division of the saccule, with its marked degree of separation from the inferior division.

The general views of the labyrinth just presented will serve as orientation for a more detailed examination of the structures of the inner ear region.

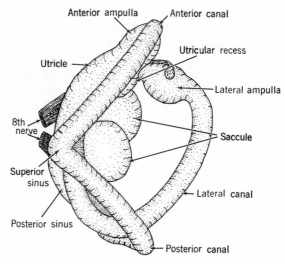

Fig. 3-14. The membranous labyrinth of *Rana esculenta*; dorsal view. Redrawn from Gaupp, 1904, p. 685; after Retzius.

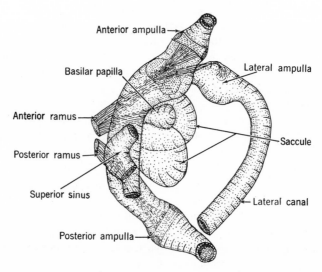

Fig. 3-15. The membranous labyrinth of *Rana esculenta*; dorsal view. Redrawn from Gaupp, 1904, p. 689; modified from Retzius by removal of anterior and posterior semicircular canals.

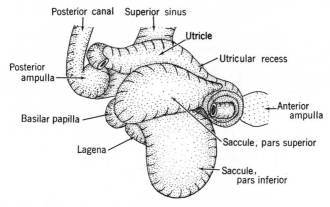

Fig. 3-16. The membranous labyrinth of *Rana esculenta* showing the divisions of the saccule after removal of most of the semicircular canals. Redrawn from Gaupp, 1904, p. 691.

These structures will be presented in a series of transverse sections in a specimen of *Rana utricularia sphenocephala* beginning anteriorly where parts of the otic labyrinth first appear and continuing by stages to the posterior end.

A schematic representation of the frog's ear is provided by Fig. 3-17. There are two main divisions, the sound conductive mechanism to the right of the picture and an inner ear portion contained in the otic capsule on the left. There is no outer ear as in higher animals, in which a pinna or auricle extends from the side of the head and serves to scatter sound waves into an inward passage, the external auditory meatus; instead the tympanic membrane lies flush with the surface of the face and receives the sounds directly.

The tympanic membrane is an area of modified skin, much thinner than ordinary skin and largely lacking the subcutaneous layers; it is usually round or broadly oval, and its edges are attached to the flaring rim of a funnel-shaped cartilage, the tympanic annulus, which itself is sustained by the squamosal bone at upper and lower surfaces. A thickened area of dense connective tissue lies in the middle of the membrane to which the outer end of the columella is attached; in most species, as in the one represented in this figure, there is also (though not shown) a cartilaginous rod, the ascending process, extending from the inner surface of the membrane, adjacent to the columellar attachment, and running inward and dorsally to fuse with the undersurface of the parotic crest. This ascending process (seen in Fig. 3-29, p. 55) probably adds stability to the tympanic membrane and protects it against undue forces.

The columella, also sometimes referred to as the plectrum, consists of three portions: the pars externa or extracolumella, which is attached to the tympanic membrane, the pars media or columellar shaft, a middle portion that

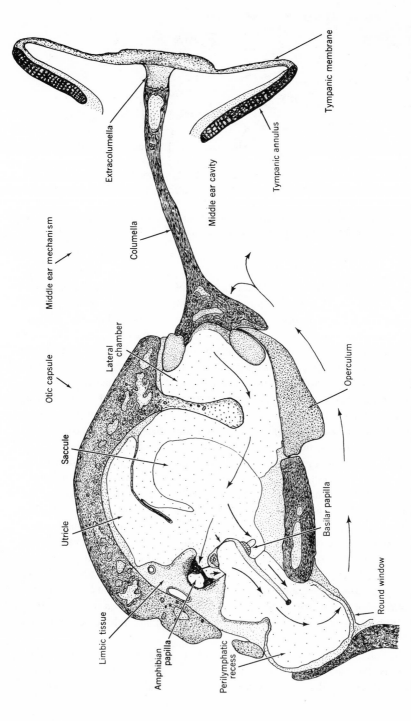

Extracolumella

Tympanic membrane

Tympanic annulus

Middle ear cavity

Columella

Middle ear mechanism

Operculum

Otic capsule

Lateral chamber

Saccule

Basilar papilla

Utricle

Round window

Limbic tissue

Amphibian papilla

Perilymphatic recess

Fig. 3-17. A schematic representation of the frog's ear. From Wever, in W. D. Keidel and W. D. Neff, *Handbook of Sensory Physiology*, 1974, vol. V-I, p. 438.

extends inward as a long, slender rod and then flares out to merge with the third portion, the pars interna, which enters an opening in the otic capsule. The outer and inner portions consist of cartilage, whereas the middle portion is osseous. The expanded inner end of the pars media together with the entire pars interna constitute the footplate of the columella, which fills the lateral opening of the otic capsule and also extends well inside a portion of the cavity, the lateral chamber, that is partially partitioned off by a bony or cartilaginous flange projecting downward from the dorsal wall of the capsule. The ventral wall of this chamber is formed by the operculum, a broad, rounded plate, usually of cartilage, but sometimes at least partially calcified.

Within the otic cavity are the six labyrinthine and two auditory endorgans; of these only the auditory ones, the amphibian and basilar papillae, are represented here.

The endorgans within the otic cavity are sustained by masses of limbic tissue, which is a form of connective tissue that evidently is peculiar to the ear; this tissue was noted by the early anatomists and often referred to as "cartilage" or "spindle cartilage," but it is a distinctive material, found nowhere else as far as is known. Two endorgans are indicated, the amphibian and basilar papillae, both in the medial region of the capsule and appearing as outpocketings of the saccular cavity. These two papillar cavities are filled with endolymph, as is the saccular cavity; but each has a small window, covered by a thin membrane, that leads into the perilymphatic sac. The perilymphatic sac or recess fills the posteromedial portion of the otic capsule and extends a little way into the cranial cavity, where it is separated from the intracranial fluid by a moderately thin membrane, the arachnoid. At the basal region of the cranium the perilymphatic sac bulges into the jugular foramen, where it is contained by a rather thick membrane, the round window membrane.

The round window lies in the posterior floor of the cranium, covered by a layer of muscle (the levator scapulae inferior) and by the epithelial lining of the roof of the mouth; below is the air of the mouth cavity. Thus this window is free to move in response to fluid pressures from within, with probably only moderate damping. When a sound wave strikes the tympanic membrane and at its positive phase produces an inward movement of the columella, a current of fluid displacement is produced as shown by the large arrows: the volume of fluid displaced by the footplate of the columella continues through the inner ear to produce a corresponding fluid displacement at the round window; and then at the negative phase of the sound wave all these actions are reversed; thus the effect of a sound is the production of a series of alternating fluid movements.

As will be noted, this fluid current in the interior of the capsule takes two outward paths, one rather direct through the basilar papilla and another somewhat more circuitous through the amphibian papilla.

The Auditory Ossicles. — The presence of an operculum as well as a columella in anurans and urodeles is a clear similarity between the two orders, but the caecilians lack this second middle ear element. The argument has been made that the caecilians also possess an operculum but it is indistinguishably fused with the columella, yet this point appears to be forced; there is no positive evidence that such is the case. It must be admitted, however, that extensive modifications of the caecilian skull to make it an effective burrowing instrument may have obscured the anatomical picture, and we must allow at least the possibility, as urged by Parsons and Williams, that all the amphibians arose from ancestors possessing an operculum.

The Auditory Papillae. — A further line of evidence of great significance is the particular character of the auditory papillae: the presence of inner ear receptors in a form entirely different from those of other vertebrates and operating by physical principles that also are distinctive. The caudates present a number of features (most notably the partial or complete loss of the middle ear) best explained as reduction occasioned by at least a partial return of these animals to an aquatic form of existence subsequent to the original tetrapod emergence. The caecilians exhibit extreme reduction and specialization of these structures also, commonly explained as attendant on their adoption of burrowing habits.

If we accept the Lissamphibian hypothesis that the three orders of living amphibians had a common origin, then the frog's ear probably represents our best available approach to the ancestral form: this ear is provided with a sound-conductive mechanism (a middle ear) consisting of a tympanic membrane, columella, and operculum, along with two (or possibly three) middle ear muscles that actuate a remarkably effective sound-control mechanism. Frogs present also an inner ear containing two distinct receptor organs, the amphibian and basilar papillae, both of a form appearing only in amphibians: nowhere else in the vertebrate series is to be found a structure of its kind. Many caudates possess two separate papillae as well, though a few have only one of them, and all the caecilians examined in the present study were found to have two papillae.

These papillae are alike in their essential features: all present hair cells of a similar kind contained in a framework of supporting cells, also of a standard form, with the ciliary tufts of the hair cells usually extended into the ends of little tubules in a sort of honeycomb structure, the tectorial body. These basic features are present in the equilibrium organs of all vertebrates from fishes to mammals.

DETAILED ANATOMY OF THE ANURAN EAR REGION

An orientation view of the region of the ear in the leopard frog, *Rana pipiens*, is given in Fig. 3-18, showing the tympanic membrane on the right

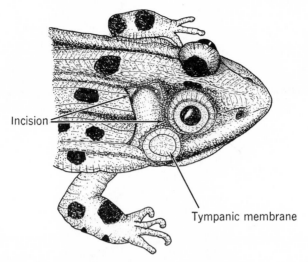

Fig. 3-18. The head region of a specimen of *Rana pipiens*, with an exposure of the prootic bone. Natural size.

side as it appears in a lateral and somewhat dorsal view. The skin was re-moved above and medial to this membrane to expose the prootic bone and its lateral cartilaginous extension as the parotic crest, to which the tympanic annulus and tympanic membrane are attached. A dark streak deep in the in-cision represents the course of the posterior and anterior semicircular canals where they come close to the bony surface. The small circle near the ante-rior end of this dark area marks the location of the anterior semicircular canal where a minute hole was commonly drilled for the placement of an electrode for the recording of responses to sounds.

The view that was presented in Fig. 3-18 will serve as orientation for Fig. 3-19. Here the surface at the side of the head behind the tympanic mem-brane has been cleared of muscle and connective tissue, and two sutures (s) are seen, one between the bony prootic and the cartilaginous parotic crest and another between prootic and the frontoparietal bone high on the head. The internal jugular vein, deep beneath the skin, roughly follows the course of this second suture. A depression in the anteromedial portion of the prootic was occupied by the attachment of the temporalis muscle; a star lies over a bony ridge that is the usual point of electrode placement as already indicated in Fig. 3-18.

The bone of the prootic was then opened to reveal the course of the three semicircular canals and the locations of their ampullae, as represented in Fig. 3-20.

Numerous muscles are found in the vicinity of the ear and need to be rec-

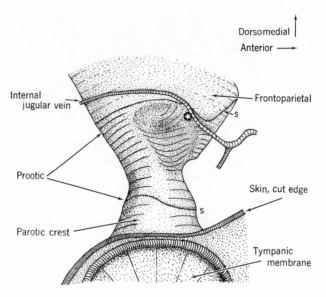

Fig. 3-19. The prootic region in *Rana pipiens*; *s, s* represent sutures. Scale 5X.

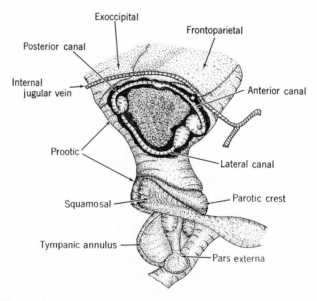

Fig. 3-20. The prootic bone opened to expose the three semicircular canals and their ampullae. Scale 5X.

ognized in a surgical approach and study of its structures. The more superficial of these muscles are indicated in Fig. 3-21 after the skin was removed in areas above and behind the tympanic membrane as far as the dorsal midline. Immediately adjacent to the ear structures are the depressor mandibulae and temporalis muscles, which are used in jaw movements.

The depressor mandibulae arises in two parts. The main part, shown in Fig. 3-21, is anchored to a nearly transparent sheet, the dorsal fascia, which extends upward from the level of this muscle to the midline. A second part, hidden in this view, arises from the posteroventral quadrant of the tympanic annulus and extends in a posteroventral direction to a small ligament that joins the main one on the tapering end of the part shown here, and then the two insert together on the quadrate cartilage at the extreme posterior tip of the lower jaw (see Fig. 3-27 below). The depressor mandibulae is the only muscle serving to open the mouth.

The temporalis muscle also arises in two parts; one part comes from a ridge far back on the prootic at its junction with the exoccipital bone, and the other part comes from the posterior arm of the squamosal; the lower end of this muscle runs to the lower jaw where it inserts on the coronoid process of the angular (shown in Fig. 3-27 below).

The next figure, Fig. 3-22, shows the upper edge of the suprascapula, a broad fan-shaped piece of cartilage and bone that is an upward extension of the shoulder blade serving for the attachment of a number of muscles, mainly those concerned with movements of foreleg and shoulder.

In this figure many of the superficial tissues in the region have been dissected away, and the depressor mandibulae muscle has been cut off to expose the whole of the dorsalis scapulae, an arm muscle covering the larger surface of the suprascapula. Now seen is the cucullaris, used in shoulder and head movements, running obliquely between suprascapula and tympanic membrane from an anchorage on the posterior underedge of the prootic. Also the temporalis is further revealed as it runs forward and laterally over the surface of the prootic on its way to the lower jaw.

Still deeper lying muscles are shown in Fig. 3-23, in which many of the muscles already identified have been removed or bent back, and a window has been opened in the middle of the suprascapula to reveal the muscles beneath it. Here may be seen the serratus superior, serratus medius, and levator scapulae inferior, all concerned in movements of the shoulder blade. Also indicated is what will be referred to as the middle ear muscle complex, a group of three muscles arising from the underside of the suprascapula at its anterodorsal corner as indicated by the dashed outline. This muscle group has conventionally been referred to as the levator scapulae superior and regarded as pulling the suprascapula forward and downward; actually it is now found to be mainly concerned with the middle ear mechanism, as will be brought out presently.

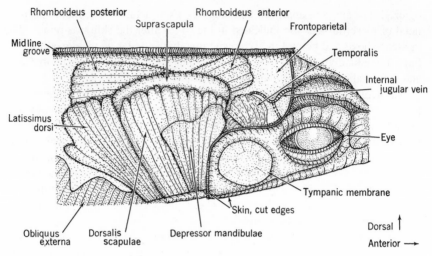

Fig. 3-21. Superficial muscles in the region of the ear. Scale 4X.

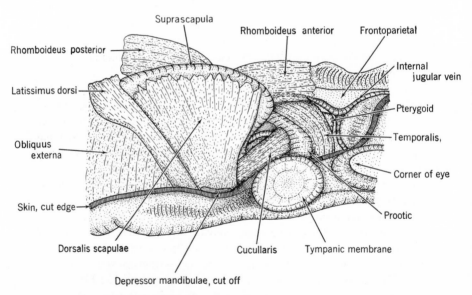

Fig. 3-22. Additional muscles of the ear region. Scale 4X.

Figure 3-24 shows structures at a still deeper level, after removal of the parotic process along with the tympanic membrane and columella, and clearing away all muscles except those of the middle ear muscle complex, now seen coming from the underedge of the suprascapula (which is displaced to the left and shown with the scapula attached). The otic capsule, exposed immediately lateral to the prootic, contains a broad opening, the oval window, whose posterior portion remains filled by the operculum, and an anterior area from which the footplate of the columella has been extracted. Forward on the ventral wall of the otic capsule is an attachment of the hyoid process, a long cartilaginous rod, usually called the cornu principale, extending from the hyoid plate that lies above the floor of the mouth and serving to support the tongue. The endpiece as represented here is often broken off during dissection and has sometimes been mistaken for an independent "styloid process" on the capsule wall.

Shown below in this figure are the squamosal and pterygoid bones that form the suspensory framework for the jaws. The maxillary is attached by way of the quadratojugal to the posterior ends of squamosal and pterygoid and is held firmly by cartilage and tendinous tissue. The mandible is hinged to the quadrate and quadratojugal surfaces.

The masticatory muscles in the frog consist of two groups, each containing two muscles. Those of one group, the masseters, insert on the angular outside of Meckel's cartilage (Fig. 3-25). The masseter major arises in two regions, from the ventral surface of the anterior (zygomatic) process of the squamosal and from the anterior and ventral portions of the tympanic annulus, and inserts as a broad apron along the lateral edge of the angular. In the figure the anteroventral edges of tympanic membrane and annulus have been turned up to show the muscle attachments on their underside. The posterior portion of Meckel's cartilage is represented as showing through the lower portion of the masseter major, but in reality this part of the cartilage is completely hidden by the muscle mass. A small part of the masseter minor may be seen in front of the jaw articulation; the main portion of this muscle lies deep below the masseter major.

The masseter minor muscle is more completely represented in Fig. 3-26 in which the masseter major has been removed; as shown, its principal portion lies deep in the area between squamosal, mandible, and elements of the jaw joint. A branch of the trigeminal nerve (t), which runs between the layers of the two masseters, is thus exposed. The masseter minor arises from the ventral edges of the squamosal and from the quadratojugal and inserts broadly on the lateral surface of the angular a little above and behind the insertion of the masseter major.

The second two masticatory muscles, the pterygoid and temporalis, are more deeplying, though portions of them have already been seen in Fig. 3-22; they are represented fully in Fig. 3-27. The temporalis is a powerful

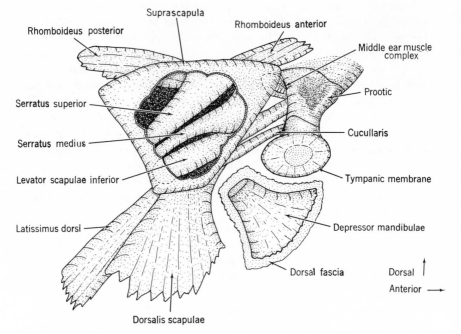

Fig. 3-23. Deeper muscles of the ear region. Scale 3X.

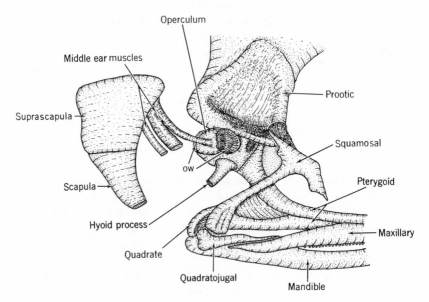

Fig. 3-24. Further structures of ear and jaw regions. *ow*, oval window. Scale 5X.

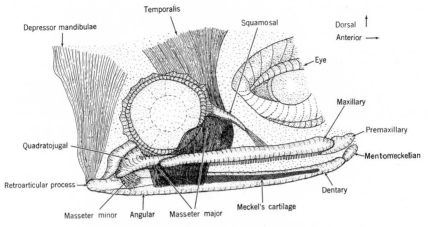

Fig. 3-25. The jaw and masticatory muscles. Scale 3X.

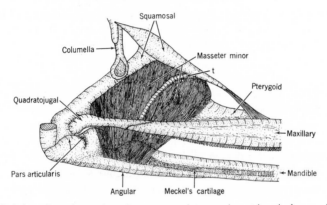

Fig. 3-26. A deep dissection to show the masseter minor muscle. *t*, trigeminal nerve. Scale 5X.

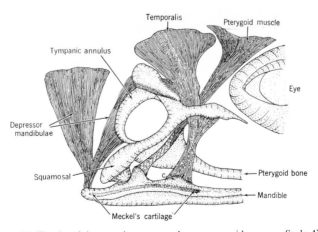

Fig. 3-27. The deep-lying masticatory muscles. *c*, coronoid process. Scale 4X.

muscle occupying most of the space between prootic and eyeball; it arises mainly along the upper part of the prootic from a ridge representing the bulge of the posterior semicircular canal and also has a few fibers from the anterior part of the posterior arm of the squamosal. It runs beneath the zygomatic process of the squamosal and inserts on the coronoid process of the mandible. The pterygoid muscle lies medial to the foregoing, close to the eyeball, and arises from the frontoparietal bone, with a few fibers from the prootic. It goes over into a long flat tendon and inserts on the angular immediately in front of the jaw joint.

The relations of these and other head muscles to one another and their general locations in the head are indicated in the cross section of Fig. 3-28. In this view the origins and insertions of the masseter muscles are clearly evident and also the broad origins of the pterygoid muscle on the frontoparietal and the side of the otic capsule. The origin of the temporalis is only generally indicated but can be better understood by reference to Fig. 3-19 above; it mainly attaches within the deep depression in the prootic shown there, and from the bordering portion of the frontoparietal.

THE MIDDLE EAR

When the muscles at the side of the head are dissected away, a picture of the oval window and ossicular structures is obtained as in Fig. 3-29. Here is seen the pars externa of the columella, a cartilaginous process with an expanded plate at its outer end that attaches to the middle portion of the tympanic membrane and a dorsomedial process that connects with the lateral end of the pars media. It sends off also a cartilaginous ascending process that runs inward below the pars media to attach to the undersurface of the parotic crest as indicated. The pars media runs inward and then expands rapidly and goes over without any clear discontinuity into the complex inner portion of the columella (pars interna). As shown here, the pars interna embraces the lateral portion of the operculum and also, as will be represented later, it expands inward and nearly fills the lateral chamber of the otic capsule. The footplate of the columella may be regarded as including the inner expanded portion of the pars media and the entire pars interna.

The Operculum. — The presence of two elements, columellar footplate and operculum, in the oval window of the frog's ear has evoked continued comment and conjecture since the early development of knowledge in this area without any reasonable hypothesis being offered about the presence of the second one. The general implication has been that the operculum takes part in the vibratory motions conveyed inward from the tympanic membrane along the columella, but does not itself play any special role. Gaupp (1904) gave this problem serious consideration, but after a detailed and largely accurate description of the anatomical relations between operculum and columella (which he called the plectrum) as well as the connections to the other

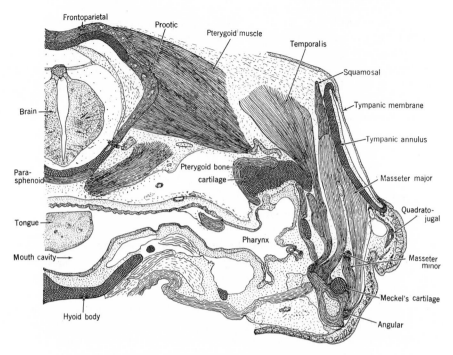

Fig. 3-28. General relations of the ear and jaw structures. Scale 10X.

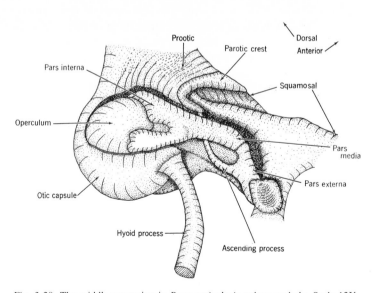

Fig. 3-29. The middle ear region in *Rana utricularia sphenocephala*. Scale 12X.

structures in this region, he was unable to offer any specific explanation of its role. He suggested merely that the operculum through its manner of connection to the pars interna might convert the in-and-out displacements of the plectrum in response to sounds into a rotational motion, but what advantage this might have in the process of hearing he was unable to say. He was noncommittal also as to the function of a muscle attached to the outer surface of the operculum, called the "pars opercularis" muscle, which he described in the then accepted way as a derivative of the levator scapulae superior. The opercular muscle had been described in urodeles also, and in these animals it was accorded a curious role as a path for the transmission of sound vibrations from the substrate. It was supposed that sounds come from the ground to the forelegs, pass along them through scapula and suprascapula to this muscle, and finally reach the operculum and inner ear (the theory of Kingsbury and Reed described elsewhere).

The situation just outlined prevailed until recently when new observations described below were carried out that showed that the operculum in the frog's ear has a special function as a part of a mechanism to protect the ear against damage from overstimulation. This protection is urgently needed when the frog itself croaks loudly or is in the company of other frogs that are vocalizing individually or in chorus. The protective mechanism is actuated by antagonistic muscles whose contractions determine the mobility of the columella and thereby control acoustic transmission to the inner ear receptors.

The Search for an Acoustic Control Mechanism. — The presence of a control mechanism in the ear of the frog was first suspected as the result of repeated observations of large variations in sensitivity in these animals. In the course of measurements of inner ear potentials in response to sounds, it was found that striking changes in output could occur in response to particular sounds when tests were repeated under conditions that so far as possible were kept constant. The animals were maintained at a uniform temperature, the oxygenation was steady as provided by artificial respiration, and the skin was kept regularly moist, yet a series of sensitivity measurements that followed immediately upon another might give results differing by large amounts, often of the order of 30–40 db, and sometimes even greater, up to 60 db or more. The larger variations were found in lightly anesthetized animals or those that were first anesthetized, whereupon the operation for electrode placement was carried out, and the animal then was allowed to recover from the anesthesia.

The more moderate variations were encountered in animals that were anesthetized to a level at which spontaneous activity ceased and surgical procedures could be carried out without any obvious movements.

The results of such measurements on a specimen of *Rana utricularia sphenocephala* in which the more moderate variations were found, without any intentional change of conditions, are presented in Fig. 3-30. The first curve (solid line) shows a region of best sensitivity in the high tones between 1200 and 2500 Hz, and a secondary region of good sensitivity in the low frequencies, improving from 300 Hz downward and reaching −2 db at the lowest point tested at 50 Hz. The second series of tests, which followed immediately after the first, gave the broken curve of this figure, in which there is a general improvement in sensitivity except for the highest tones. Here the maximum point is at 1500 Hz where −38 db is reached, and a substantial improvement appears in the low-tone region, with notable changes around 290 and 400 Hz and also near the low end of the curve. The change at 290 Hz is 31 db and at 70 Hz is 26 db.

Variations occurred likewise in another specimen of the same species as shown in Fig. 3-31. In this animal the first series of tests gave the solid curve, with maximum sensitivity at 1200 Hz where −42 db was reached, and with a minor maximum in the low-tone region at 150 Hz where a reading of −6 db was obtained. The second series of tests gave the broken curve of this figure, in which the sensitivity showed a general loss, amounting to about 43 db on the average, but reaching as much as 52 db in the high tones (at 1200 Hz) and 37 db in the low tones (at 200 Hz). A third run in this same animal was made somewhat later, in which the readings returned almost to the level seen in the first one.

Still more extreme variations were sometimes found, though usually over a more limited region of the frequency scale as seen in Fig. 3-32. Here the first series of tests gave a rather poor sensitivity curve as shown by the solid line, in which the best response in the upper frequencies was +15 db at 900 Hz, and only slightly better sensitivity was shown in the low range, where +14 db was recorded at 100 and 200 Hz. The second series of measurements gave the results represented by the broken line, which indicates great improvement for all but the lowest tones with the most striking difference at 1200 Hz where it amounted to 72 db.

As already mentioned, the sensitivity did not always vary between tests; indeed in well-anesthetized specimens a high degree of stability was the rule. Figure 3-33 gives results in which only minor variations, of the order of 3-5 db, were found between two successive tests. Here the sensitivity in the high frequencies reached −36 and −42 db at 1500 Hz and in the low frequencies showed a minor maximum of −16 and −18 db at 150 Hz, with good agreement in the form and level of the two functions everywhere.

In general, in the systematic study of frogs of various species the tests were repeated until good agreement was found between successive series, and the anesthesia was made deeper as necessary to satisfy this condition. It must be pointed out, however, that the presence of the protective mechanism in the middle ear imposes a certain limitation on the interpretation of the sensitivity measurements; the animal is able to hear at least as well as the best curves indicate, but there is always the possibility that it could hear bet-

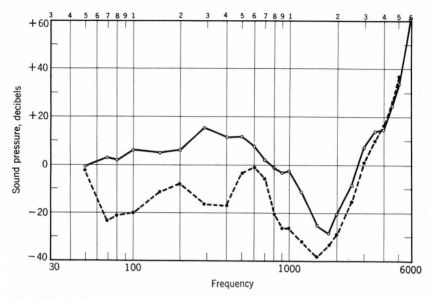

Fig. 3-30. Sensitivity curves for a specimen of *Rana utricularia sphenocephala*, representing two series of measurements made in close succession. Shown are the sound pressures, in decibels relative to a zero level of 1 dyne per sq cm, required for a response of 0.1 microvolts.

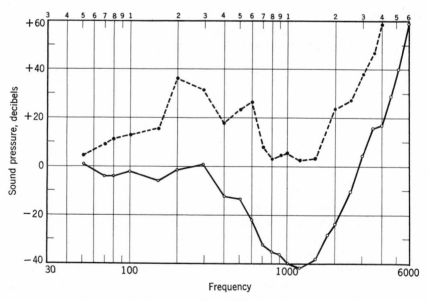

Fig. 3-31. Sensitivity curves for a second specimen of *Rana utricularia sphenocephala*, with conditions as for the preceding. Again 0 db = 1 dyne per sq cm.

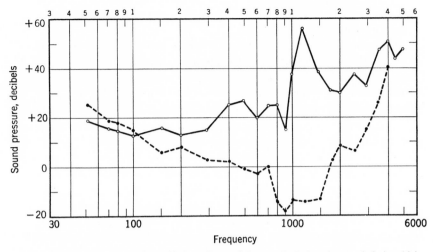

Fig. 3-32. Sensitivity curves for a third specimen of *Rana utricularia sphenocephala* in which the two series of measurements were made in close succession. As before, zero level is 1 dyne per sq cm.

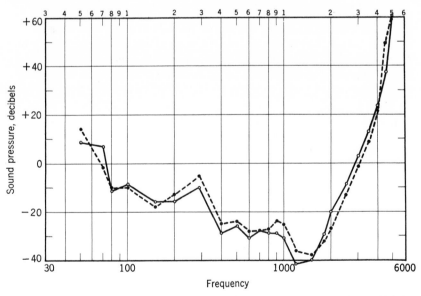

Fig. 3-33. Sensitivity curves in a specimen of *Rana utricularia sphenocephala* in which the responses were relatively stable. Zero level is 1 dyne per sq cm, for a response of 0.1 micro-volts.

ter if no restraint at all were imposed on the columellar movements by the control mechanism.

The control mechanism referred to here was at first assumed on a basis of the evidence of variability like that just presented, but the identification of the mechanism was made only after a deliberate search.

The search for a specific sound-control mechanism in the frog's ear required several months, though after this structure was found it was difficult to explain why it took so long, and also why others had not seen the structure earlier. This is perhaps a comment on the nature of perception: we note what we are disposed to recognize; but a part of the difficulty in this instance was due to the complexity of the mechanism and its orientation in the head, which is such that the planes in which serial sections are most commonly cut give relations that are difficult to interpret without foreknowledge. The use of direct dissection, which ordinarily provides the most comprehensive view of a complex structure, was at first of limited benefit and became successful only with much practice and after the study of serial sections had revealed the significant features.

Most generally useful for this sectional study is the frontal plane, the plane perpendicular to the dorsoventral axis as indicated in Fig. 3-34. A section cut in this plane through the ear region in a specimen of *Rana catesbeiana* is represented in Fig. 3-35. The otic capsule, with heavy bony walls, encloses the otic cavity which consists of two connected spaces, a larger one which is the otic cavity proper and another of about a fourth that size which is here referred to as the lateral chamber. At the level shown these two spaces are incompletely separated by a heavy bony partition; at more dorsal levels

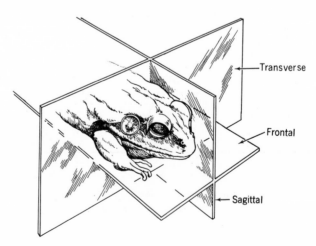

Fig. 3-34. Planes of sectioning used for the amphibian ear.

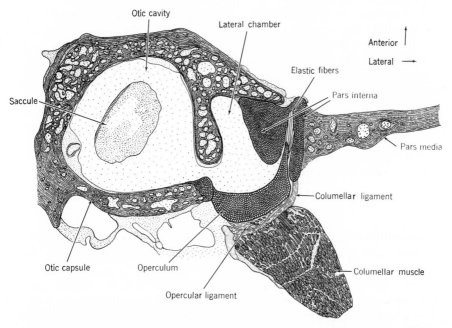

Fig. 3-35. A section through the ear region of *Rana catesbeiana*, cut in the frontal plane. Scale 10X.

the division appears complete. Previously this lateral chamber was hardly recognized as a significant part of the otic cavity, but has been referred to simply as the oval window. Gaupp (1896) gave this region the more distinctive name of "fossa fenestrae ovalis," but it is obviously much more than a simple "pit" for the oval window: it is a distinct antechamber with a special function of containing the sound-control mechanism.

The main cavity as seen here contains the saccule lying in the middle of the large perilymph space, and the lateral chamber includes two skeletal elements, the inner portion of the columella and the operculum (which indeed forms a large portion of the posterolateral wall of the chamber).

The osseous part of the columella, or pars media, extends inward from its attachment to the pars externa and then expands somewhat as it goes over into the inner cartilaginous portion, which is the pars interna. This inner portion is of complex form; it continues the expansion of the pars media and develops as a large mass of cartilage that here nearly half fills the lateral chamber but at a still more ventral level takes up almost the whole space. As shown here it presents a deep notch in its posterior surface into which is inserted an ascending wedge-shaped portion of the operculum.

The operculum is an oval disk of cartilage lying over and extending within the opening in the otic capsule. As seen in frontal sections as in Fig. 3-35 it appears as a flap hinged to the cartilaginous border of the medial wall of the otic capsule and is held here by a heavy ligamentous thickening along its inner surface. More ventrally the operculum becomes fused to this wall, yet it retains a considerable degree of mobility by virtue of the pliability of the cartilage.

In some species the cleft into which the anterior border of the operculum extends is not within the pars interna as shown here, but lies a little more laterally, between the pars interna and the bony pars media.

The free edge of the operculum does not extend all the way into the depths of the columellar notch, but a thick bundle of elastic fibers continues from this edge and serves to hold the operculum snugly in the notch.

THE MIDDLE EAR MUSCLES

The middle ear muscles further control the closeness of contact between the columellar footplate, operculum, and otic capsule, and thereby determine the ear's sensitivity to sounds. As already shown in Fig. 3-23, there are three of these muscles running close alongside at their origins on the suprascapula, and then they separate at their insertions deep in the middle ear region.

The Opercular Muscle. — One of the muscles of the middle ear, which has long been known, is the opercular muscle, arising at the anterodorsal corner of the suprascapula and running ventrolaterally and somewhat anteriorly to insert on the dorsal surface of the operculum. Its course is schematically shown in Fig. 3-36.

This figure is based upon a dissection of a specimen of *Rana pipiens* in which the suprascapula was cut away to reveal the courses of the muscles along its undersurface up to the outer rim, which remains. The attachments of the ends of these muscles on the cartilaginous plate are adjacent and actually in close contact at their upper ends, but for clarity in presentation these muscles were artificially separated. This was accomplished by inserting a fine needle between two of the bundles at a lower level where they were more clearly separated and then drawing the needle upward so as to part the fibers all the way to the suprascapula. This was easily done, and it was evident that the fiber bundles were merely peeled apart, with tearing only of the delicate interfascicular tissues, without breaking the muscle fibers themselves.

The opercular muscle will be recognized as a distinct bundle and in no sense is to be regarded as an offshoot of the levator scapulae superior as commonly described and as the name of "pars opercularis" formerly given to it connotes. Indeed, it bears no special relation to the levator scapulae superior; both origins and insertions of these two muscles are distinct, and throughout their courses they are well separated by the columellar muscle. The origin of the opercular muscle is on the anteromedial undersurface of the suprascapula, and its fibers run directly to its insertion on the operculum.

Furthermore, the fibers of the opercular muscle have a certain distinctive

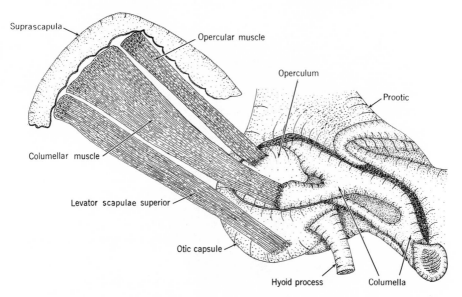

Fig. 3-36. The middle ear and its muscles in a specimen of *Rana pipiens*. Scale 12X.

appearance both in direct visual examination and as seen in serial sections. As examined under the dissection microscope they present a peculiar texture and a translucent quality that distinguishes them from the other two muscles. In sections it is immediately noted that these fibers are smaller in cross section than the others (see Fig. 3-37), and often they stain differently. Thus in most preparations the opercular muscle can be identified at a glance.

The Columellar Muscle. — Shown in Fig. 3-36 and already pictured in cross section in Fig. 3-35 is a second muscle of considerably larger size, arising from the suprascapula immediately adjacent to the opercular muscle and sending a broad, flat ligament anteriorly to an insertion on the footplate of the columella. This insertion broadly covers the medial rim of the posterior arm of the footplate, which includes in this species both cartilage (as Fig. 3-35 shows) and an extended portion of the osseous pars media (seen in Fig. 3-37).

Figure 3-37 shows the columellar muscle cut obliquely, with its ligament extending posteriorly to the ligament of the opercular muscle and then continuing anteriorly alongside the operculum to its insertion. This ligament follows closely the rounded contour of the operculum, and it seems likely that when it is straightened through the contraction of its muscle it tends to displace the operculum inward, adding to the immobilization of this element in its contact with the footplate. This action is suggested by the large arrow in Fig. 3-37.

The Levator Muscle. — The third muscle in this complex will continue

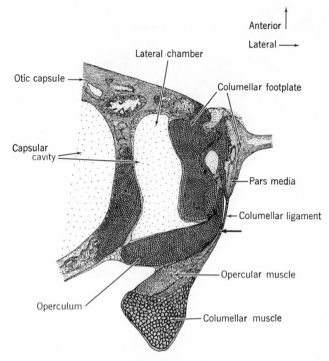

Fig. 3-37. A section through the middle ear region of a specimen of *Rana pipiens* showing opercular and columeller muscles, and the operculum in relation to the columeller footplate. Scale 20X.

to be referred to as the levator scapulae superior, though there is considerable doubt that it has the function that this traditional name implies; its action appears to be in some degree supplementary to that of the columellar muscle. This bundle arises on the undersurface of the suprascapula in a region posterior to the columellar muscle (shown in Fig. 3-36) and runs alongside that muscle, partially covering it on the medial side. It inserts mainly on the lateral surface of the otic capsule, but with many of its fibers attaching to a small ligament that follows closely the course of the columellar ligament and inserts on the same site on the columella. This course is not represented in Fig. 3-36, but appears in Fig. 3-38.

THE LOCK MECHANISM

The muscles of the middle ear just described constitute a mechanism for the control of sound transmission in the frog's ear that protects against the damaging effects of excessive sounds, no doubt primarily the animal's own vocalizations.

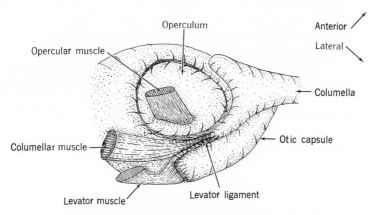

Fig. 3-38. Insertion of the levator muscle in *Rana catesbeiana.*

The opercular and columellar muscles form an antagonistic pair, with the levator perhaps adding some force to the columellar contraction. When the opercular muscle is relaxed and the columellar muscle contracts, it pulls the footplate posteriorly, driving the end of the opercular wedge deeply into the notch in the footplate, which it fits closely, as pictured in Fig. 3-35. This locks the pars interna and operculum together, so that these elements as a whole are rigidly held between anterior and posterior walls of the otic capsule. A back-and-forth motion of the columella as produced by a sound acting on the tympanic membrane then encounters a large resistive force. Such a movement tends to displace the otic capsule as a whole and thereby the entire head, and the relative motion between columella and capsule, which is essential for producing the fluid pulsations that would stimulate the auditory papillae, is greatly diminished. As already suggested, the levator probably contributes to the immobilization, though its effect is probably minor compared with that of the columellar muscle.

A contrary operation of these muscles reverses the action: a relaxation of the columellar and levator muscles and a contraction of the opercular muscle pulls the wedge end of the operculum out of the notch and leaves the columella free to produce its vibratory responses to sounds.

Here is an acoustic regulatory mechanism that has no counterpart elsewhere among animals. Many vertebrate ears contain devices that provide some measure of protection against excessive sounds, but none approaches this one in degree of effectiveness. The mammals in general possess middle ear muscles, the malleus and stapedius muscles, which contract reflexly in the presence of loud sounds and reduce vibratory transmission to the cochlea. This reduction is produced by the application of the muscle force in a direction essentially perpendicular to the normal motion of the ossicular chain, thereby adding friction and stiffness, but the effect is only moderate, of the order of about 30 db at most, which is an amplitude reduction of about 30-fold. This effect is slight in comparison with the 60 db effect readily observed in the frog's ear, which corresponds to a 1000-fold reduction in amplitude. Some other animals, such as bats, are able to supplement their middle ear muscle action with a closure of the external auditory meatus, and thereby attain a more significant reduction, but even the combination of these muscular and occluding measures falls far short of what the frog's ear is able to do in its own protection.

An examination of a series of specimens of a given species revealed some minor variations in the courses and distribution of the fibers of these muscles. Sometimes the opercular muscle was in two parts, with separate but closely adjacent attachments on the suprascapula. In one specimen the levator muscle also was clearly divided into two bundles, as shown in Fig. 3-39.

Some Experimental Tests. — A number of experiments were carried out in an attempt to study further the functional characteristics of the middle ear muscles. With careful dissection a portion of the upper ends of these three

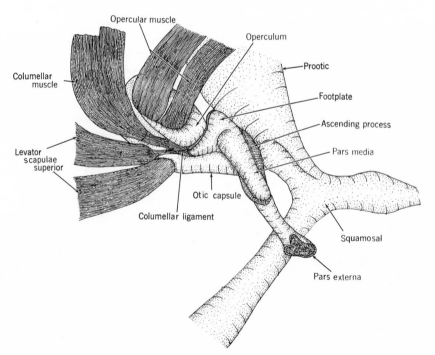

Fig. 3-39. The middle ear muscles in a specimen of *Rana utricularia sphenocephala* in which these muscles show a degree of duplexity. Scale 10X.

muscles was exposed as indicated in Fig. 3-23. Usually three distinct portions of the total mass could be made out; the upper bundle is the opercular muscle, the lower one is the levator, and the columellar portion lies between, partially covered by the opercular. Careful manipulation showed the boundaries of the muscles and the lines of separation between them.

While the ear was exposed to a continuous tone, the effects of applying tension to these muscles was investigated in terms of the electrical responses of the ear. It was observed systematically that tension applied to the columellar bundle reduced the responses and tension applied to the opercular bundle increased them. Thus the control function of this mechanism was fully substantiated.

Other experiments in which the muscle group as a whole was put out of action by the application of Xylocaine, or different portions were sectioned, were less successful. The use of a local anesthetic probably leaves the mechanism in whatever locked or unlocked position it happens to have at the moment the drug takes effect, and thus the results are variable. The sectioning experiments in preliminary trials gave uncertain results also, and these experiments were interrupted before they could be carried to completion; they ought to be resumed at some future time. A systematic use of sensors in the individual muscles should also be attempted to discover the relations between performance of these muscles and other bodily activity: it seems likely that contraction of the columellar muscle to protect the ear will be found to precede the animal's own vocalizations.

The Anuran Inner Ear

A comprehensive picture of the inner ear and the relations to neighboring structures is given by a series of 20 sections for the species *Rana utricularia sphenocephala* in which the sectioning was transverse, from anterior to posterior ends.

1. The first figure in this series (Fig. 3-40) presents an entire section passing through the head at the level of the anterior semicircular canal and its ampulla; all remaining sections will be limited to the region of the right labyrinth.

As will be noted, the skin and subcutaneous tissues had been removed over the dorsal surface prior to the sectioning, and the temporalis muscle had been trimmed away on the right side. This was done because the animal was used initially for the recording of inner ear potentials in response to sounds, and in this procedure the bone over the anterior semicircular canal region was exposed and a minute hole drilled through the capsule for the insertion of a recording electrode. This procedure produced a noticeable distortion of the ampullar wall and an infiltration of blood and tissue fragments into the ampullar cavity as the picture shows.

The labyrinth is enclosed above by the prootic bone with its parotic car-

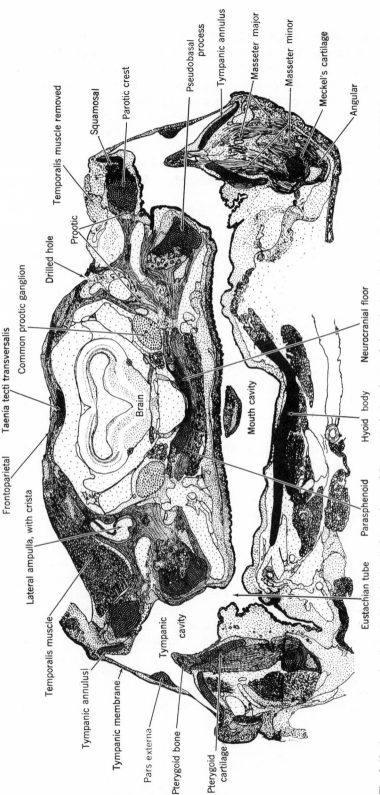

Frontparietal Taenia tecti transversalis Common proötic ganglion Drilled hole Temporalis muscle removed Squamosal Parotic crest Pseudobasal process Tympanic annulus Masseter major Masseter minor Meckel's cartilage Angular

Lateral ampulla, with crista

Proötic

Brain

Temporalis muscle

Tympanic annulus|

Tympanic membrane

Pars externa

Pterygoid bone

Pterygoid cartilage

Tympanic cavity

Mouth cavity

Neurocranial floor

Hyoid body

Parasphenoid

Eustachian tube

Fig. 3-40. A transverse section through the entire head of a specimen of *Rana utricularia sphenocephala* at the level of the tympanic membranes. Scale 7.5X.

tilaginous extension and below by the pseudobasal process of the quadrate; a deep cleft between these two elements serves for the passage of nerves and blood vessels. The tympanic membrane appears in cross section as a thinned-out extension of the surrounding skin and is supported principally by the outer flange of the tympanic annulus, here cut through at two places. Near the center of the tympanic membrane is the end of the pars externa of the columella.

At the level shown the middle ear space, filled with air, passes completely through the head and includes middle ear cavities, pharyngeal spaces, and the cavity of the mouth. There are diverticula also into the tissues of the lower jaw medial to the masseter muscles. Below the mouth cavity is the hyoid body, a plate of cartilage supporting the muscles of the tongue.

2. The next figure (Fig. 3-41) shows only the labyrinthine region on the right side and includes the anterior and lateral semicircular canals, the utricular recess with its macula, and below a small portion of the saccule appearing in the large perilymphatic cistern. The utricular macula is an elongated mound covered with hair cells, over whose ciliary tufts lies a mass of otolithic crystals (otoconia). The crista of the lateral canal is seen extending transversely across the lumen of its ampulla.

3. The section in the previous figure only grazed the anterior wall of the saccule; the next figure (Fig. 3-42) approaches the main body of this endorgan with its broad plate of hair cells surmounted by a large otolithic mass. A slight constriction separates the utricular cavity into a medial portion, the utricle proper, and its recess which contains the macular organ. Within the lateral chamber may be seen the inner end of the pars interna of the columella, a cartilaginous plate that has here a firm attachment to the roof of the chamber.

4. In the next figure (Fig. 3-43) the saccule continues, and its macula is seen to receive a rich nerve supply from the anterior branch of the eighth nerve located above and to the left. Above the utricular recess (in which the utricular macula is still to be seen) is a narrow slit, which represents the most anterior portion of the superior division of the saccule. The pars interna of the columella is more prominent here and shows its attachment to the bony shaft, which is the pars media.

5. The utricular recess as seen in Fig. 3-44 has almost come to an end, but the utricle itself continues. The superior part of the saccule has expanded somewhat and its connection to the main part is now evident.

6. The next section, shown in Fig. 3-45, lies beyond the main shaft of the columella, but the footplate remains, consisting of both bony and cartilaginous portions. Its dorsomedial end is here attached to the prootic bone only by loose connective tissue. The lateral chamber is large, but is still separated from the main cavity of the otic capsule by a thin web of bone.

At this level the inferior saccule has reached about its maximum dimen-

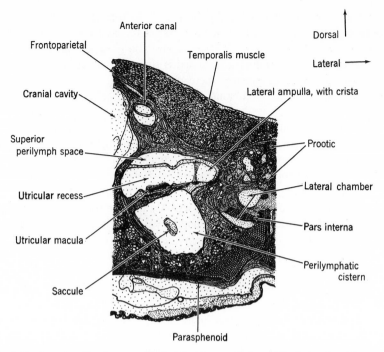

Fig. 3-41. A section like the preceding but farther posteriorly and limited to the right ear region. Scale 10X.

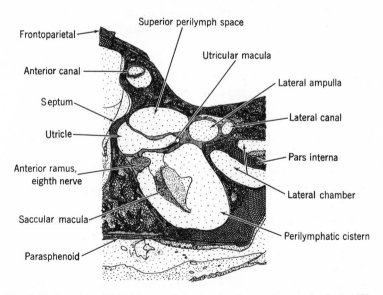

Fig. 3-42. A section still farther posteriorly showing the saccular macula. Scale 10X.

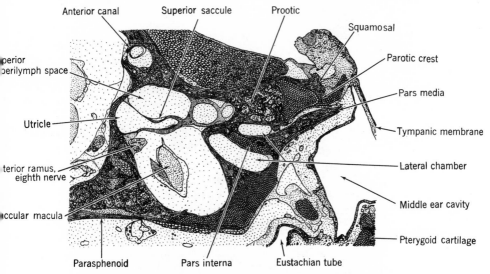

Fig. 3-43. A section still farther posteriorly showing the saccular macula. Scale 10X.

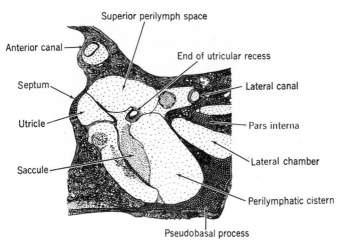

Fig. 3-44. A section near the end of the utricular recess. Scale 10X.

sions and is nearly filled by its large otolithic mass. Its superior portion makes connection with the utricle through the utriculosaccular duct, which is partly occluded by a fold of the membranous wall between the utricle and the superior perilymph space, a fold sometimes referred to as the utriculosaccular valve (Bast, 1928). This "valve" was first described in human ears and has

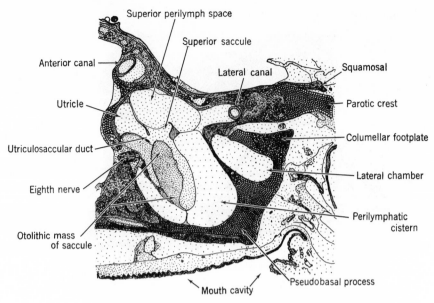

Fig. 3-45. A section farther posteriorly, through the middle of the saccular otolith. Scale 10X.

been studied chiefly in mammals. Bast, its discoverer, believed that it served to maintain fluid pressure in the utricle after damage to the saccule, but the form of the structure does not support this hypothesis. There are no controlling muscles, and the flap of tissue must respond passively to pressure changes in the surrounding fluid cavities; from the configuration as shown it appears that excess pressure in the utricle should hold the utriculosaccular passage open, whereas excess pressure in the saccule should tend to close it. This action is just the reverse of that proposed by Bast (1934). Further investigation is needed to determine whether this structure has any special function.

The columellar footplate as seen at this level consists of bone in its lateral portion and cartilage at its medial end. The bony portion is fused to the large mass of cartilage in the pseudobasal process below, whereas the medial end is only loosely attached by connective tissue to the bone of the otic capsule.

7. In the next figure, Fig. 3-46, a fold appears in the membrane along the floor of the utricle just below the utriculosaccular foramen, which in further sections will be seen to form the mouth of the endolymphatic duct. This duct runs dorsomedially to the otocranial septum, where it is seen in Fig. 3-47 as about to enter an opening in the septal wall; it then passes through into the cranial cavity as seen in Fig. 3-48. Figure 3-47 shows a plate of limbic tissue with an opening in the middle along the otocranial septum; far-

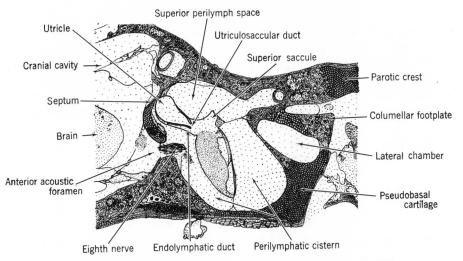

Fig. 3-46. A view in which the anterior acoustic foramen appears. Scale 10X.

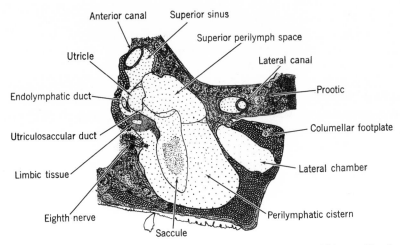

Fig. 3-47. A level showing limbic tissue from which the shelf for the amphibian papilla arises. Scale 10X.

ther posteriorly it will be seen that this plate forms a shelf for the suspension of the amphibian papilla, and the opening just noted is the entrance into its recess (Fig. 3-48). The limbic shelf for the amphibian papilla is evident in Fig. 3-48, but the hair cells belonging to this receptor organ begin a little farther posteriorly.

8. In Fig. 3-49 the anterior semicircular canal runs into the superior sinus. At this level there is still an enchondrotic connection between the columellar footplate and the pseudobasal cartilage, but the wall between the lateral chamber and the perilymphatic cistern has become exceedingly thin.

9. At the level shown in Fig. 3-50 the columellar footplate loses its direct connection with the ventral border of the window (the dorsolateral extension of the pseudobasal process of the quadrate), and a small piece of cartilage lies between; this intervening cartilage is the operculum and becomes increasingly prominent in subsequent sections. Here also is the most anterior appearance of hair cells in the amphibian papilla, which more posteriorly form an extended layer on the underside of the limbic shelf. The anterior wall of the lagena comes into view in this section also.

10. In Fig. 3-51 the amphibian papilla is well developed, and the lagenar recess shows its macula facing dorsolaterally. There is a wide opening between the lateral chamber and the main cavity of the otic capsule, and the operculum begins to intrude between the columellar footplate and the extended end of the pseudobasal cartilage.

11. A little farther posteriorly, in Fig. 3-52, the amphibian papilla exhibits a clear division into two parts, with well-separated patches of hair cells and tectorial strands from each that are united in a single mass below.

At this level a marked constriction separates the superior portion of the saccule from an inferior portion that is comparatively small. The lagena opens near the junction of these two divisions.

12. Figure 3-53 shows the posterior semicircular canal beginning to come out of the superior sinus; in subsequent sections this canal will take the position that anteriorly was occupied by the anterior canal. The lateral canal remains in its former position on the right side.

The amphibian papilla continues to show two patches of hair cells, and from the lateral one the tectorial tissue runs into a sensing membrane that is anchored to the floor of the papillar recess.

The inferior saccule at this level is much reduced and appears to be independent (though of course confluent with the superior division anteriorly). The lagena is still present, here connected with the superior division of the saccule.

In Fig. 3-54 the posterior semicircular canal has separated from the superior sinus, and a posterior sinus now appears below it. The inferior division of the saccule has ended, but the superior division remains, to which both amphibian and lagenar recesses are connected.

13. At the level shown in Fig. 3-55 the posterior canal continues, as does the lateral canal, and a return path for the latter, known as its "canalis," is present also. The sensing membrane of the amphibian papilla has ended, and this papilla continues in its extended posterior portion. The superior saccule is much reduced, and the posterior end of the lagena has been reached. The

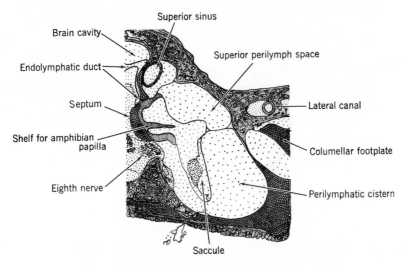

Fig. 3-48. The limbic shelf in more complete form. Scale 10X.

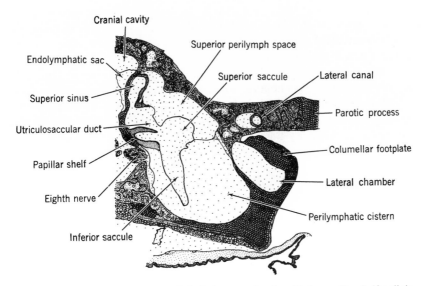

Fig. 3-49. *Rana utricularia sphenocephala* farther posteriorly with the papillar shelf well defined. Scale 10X.

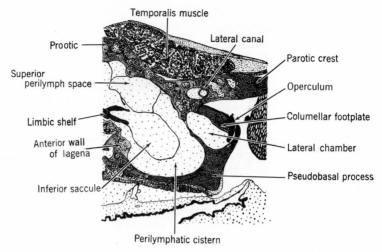

Fig. 3-50. The operculum appears. Scale 10X.

lateral chamber is now fully opened to the main cavity of the otic capsule and is covered laterally by the operculum, to whose lateral surface are attached a few fibers of the opercular muscle.

A little farther posteriorly, as shown in Fig. 3-56, the posterior portion of the amphibian papilla continues, with the sensory structure much reduced. It is important to note that anteriorly to this point, from the beginning of the posterior division of this papilla to the present level, the papillar cavity has been fully enclosed by limbic tissue (as seen in Fig. 3-55), but here and beyond, almost to the posterior end of this papilla, the lateral wall of this cavity is formed by an exceedingly thin membrane that is bounded laterally by a perilymph space of considerable size. This inferior perilymph space leads finally to the round window and serves as the escape path for vibratory pressures.

A recess of the superior saccule belongs to the basilar papilla, which will appear farther on. The operculum has greatly thickened, and its muscle is evident.

Figure 3-57 represents the posterior end of the amphibian papilla, in which a good many hair cells may be seen, but no longer in an orderly row. Soon hereafter comes the blind termination of this papilla.

In this section the basilar papilla lies in the middle of a heavy plate of limbic tissue, with a thin membrane at its ventromedial end that separates it from a relatively large perilymphatic cavity. Dorsolateral to this papilla is the remaining end of the superior saccule. A small nerve bundle (n) runs in the dorsal edge of the limbic plate between basilar papilla and superior sac-

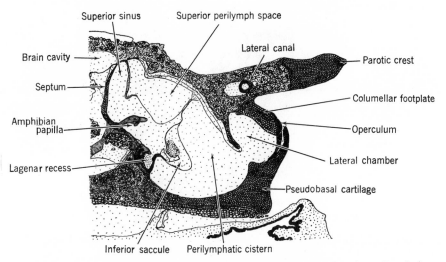

Fig. 3-51. The same specimen with the lateral chamber opening freely into the perilymphatic cistern. Scale 10X.

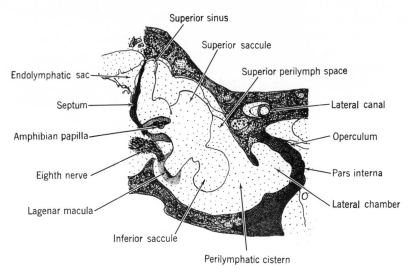

Fig. 3-52. The amphibian papilla appears. Scale 10X.

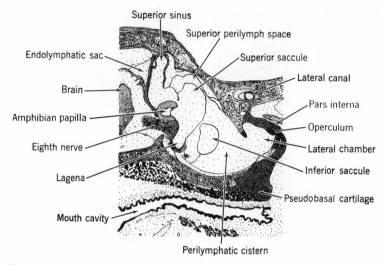

Fig. 3-53. The amphibian papilla is duplex and shows a sensing membrane. Scale 10X.

cule; this bundle extends laterally to supply the crista of the lateral semicircular canal.

The early anatomists, who sought to identify the auditory endorgans by following to the periphery the ramifications of the eighth nerve, mistakenly regarded this nerve bundle as terminating in the area seen here and specifically considered it to serve the limbic plate enclosing the papillar cavity to which had been given the name of "tegmentum vasculosum." Accordingly they regarded this structure as an auditory receptor, unaware of the fact, as serial sections like the present ones ultimately disclosed, that this "tegmentum" is completely devoid of sensory structures.

The thin separating membrane between the basilar cavity and its recess was named "the basilar membrane" and considered to be homologous with this membrane in higher vertebrates, an erroneous conception that had a long period of unquestioning acceptance, and traces of which remain even today.

The basilar cavity and its recess continue in Fig. 3-58 and then a little farther posteriorly the cavity ends blindly. The recess then continues alone and becomes covered on its ventromedial side by a relatively thick membrane, as Fig. 3-59 shows. This membrane is an extension of the arachnoid, and below it, covering a relatively large opening in the cranial capsule, is the round window membrane. Here is the principal place of discharge of vibratory pressures exerted upon the labyrinthine fluids by sounds transmitted to the columellar footplate; the auditory hair cells contained in amphibian and basilar papillae lie in the path between and are stimulated by the fluid surgings.

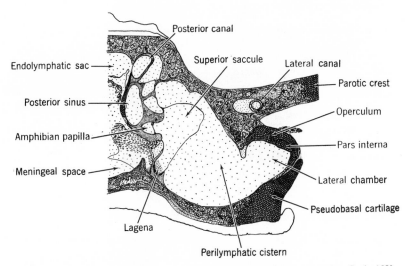

Fig. 3-54. The amphibian papilla is reduced; a sensing membrane remains. Scale 10X.

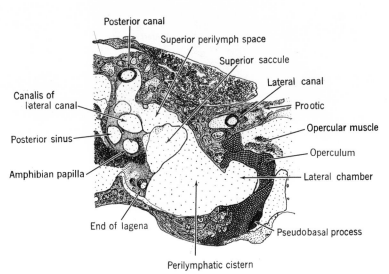

Fig. 3-55. The posterior division of the amphibian papilla begins. Scale 10X.

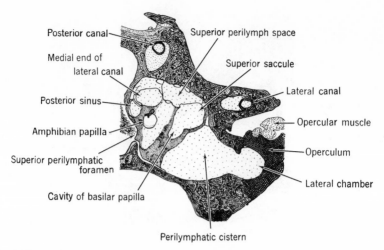

Fig. 3-56. *Rana utricularia sphenocephala*, showing the most posterior portion of the posterior division of the amphibian papilla. Scale 10X.

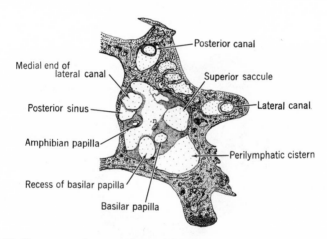

Fig. 3-57. The same specimen near the posterior end of the amphibian papilla. Scale 10X.

The Amphibian Papilla in the Anuran Ear. The series of transverse sections through the labyrinth of the frog *Rana utricularia sphenocephala* as just displayed has revealed the changing form of the amphibian papilla as it is viewed successively along its anterior-posterior axis. These views are on a small scale, however, and the papillar structure needs to be examined in further detail, as provided by the series now to be presented.

As already seen, this papilla is located in a recess of the saccular cavity

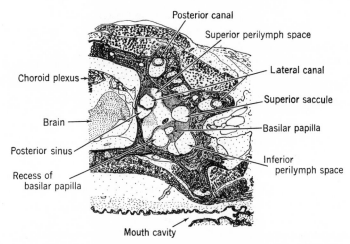

Fig. 3-58. The same specimen where the amphibian papilla has ended and the basilar papilla continues. Scale 10X.

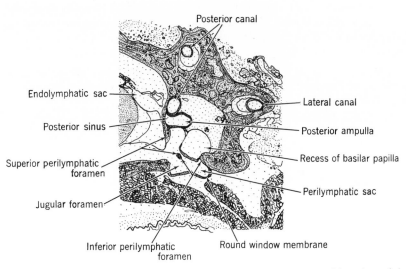

Fig. 3-59. The same specimen; the round window forms the posteroventral boundary of the perilymphatic sac. Scale 10X.

formed by a shelf of limbic tissue attached to the septum separating cranial and otic cavities. The location of this limbic shelf on the lateral wall of the otocranial septum is indicated in the transverse section of Fig. 3-60, and its position with respect to the other ear structures is shown in more pictorial fashion in Fig. 3-61. Here the otic capsule has been opened to give a direct dorsal view into its cavity, with the lagena at the deepest point and the amphibian and basilar papillae on either side.

The mass of limbic tissue enclosing the amphibian papilla has its opening into the saccule facing lateroventrally, and this opening is partially obstructed by a pillar transmitting the terminal end of the papillar nerve.

The basilar papilla is situated a little below and likewise is enclosed in limbic tissue. This tissue forms a tube running obliquely into the perilymphatic recess and contains the sensory receptive apparatus at its upper end, as Fig. 3-17 has shown.

The action of sounds on the ear may be followed in Fig. 3-17 just referred to. Aerial waves striking the tympanic membrane produce oscillatory motions that are transmitted through the columella to the fluid of the lateral chamber and thereafter follow two paths through the otic capsule as indicated by the arrows. The positive pressure exerted on the inner ear fluid by an inward thrust of the columella is able to produce a displacement of a quantity of fluid because of the presence of the round window as a yielding area.

This window is a round opening (about 1 mm in diameter in *Rana utricularia*) in the occipital bone located at the back of the mouth cavity just posterior to the cross arm of the parasphenoid bone and is covered by a thin membrane. Further blanketing of this area is produced by the flat end portion of the levator scapulae inferior muscle, and over this lies the lining of the roof of the mouth. This tissue covering is sufficiently flexible to permit the displacements required in the action of sounds.

Operation of the Amphibian Papilla. — In the advanced frogs the amphibian papilla consists of three portions, with two different modes of operation. As shown in Fig. 3-17 this structure occupies a deep notch in a mass of limbic tissue lying on the septum separating otic and cranial cavities, with its opening facing ventrolaterally. Within the notch, as shown in detail in Fig. 3-62, a large rounded mass of hair cells borne by a compact tectorial body occupies the anterior portion of the cavity, and extending posteriorly from this is a long arm bearing several rows of hair cells. Here we distinguish a middle portion and a posterior end portion that is largely out of view in this picture. Then lateral to these two portions is a thin web of tissue, the sensing membrane, that stretches from a pillar in a dorsolateral direction to attach to a small group of hair cells making up one edge of the main mass. This pillar consists of limbic tissue and transmits the papillar nerve, as the figure indicates.

Stimulation of these three portions of the sensory structure involves currents of vibratory flow as suggested by the curved arrows of Fig. 3-17. These currents enter the lateral opening of the limbic notch, billow through the papillar structure, and exit along two paths, one indicated by a short arrow

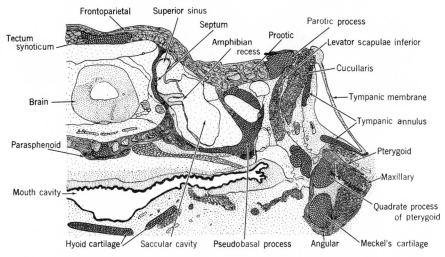

Fig. 3-60. The ear region of *Rana utricularia sphenocephala* in a transverse section, right side. Scale 7.5X.

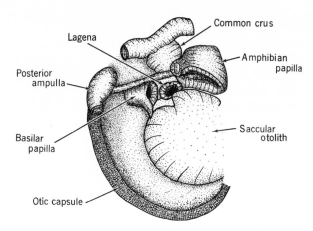

Fig. 3-61. A dorsal view into the otic capsule of *Rana pipiens* showing the two auditory papillae in relation to other structures.

in Fig. 3-17 going into the perilymphatic duct that passes by way of the perilymphatic recess to the round window and another path that is not shown, but runs along the space that contains the posteromedial arm of this papilla and also finds its outlet through the perilymphatic recess.

Within the papillar cavity the sound waves act upon the main mass of the hair cells directly through the tectorial filaments that connect to their ciliary

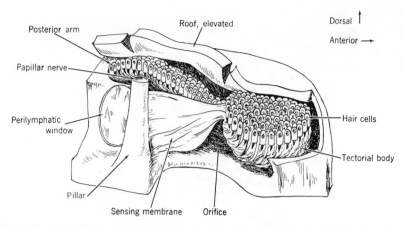

Fig. 3-62. A pictorial view of the amphibian papilla, opened laterally. (Drawing by Anne Cox; from *J. Morphol.*, 1973, 141, p. 467)

tufts and also reach a limited number of these hair cells through the sensing membrane. This latter mode of stimulation clearly adds greatly to the sensitivity for these particular hair cells, because for these the vibratory motion is summated over the membrane surface. The hair cells thus favored in the stimulation process make up about 10 percent of the total in this principal mass.

The hair cells contained in the posteromedial arm, and especially the ones toward its farther end, are much less favorably situated to receive this stimulation: the movements must pass along a comparatively narrow tube, encountering considerable resistance in the path to the perilymphatic sac and round window. Thus there is an extensive dynamic range in the operation of this receptor: those hair cells that are contacted by the sensing membrane serve for reception of the faintest sounds, and those distant along the posteromedial arm of the papilla will be responsive only to the more intense ones.

The Mass Movement Mode. — The action of the frog's amphibian papilla will now be examined further in its anterior portion. The hair cells are attached to the roof of the papillar cavity, with their cell bodies clasped between the extended processes of supporting cells, shown in Fig. 3-62 but now represented schematically in Fig. 3-63. Below this hair-cell layer is a mass of tectorial tissue in the form of a meshwork closely resembling a honeycomb, made up of a number of closely packed canals or tubules. Each hair cell has a canal fitted over its end so that the ciliary tuft extends into the opening. In many instances it has been observed that the tip of the ciliary tuft is attached to the inner wall of the canal near its mouth, and this is prob-

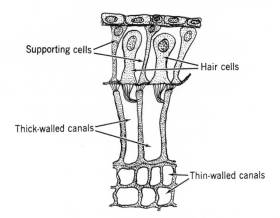

Fig. 3-63. Hair cells of the amphibian papilla in relation to the tectorial canals.

ably the general rule. The canals that cover the hair cells in this way usually have relatively thick walls.

Then there are numerous other canals and vesicles of various sizes, mostly out of direct contact with the hair cells, that have thin walls. These constitute an extended mass that hangs down in the cavity so as to form a partial barrier to the vibratory fluid stream.

A somewhat different structural arrangement has been seen occasionally in specimens in which the plane of sectioning was oblique and the staining was particularly effective. Here is seen a perforated membrane lying close beneath the middle rows of hair cells and leading to thin-walled canals and vesicles below. It has not been possible to ascertain whether the perforated membrane is an independent structure or simply the thickened upper ends of the thin-walled canals that make up the main tectorial body. The sketch in Fig. 3-64 represents this membrane as independent, but this feature needs further study.

Figure 3-65 shows the amphibian papilla of *Rana pipiens* in the middle region at a point of transition from unitary to duplex forms and indicates the two types of canals: the thick-walled canals on the left side and the thin-walled canals elsewhere. Also shown here is the perforated membrane suspended below the middle series of hair cells. The separation between hair cells and membrane is an artifact; normally these are in contact and the ciliary tufts extend fully into the membrane openings.

The thin-walled canals and vesicles that make up the main mass of the tectorial body, and always constitute its lowermost part, clearly have the function of taking up the vibratory motions of the surrounding fluid as produced by the action of sounds. The physical characteristics of this thin, fluid-filled meshwork seem well suited to this purpose: the density and elasticity

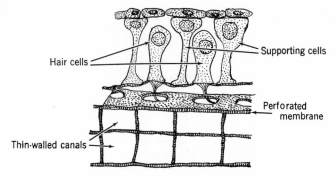

Fig. 3-64. Cellular elements of the amphibian papilla in relation to the perforated membrane.

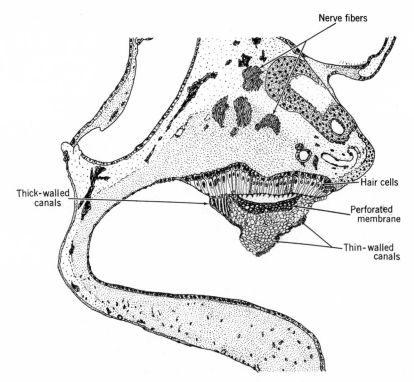

Fig. 3-65. The amphibian papilla of *Rana pipiens* in a region of transition to a duplex form. Scale 100X.

of this material should closely resemble these properties in the fluid itself, and thus there is an impedance match that facilitates the transfer of vibratory energy from fluid to tectorial body.

This tenuous material then leads usually to other canals, most often located toward the medial edge of the structure, in which the walls are thicker, often by as much as tenfold. This arrangement can be understood as having the role of a mechanical transformer: the motions taken up from the fluid by the peripheral thin-walled material is transferred to the thick-walled canals that in turn energize the ciliary tufts of the hair cells. A thick-walled canal because of its added rigidity will be more effective in transferring energy to the ciliary tuft than a thin-walled canal would be, especially if this tuft is relatively stiff and resistive to bending. Thus there appears to be an advantage in the development of thick-walled canals in direct contact with a large proportion of the hair cells.

There may also be a virtue in a degree of variation in the thickness of the walls of different canals as a way of extending the dynamic range of the receptor. If the ciliary tuft of a hair cell has an appreciable resistance that increases with the degree of deflection (as well may be the case), then a hair cell served by a thin-walled canal may be highly sensitive but subject to early overloading; another element acted upon by a thick-walled canal may not be affected by a faint stimulus but will respond to a strong one. The range of effectiveness of the receptor is thereby extended.

The posterior portion of the amphibian papilla appears beyond the sensing membrane and consists of a long, narrow mound bearing a limited number of rows of hair cells, covered by a correspondingly small tectorial body. There is then a moderate expansion of the width of the mound toward the posterior end. This posterior portion of the papilla will be examined further as the form of the structure is traced over its whole course, but it appears that the manner of functioning is essentially the same as in the anterior region, except that the level of stimulation is probably lower. It seems likely that this level declines progressively along the path through the papilla due to friction and leakage. If so, then this posterior region will serve for the reception of relatively loud sounds, and thus extends the dynamic range of the ear.

The Sensing Membrane Mode. — As already noted, in the region between anterior and posterior portions of the amphibian papilla a thin membrane lies across the path of vibratory fluid flow. This membrane, a thin net of tectorial material, is suspended between a wall of tissue that serves as an enclosure for the papillar nerve and the posterior edge of the mass constituting the anterior tectorial body (see Fig. 3-62). This membrane hangs like a curtain across the fluid path and moves with the vibratory current. The hair cells to which the lateral edge of the membrane makes connection thus receive the summated motional effects over the membrane surface in addition to what the mass in general is exposed to, and these cells are probably the ones concerned in responses to the faintest sounds. In most anuran ears these favored cells constitute something like a tenth of the total number in this papilla, but there are wide individual and species variations.

Successive Views of the Amphibian Papilla. — The varying forms of the amphibian papilla will now be followed in a specimen of *Rana utricularia sphenocephala* in a series of transverse sections from anterior to posterior ends.

At its first appearance anteriorly this papilla contains only a few small hair cells; at the beginning only three or four cells are seen, and then these increase rapidly in numbers and also somewhat in size until at the level shown in Fig. 3-66 there are about 17 rows. The cells have their ciliated ends entering the openings of a tectorial body made up entirely of parallel thin-walled canals. At the point represented here 19 canals could be counted, of which all but a few at the lateral end run continuously through the structure.

A little farther posteriorly—in this series only 60 μm from the anterior end—the picture changes with the appearance of thick-walled canals at the medial edge, and the number of these increases progressively until at the 100 μm point, as seen in Fig. 3-67, there are 13 of them. The parallel character of the canals seen in the previous figure is retained in these thick-walled canals, but is largely lost over the remainder of the tectorial body.

Farther posteriorly, as in Fig. 3-68, this progression continues, until the uppermost canals, those entered by the ciliary tufts of the hair cells, all have thick walls. Among these the longer canals are on the medial side, and the shorter and less regularly arranged are in the middle of the array. The lower middle portion of the tectorial body is made up of very small round randomly scattered vesicles.

Then almost abruptly as we continue posteriorly through the structure a bimodal character appears: the hair cells in the middle of the papilla become shorter, lose their connections with the canals, and then disappear, while the medial and lateral hair cells remain as separate groups. This condition is shown in Fig. 3-69, which represents a section 200 μm from the anterior border. Here even the supporting cells have vanished in the middle of the papilla and are replaced with simple cubic cells.

In the next section, 20 μm farther along (Fig. 3-70), a further change is seen; the two patches of hair cells are now well separated, with only squamous epithelium between, and the medial patch has lost its connection with the tectorial body. The connection for the lateral patch remains, formed by long parallel thick-walled canals that enter a complex mass below that is now joined by a sensing membrane whose ventral foot is anchored in the floor of the limbic recess. (The terminal canals normally fit closely over the ends of the hair cells, but have been pulled away as an artifact of preparation.)

A little farther posteriorly the pattern continues, but with a rapid reduction in the medial group of hair cells, and then at a point 300 μm from the an-

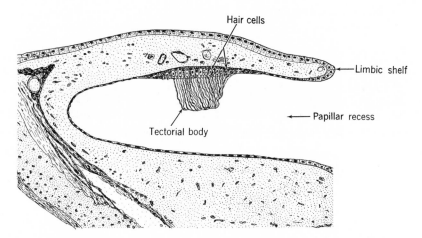

Fig. 3-66. The amphibian papilla of *Rana utricularia sphenocephala* a little beyond its anterior end. Scale 100X.

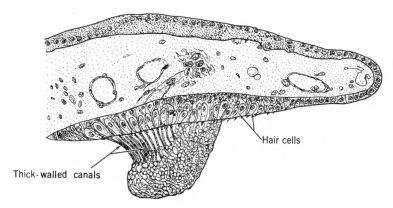

Fig. 3-67. The amphibian papilla of *Rana utricularia sphenocephala* somewhat farther posteriorly than the preceding section. Scale 200X.

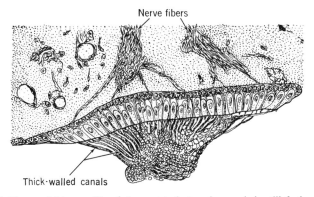

Fig. 3-68. The amphibian papilla of *Rana utricularia sphenocephala* still farther posteriorly. Scale 200X.

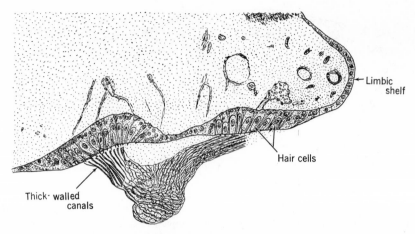

Fig. 3-69. The amphibian papilla of *Rana utricularia sphenocephala* farther posteriorly where it has divided into lateral and medial portions. Scale 200X.

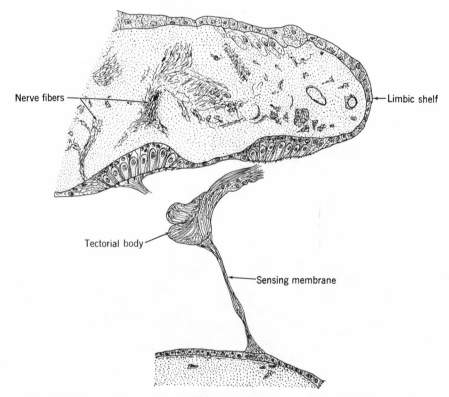

Fig. 3-70. The amphibian papilla of *Rana utricularia sphenocephala* still farther posteriorly where a sensing membrane is present. Scale 200X.

terior end this medial part of the papilla disappears, as seen in Fig. 3-71. At this level the rows of hair cells are few and the tectorial tissue is much reduced, consisting only of the sensing membrane and a small canaliculated portion that probably in the normal state received the ciliary tufts of the hair cells into their canals, but here is drawn away leaving a few connections by fine fibers.

This condition continues farther posteriorly, except that there is a moderate increase in the number of rows of hair cells and the tectorial body shows a corresponding enlargement, as Fig. 3-72 indicates. Here the number of rows of hair cells has reached about 8 at a point 460 μm from the anterior end. Then the sensing membrane terminates.

Measurements on this specimen indicated that the sensing membrane is about as wide as it is high, approaching 200 μm for each of these dimensions; the area was determined as 0.038 sq mm.

A little posterior to the level at which the sensing membrane ends the amphibian recess goes over into a cavity: the lateral edge of the roof portion of the limbic shelf extends downward and meets an elevated edge of the floor so that the space beneath the papilla becomes closed off from the saccular cavity. A section just anterior to this closure of the papillar cavity is shown in Fig. 3-73. Here three patches of nerve fibers are seen as the fibers of the papillar branch of the eighth nerve pass outward from the medulla, run beneath the papillar cavity, and then swing around it to enter the roof layer, from which they disperse downward to the hair cells. A reconstruction including several sections from this region indicates the course of the papillar nerve as shown in Fig. 3-74. More posteriorly a portion of the eighth nerve bundle indicated at p continues to the ampulla of the posterior semicircular canal (which is out of the picture to the upper right), and the part at b goes to the basilar papilla (which appears laterally and above the level shown here).

The amphibian papilla in this region is small, containing only about 7 rows of hair cells, which are served by a much reduced tectorial body as seen in Fig. 3-75. It is of interest, however, that this body maintains the form seen earlier, with thick-walled canals on its medial side and thin-walled canals elsewhere.

This form of the papilla and its tectorial body continues for a considerable distance posteriorly with no essential change, and then rather suddenly, at a point about 780 μm from the anterior end of this organ, there is a noticeable increase in hair-cell numbers. Also at this level, after the papillar branch of the eighth nerve has completed its passage around the papillar recess, the limbic tissue in which the nerve runs becomes reduced and finally disappears. Thereafter the papillar cavity on its lateral side has only a thin membrane separating it from a perilymph space, as seen in Fig. 3-76. This perilymph space leads posteriorly and ventrally to the round window and constitutes the outward path of discharge of vibratory pressures applied at the columellar footplate. Toward its very end the papilla takes a slight bend,

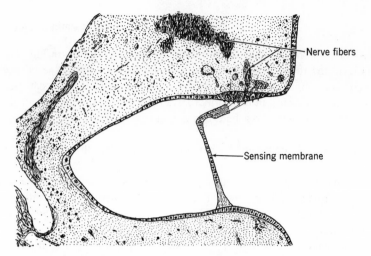

Fig. 3-71. The amphibian papilla of *Rana utricularia sphenocephala* at a level where the lateral portion only of the sensory structure is present, along with the sensing membrane. Scale 100X.

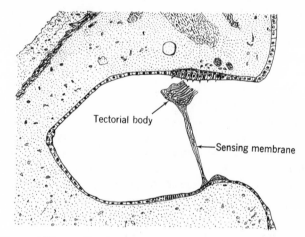

Fig. 3-72. The amphibian papilla of *Rana utricularia sphenocephala* farther posteriorly where the tectorial body is enlarged and the sensing membrane remains. Scale 100X.

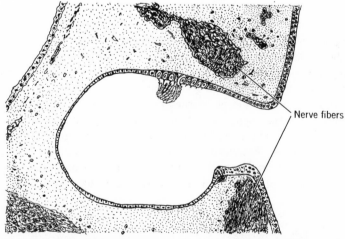

Fig. 3-73. The amphibian papilla of *Rana utricularia sphenocephala* far posteriorly where the sensing membrane has ended. Scale 100X.

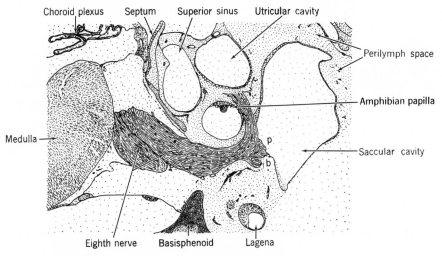

Fig. 3-74. The amphibian papilla of *Rana utricularia sphenocephala* near its posterior end showing the course of the papillar nerve. Scale 25X.

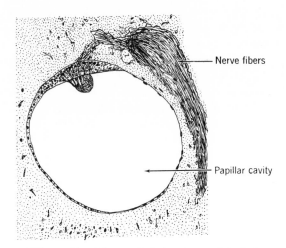

Fig. 3-75. The amphibian papilla of *Rana utricularia sphenocephala* farther posteriorly where both the rows of hair cells and the tectorial body are much reduced. Scale 100X.

so that the hair cells are seen in the sections as in double or triple rows, and then the papilla terminates. This terminal portion of the papilla is again enclosed in limbic tissue.

The Basilar Papilla in the Anuran Ear.   The transverse sections examined in the inner ear series above (Figs. 3-40 to 3-59) revealed the basilar papilla

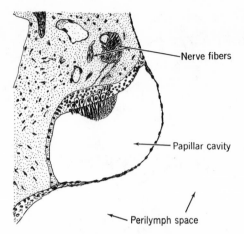

Fig. 3-76. The amphibian papilla of *Rana utricularia* at its posterior end where only a thin membrane separates it from a large perilymph space. Scale 100X.

of the frog *Rana utricularia sphenocephala* only far posteriorly, beginning with Fig. 3-56 where the papillar recess and cavity came into view as a ventromedial extension of the saccule and then cutting across this cavity in the next three figures. This sense organ consists of a thin vane of tectorial tissue of semilunar form stretching across a short tubelike passage so as partly to obstruct it and leading to a series of hair cells arranged along its outer edge.

The relation of this papilla to adjoining structures is shown in the frontal section of Fig. 3-77. The columella appears with its footplate extending into the lateral chamber, which at the level represented is only a recess separated by a bony wall from the perilymphatic cistern, but more dorsally is seen as a true chamber with an opening into the main part of the cistern. Thus the vibrations of the columella are able to set up a train of fluid movements across this chamber and the perilymphatic cistern, then through the saccule and basilar cavity into the inferior perilymph space, with the final outflow through the round window. Only a thin membrane, the perilymphatic window, lies in the path between the basilar cavity and the inferior perilymph space, and the basilar papilla lies over this window, partially blocking the pathway as shown in Fig. 3-78.

It has already been noted that the perilymphatic window as shown here when observed by the early anatomists was considered to be homologous with the basilar membrane in man and the other higher vertebrates because the nerve fibers that had been followed to this region were mistakenly thought to terminate on this membrane (rather than on the hair cells in the papilla near by as we now know). Thus the window was named the "basilar mem-

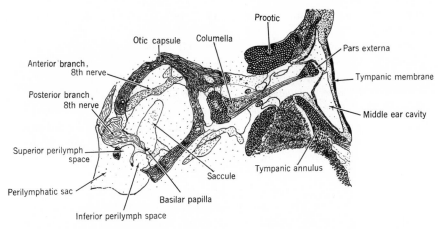

Fig. 3-77. A section through the ear region of *Rana utricularia sphenocephala* at a level showing the basilar papilla. Scale 10X.

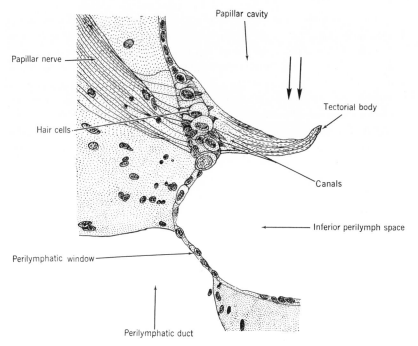

Fig. 3-78. The basilar papilla of *Rana utricularia sphenocephala* as in the preceding section, greatly enlarged. Scale 300X.

brane,'' a misconception reflected in the name still given to the papilla in this region.

Fluid movements passing over the free end of the tectorial vane set it in motion, and this motion is transmitted along the structure to the hair cells whose ciliary tufts are embedded in the base portion of the vane.

Figure 3-78 gives a detailed picture of the basilar papilla with its hair cells, oriented as in the preceding figure. Here are seen about 6 rows of hair cells, with their outer ends directed anterolaterly and the tips of their ciliary tufts attached to the undersurface of the tectorial body. These hair cells are innervated by fibers from a root of the eighth nerve coming from the dorsomedial side.

The tectorial vane as seen in Fig. 3-78 lies nearly in the transverse plane and in a suitably oriented series of sections its form can be seen as in Fig. 3-79. A single row of hair cells is represented in relation to their series of canals along the outer edge of the web of tectorial tissue.

This tectorial web is very thin; in the species pictured it measured only 12 μm in thickness, so that in the 20 μm series of sections as commonly used it appeared only once when the orientation was precisely in its plane. More often the orientation was not as fortunate, and the vane was cut across obliquely, with portions showing in three or four adjacent sections.

As may be seen, the web of tectorial tissue extends across about a third of the lumen of the tube and is supported on its inner side and through its midregion by struts formed by thick bundles of fibers. These struts reinforce the membrane along its inside edge and also extend from the lateral border to a common point at one end.

Around the inner end of the structure the web forms a series of short canals or cups, each of which encloses a hair cell. Only a single row of these canals is shown, in which they number 20, but by varying the focus of the microscope (or by examining a section cut at right angles to this one) it could readily be seen that there were 5 or 6 series of these canals through the depth of the web, corresponding to the rows of hair cells. The total number of canals, and of hair cells covered by them, was counted as 69 in one specimen and 75 in another.

There is reason to believe that the end of the ciliary tuft of each hair cell is firmly attached to the side wall of its canal, because in some specimens, such as the one illustrated here, the ciliary tufts in one region were found to be torn away from their hair cells and appeared in isolation in the vicinity of the canals. No doubt in such instances there had been an unusual stress, possibly caused by tissue deformation that produced the disruption in this structure sometime in the course of histological preparation. It is notable that the attachment of the ciliary tuft to the canal wall was stronger than the anchorage to its hair cell. This disruption was exceptional; in most prepara-

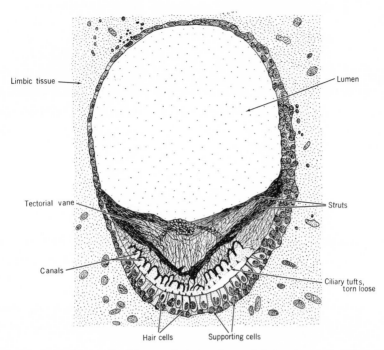

Fig. 3-79. The basilar papilla of *Rana utricularia sphenocephala* in a transverse section. Scale 300X.

tions the ciliary tufts remained in their normal positions at the ends of the hair cells.

These basilar hair cells resemble the ones found in the amphibian papilla and have large U-shaped nuclei that nearly fill the lower portion of the cell. The cells lie close together, as seen in Fig. 3-78, separated by thin columns formed by supporting cells that lie in a regular row beneath the hair cells. The supporting cells have well-defined cell walls and ovoid nuclei that fully fill the bases of their cells so as to leave only a thin layer of cytoplasm on the outside. The supporting cells rest on the smooth cavity wall formed by limbic tissue. In these features the papillar structure conforms closely to that found in all other auditory receptors.

PART II. THE ANURANS

4. THE PRIMITIVE FROGS:

THE ASCAPHIDAE AND DISCOGLOSSIDAE

INTRODUCTION

This chapter begins a systematic examination of the anurans, with a consideration of the ear structures in all species for which specimens were available for study. This amphibian order is divided into 21 families, one of which is known only from fossil remains. It was possible to obtain representatives of all but 4 of the 20 families of living species.

The systematic arrangement of these forms has presented a continuing problem. No one has any hesitation in identifying a given animal as a frog: the peculiar body structure and the great uniformity among members of the group insure their immediate recognition. Yet these very uniformities have made difficult an orderly arrangement of the species, and it becomes necessary to look for subtle differences.

The unique character of the anurans was early recognized, and Laurenti in 1768 gave them the distinctive name of Salientia, a designation still in frequent use, but it was not until nearly a century later that serious attempts were made to delineate a series of subdivisions within this group that would exhibit orderly relationships and ideally would reflect phylogenetic developments as well.

The earlier of these systematic efforts depended upon the use of such features as the form of the tongue, the presence of teeth and their locations on jaw and roof bones, and the presence or absence of adhesive disks on fingers and toes. The results of these efforts were not very salutary either because the features utilized were too general, occurring in a large number of forms that were clearly different in other respects, or because they split off only a small group from the general class.

A forward step was taken by Cope in 1864 in the use of the shoulder girdle by which he separated the toothed frogs into two groups, the Arciphera, with separate and overlapping epicoracoid cartilages and divergent clavicles and coracoids, and the Raniformia, with closely adjacent clavicles and coracoids and fused epicoracoids. Similarly a little later (1867) Cope subdivided the toothless Bufonidae by use of the same criteria. Boulenger thereafter (1882, 1897) devised a scheme that used the same shoulder girdle features, but without reference to dentition, with the two principal divisions referred to as Arciphera and Firmisternia. This system was widely accepted for a number of years, until Noble (1922, 1931) noted that an intermediate form of the pectoral girdle was to be found in the species *Sminthillus limbatus* and considered this to obscure the distinction between these two divisions. Noble thereupon presented a system containing four suborders (to which a fifth was soon added) based upon characteristics of the vertebral column as earlier suggested by Nichols (1916). Noble also gave weight to the form of the thigh musculature and its tendinous connections.

Noble's system of classification replaced Boulenger's and was accepted widely over more than three decades, perhaps in part because it was presented in detail in his highly successful textbook, *The Biology of the Amphibia* (1931). Only comparatively recently have serious questions been raised about the relationships assumed by his categories. Griffiths in 1963 suggested major modifications of this system, and soon thereafter a number of diverging opinions were indicated in a symposium on the phylogeny of frogs held in Kansas City in 1970 (see Vial, 1973). In the symposium discussions the evolutionary patterns of anurans were emphasized and several attempts were made to derive phylogenetic relationships from the fossil evidence. A number of different classification schemes were suggested, though not formally developed, in the treatment of early fossil history, in the examination of relationships among larval forms, and in the study of the geographical distribution of species.

This situation led Duellman (1975) to a new classification of the anurans, which was further developed in a monograph presented by Dowling and Duellman in 1978. The present treatment follows this new system, beginning with the primitive frogs in the family Ascaphidae.

THE FAMILY ASCAPHIDAE, THE RIBBED FROGS

The Ascaphidae, sometimes referred to as the Leiopelmatidae, include only two genera: *Ascaphus* with the single species *truei*, found in limited areas in the western United States and a border region of Canada, and *Leiopelma* with three species (or perhaps only two) that comprise the whole native frog population of New Zealand. These are generally regarded as the most primitive of existing frogs. Noble in 1931 recognized the similarities of these two groups and also their primitive characters, pointing to such features as their retention of ribs in the adult stage and the presence of tail-wagging muscles even after the embryonic tail is lost. At first he followed Stejneger in placing these two genera within the Discoglossidae, but later in recognition of their primitive status he accorded them a family of their own, as first suggested by Fejérváry (1923). This arrangement is now generally accepted, though some, like Savage (1973), in consideration of the wide separation of the two areas where these frogs are now found, have recognized separate families of Ascaphidae and Leiopelmatidae.

Ascaphus truei was discovered in 1897 in a cold mountain stream in northern Washington, and since has been collected in small numbers from adjoining areas as far away as Oregon, Montana, and British Columbia.

The adults as well as larvae are found in shallow, rock-bedded mountain streams, especially the smaller ones that contain few fish predators. In the larger streams the adults usually are well hidden in pockets under stones. At night, sometimes by day in wet, cool weather, some of them leave the stream and wander onto the banks near by.

Probably all mating occurs in the water, and fertilization is internal: the sperm are introduced into the cloaca of the female by use of the so-called "tail" of the male, which is an extension of the cloaca that serves as an intromittent organ. The mating activity was thoroughly studied by Noble and Putnam (1931) who also reported that the species is voiceless. A dorsal view of an adult specimen is represented in Fig. 4-1.

Stejneger (1899), who examined and described the first known specimen of *Ascaphus*, noted the absence of a tympanic membrane and columella; this animal has only an operculum in the oval window. De Villiers (1934) found a muscle running from the dorsolateral surface of the operculum to the su-

Fig. 4-1. A specimen of *Ascaphus truei*. Drawing by Anne Cox.

prascapula, which he identified as a part of the levator scapulae superior, as commonly accepted for other frogs.

In his general description of the ear region of *Ascaphus*, De Villiers indicated the presence of an amphibian papilla (which he referred to under the old name of pars neglecta), but he failed to find a second papilla. Shortly thereafter, however, Wagner (1935), working in the same laboratory and presumably with much the same materials, was able to identify this second auditory receptor (the basilar papilla) and also located its nerve supply.

The Ear of *Ascaphus*.   Two specimens of *Ascaphus truei* were available for the present study. Both of these were first tested in terms of electrical potentials with an electrode introduced into the anterior semicircular canal and thereafter were prepared for histological examination. One of these specimens was sectioned transversely and the other in the frontal plane.

A transverse section through the right side of the head of *Ascaphus* is pictured in Fig. 4-2 and shows the amphibian papilla in its usual location in a recess dorsal to the saccule, hanging from a limbic shelf that extends from the cartilaginous septum separating cranial and otic cavities. Above this shelf is the superior sinus, partially separated from the utricle, which leads dorsally to the anterior semicircular canal. Lateral to the papillar recess and the saccule is the large perilymphatic cistern. An opening in the lateral cartilaginous wall is the oval window, filled at this level by somewhat loose connective tissue. All these structures are enclosed by the cartilage of the otic capsule, which on the dorsolateral side becomes very thick and contains the lateral semicircular canal. A thin membranous window in the medial wall of the papillar recess leads to a perilymphatic duct, which at another level passes through a foramen into the brain cavity. At a more posterior level, represented in Fig. 4-3, the oval window opens widely and is largely filled by the disk-shaped operculum.

The basilar papilla is present at the level shown here in its usual position dorsal to the lagena. Its cavity connects through a thin membranous window with the perilymphatic duct.

Still farther posteriorly, as shown in Fig. 4-4, the oval window becomes closed but the operculum continues, now extending outward from a loose attachment on the prootic wall. From its dorsolateral surface a bundle of fibers constituting the opercular muscle runs dorsally and a little medially to attach to the undersurface of the suprascapula.

Large muscle masses lie between the otic capsule and the skin at the side of the head, as may be seen in the frontal section represented in Fig. 4-5. No ossicular structures are present to assist the transmission of vibrations from the air or water outside, and this transmission must be carried out largely by the bulk movement of the peripheral tissues. The prootic cartilage is ex-

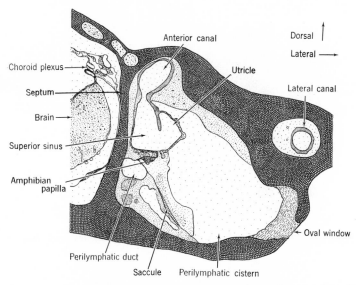

Fig. 4-2. A transverse section through the ear region of *Ascaphus truei*. Scale 25X.

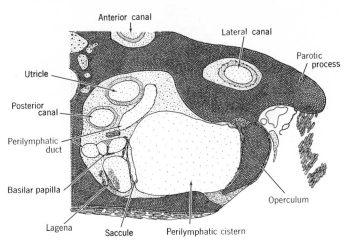

Fig. 4-3. A transverse section in *Ascaphus truei* far posterior to the preceding one. Scale 25X.

tended laterally as the parotic crest so as to lie close to the surface, and this cartilage may assist in the inward conduction of vibrations.

The Amphibian Papilla. — A detailed representation of the amphibian papilla of *Ascaphus* is given in Fig. 4-6. As shown, there are 14 rows of hair

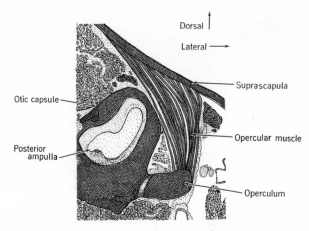

Fig. 4-4. The operculum and opercular muscle in *Ascaphus truei*. Scale 20X.

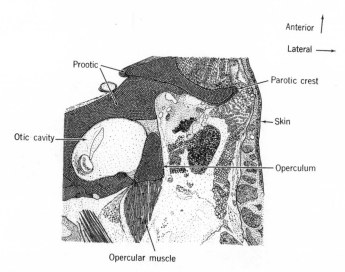

Fig. 4-5. The ear structures in relation to the exterior, in a frontal section in *Ascaphus truei*. Scale 15X.

cells extending downward to the complex tectorial body, into whose canal-like openings the ciliary tufts of these cells are inserted. At this level the canals are mainly thick-walled; those at the lateral side are short and end abruptly, and those toward the medial side are longer and run obliquely into a large globular mass made up of numbers of small vesicles.

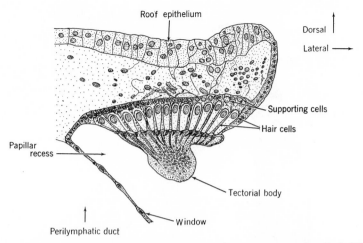

Fig. 4-6. The amphibian papilla of *Ascaphus truei* in a transverse section. Scale 250X.

The amphibian papilla as seen here has the same general appearance as the anterior division of this papilla in the advanced frogs such as the ranids, shown in Fig. 3-67 above. As this papilla is followed farther posteriorly, however, an important difference is found: this papilla in *Ascaphus* extends posteriorly only a short distance (140 μm in this specimen) and then ends abruptly; the long posterior division seen in most other frogs is entirely lacking. Lacking also is the sensing membrane that in the others extends from the posterior edge of the main mass of hair cells to an anchorage on the portion of the limbus that contains the papillar nerve and that partially blocks the aperture into this papilla (as may be seen in Figs. 3-70 and 3-71 above).

A more comprehensive view of the amphibian papilla in *Ascaphus* is given in Fig. 4-7. This picture represents a single section in a series cut frontally and shows the hair cells radiating about a central tectorial plate that takes the form of a perforated membrane. The array of hair cells forms an incomplete circle, and if the plane of sectioning were a little different this circle would probably be complete or nearly so since a small patch of hair cells is present at the appropriate place in the next adjacent section. Although in this view the tectorial structure appears as a perforated membrane, it in reality has much the form of a honeycomb, with peripheral openings that are the outer ends of little canals—the ''cells'' of the honeycomb. A considerable bundle of nerve fibers is seen on the posterior side of the papilla, and these fibers ramify around the fringes of the array. By use of the highest powers of the light microscope many of these fibers could be followed to the bases of the hair cells.

Thus it appears that the amphibian papilla of *Ascaphus* is much simpler

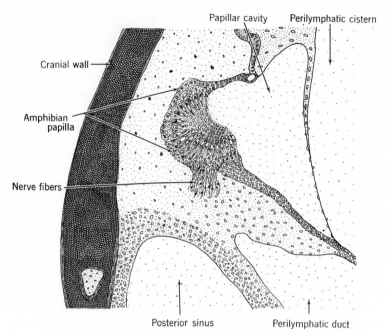

Fig. 4-7. The amphibian papilla of *Ascaphus truei* in a frontal section. Scale 100X.

than the one found in the advanced frogs, without the posterior segment of the papilla and thus with fewer hair cells, and also without the special aid to sensitivity produced by the sensing membrane.

The Basilar Papilla. — The basilar papilla closely resembles that found in *Rana* and others and is represented in Fig. 4-8. Here a relief path for vibratory pressures exerted at the operculum (which is anterolateral and ventral to the level shown) is provided by the round window.

Papillar Stimulation. — The mode of stimulation of the two papillae appears to be much the same as conceived for other frogs. Acoustic vibrations transmitted through the tissues at the side of the head enter the perilymphatic cistern and find a path through the two papillar structures to the perilymphatic duct and then pass through this duct to the round window area in the pharyngeal region. At a level a little ventral to the one shown in Fig. 4-7 the perilymphatic duct extends farther anteromedially, so that alternating pressures exerted in the cavity of the amphibian papilla can be relieved by displacements of the cellular structures in and out of this duct.

For the basilar papilla also a branch of the perilymphatic duct provides a relief path for vibrations transmitted through the perilymphatic cistern to the semilunar sensing membrane stretched across the lumen of this papilla. When

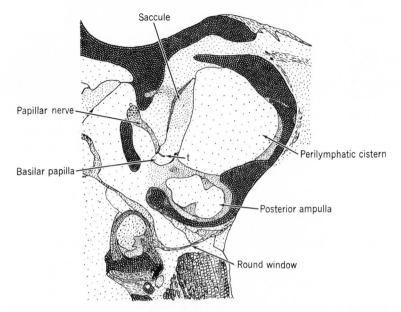

Fig. 4-8. A frontal section through the ventral region of the ear of *Ascaphus truei* showing the basilar papilla. Scale 20X.

the ciliary tufts of these papillae are set in motion by the tectorial tissues attached to them, the bodies of these cells remain relatively stationary; and through this relative motion the hair cells are stimulated.

This system thus follows the standard plan of stimulation for amphibian ears, and the condition in *Ascaphus* differs from that in other frogs only in the degree of effectiveness with which the sounds are transmitted and their energies are utilized. The absence of a tympanic membrane and columella in *Ascaphus* reduces the effectiveness of sound conduction to a level that bulk transmission by the tissues at the side of the head can provide, and the number of hair cells in this papilla is smaller than in many other ears.

The *Ascaphus* specimen sectioned in a transverse plane was suitable for the counting of hair cells, and the amphibian papilla on the left side was found to contain 173 hair cells, while that on the right contained 171 hair cells. In this same specimen there were 80 hair cells in the basilar papilla on the left and 90 hair cells in the one on the right. Thus the amphibian papilla in this animal has about twice as many hair cells as the basilar papilla.

The total complement of hair cells in *Ascaphus* is somewhat meager, and this condition combined with the poor transmission suggests a low level of auditory sensitivity.

Cochlear Potential Sensitivity. — The results of the cochlear potential tests for the two animals are shown in Fig. 4-9. These curves indicate only a moderate level of performance, with the best sensitivity in the region of 290 to 1000 Hz. In one animal there are peaks approaching 0 db at two points in this range, and in the other animal there is a region at 500–700 Hz where the sensitivity reaches +10 db. This poor level of sensitivity to aerial sounds is consistent with the absence of tympanic membrane and columella, and with the limited population of hair cells. In the water, however, this ear would be expected to operate noticeably better than Fig. 4-9 shows for aerial transmission, because of a better impedance match with the tissues of the ear region.

The Ear of *Leiopelma*.   The genus *Leiopelma* was first described by Fitzinger in 1861, but only much later, as already noted, was it assigned to a separate family. Usually three species are recognized, which are *L. archeyi, L. hamiltoni*, and *L. hochstetteri*, but E. M. Stephenson (1960) on the basis of a study of skeletal features concluded that the first two belong to a single species: that *L. archeyi* is a neotenic form of *L. hamiltoni*, which under certain conditions reaches sexual maturity at a relatively early age and when small in size. The present study was carried out on *L. hochstetteri*, the most available of these rare types of frog.

The general characteristics of the skull of *L. hochstetteri* were described by Wagner in 1934, who a little later (1935) gave particular attention to the structure of the inner ear and made comparisons with *Ascaphus* to obtain further evidence on the question whether these two genera should be brack-

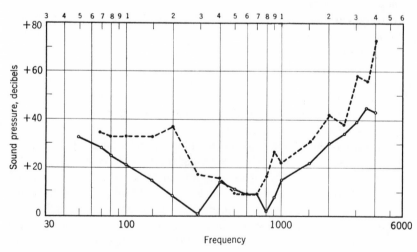

Fig. 4-9. Aerial sensitivity in two specimens of *Ascaphus truei.* Shown is the sound pressure, in decibels relative to 1 dyne per sq cm, required for an inner ear potential of 0.1 μv.

eted in the same family. Many similarities were found, and a close relationship between the two forms was supported.

Both amphibian and basilar papillae are present in these labyrinths. The present observations extend the earlier ones of Wagner to more precise details.

The Amphibian Papilla. — The amphibian papilla of *Leiopelma hochstetteri* as seen in a transverse section is represented in Fig. 4-10. At the level shown there are 14 rows of hair cells with their ciliated ends extending into canal-like openings in a complex tectorial body. The resemblance to the structure shown for *Ascaphus* in Fig. 4-6 is extremely close.

A more comprehensive view of this papilla in *Leiopelma* is presented in Fig. 4-11. This is a frontal section and reveals almost the whole array of hair cells cut across in the plane of their long axes. These cells form a complete circle, radiating about the tectorial body into which they extend their ciliated ends.

In one of the specimens examined there were 247 hair cells in the amphibian papilla on the left side and 224 hair cells in the one on the right. In another specimen, sectioned in the frontal plane and somewhat less satisfactory for the counting of hair cells, these numbers were 204 on the left and 232 on the right.

The Basilar Papilla. — The basilar papilla in *Leiopelma hochstetteri* is represented in Fig. 4-12. This is a frontal section and shows the structure cut across perpendicular to the plane of its tectorial body. This view assists in a determination of the path of sound transmission: sounds applied from the outside to the operculum set up vibratory motions of the fluid of the perilymphatic cistern that have an outward path through the membrane bounding the common chamber of saccule and papillar recess, then through a thin

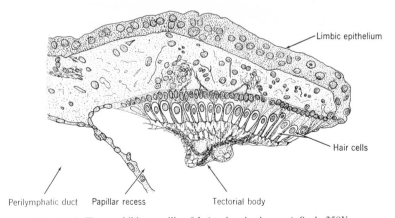

Fig. 4-10. The amphibian papilla of *Leiopelma hochstetteri*. Scale 250X.

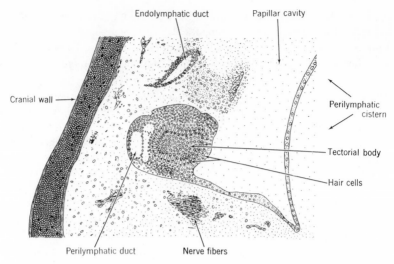

Fig. 4-11. The amphibian papilla of *Leiopelma hochstetteri*, in a frontal section. Scale 100X.

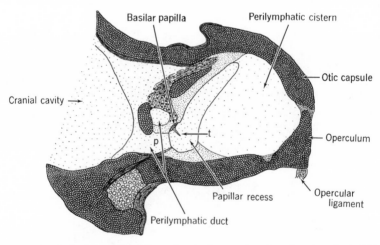

Fig. 4-12. The basilar papilla of *Leiopelma hochstetteri* in a frontal section. Scale 20X.

membrane into the perilymphatic duct adjoining the brain cavity (to point *p* in Fig. 4-12), and finally, as seen at a more dorsal level in Fig. 4-13, passing through the jugular foramen alongside nerves IX and X to fluid spaces posterior to the otic capsule. This region is somewhat encumbered with nerves and muscles masses but leads outward to the surrounding skin and carries out its essential service of completing the pathway of vibratory fluid displacement.

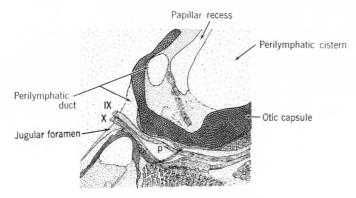

Fig. 4-13. A section at the level of the jugular foramen of *Leiopelma hochstetteri*. Scale 20X.

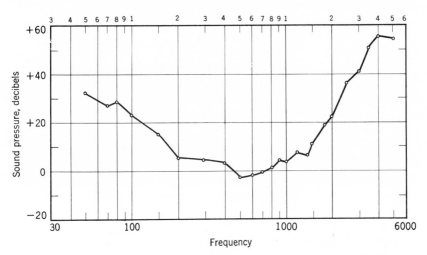

Fig. 4-14. An aerial sensitivity function for a specimen of *Leiopelma hochstetteri*. Shown is the sound pressure, in decibels relative to 1 dyne per sq cm, required for an inner ear potential of 0.1 μv.

The basilar papilla in one of these specimens contained 76 hair cells on the left side and 70 on the right. The second specimen gave corresponding numbers of 65 and 71.

The average hair-cell populations for the four ears studied are 226.7 for the amphibian papilla and 70.5 for the basilar papilla, a ratio of a little over threefold.

Sensitivity in *Leiopelma*. — The sensitivity of the ear in terms of cochlear potentials is presented for one specimen of *Leiopelma hochstetteri* in Fig. 4-14. This curve shows the most favorable region around 200-1500 Hz, in the middle of which a maximum of -3 db is attained. These results are similar to the ones for *Ascaphus* and indicate a rather poor response to aerial sounds.

The simple form of the amphibian papilla as found in *Ascaphus* appears likewise in *Leiopelma*: in both genera this papilla resembles the anterior division of this organ in the advanced frogs; it lacks the sensing membrane and the long posterior extension commonly found. Here is further evidence in support of the view that these two species of frogs, now found only in widely separated regions of the earth, in fact are closely related and are properly included in a common family.

THE DISCOGLOSSIDAE

Among the primitive frogs are the Discoglossidae with four genera: *Alytes*, *Barbourula*, *Bombina*, and *Discoglossus*. Of these only *Alytes* and *Bombina* species were available for study.

Alytes obstetricans. The species *Alytes obstetricans* is widely distributed over central and southern Europe; its specific and common names (the midwife toad) reflect the peculiar practice of the male in caring for the eggs. This frog is largely terrestrial in habit, living in burrows, but in the mating season the male comes forth and engages in active calling. If a female appears and mating occurs, the male takes possession of the fertilized eggs, which are embedded in a strand of gelatinous material like a string of beads. He winds this string around his hind legs and carries it about for a period of several weeks until the eggs are ready to hatch. It is said that he moistens the eggs from time to time to prevent their drying, and then as they begin to hatch he seeks out a pool and partially immerses his body so that the emerging larvae enter the water, where they grow to maturity.

Auditory Structures. — A tympanic membrane is present, readily seen as an oval disk on either side of the head, and in a specimen of about average size, with a body length of 38 mm, it measured 3.0×3.8 mm, with the longer axis running anterolaterally. When this membrane was removed, as in Fig. 4-15, the pars externa of the columella could be seen suspended in an opening in the tympanic annulus.

A frontal section through the ear region of a specimen of *Alytes* is presented in Fig. 4-16. Shown is the columella with an attachment to the tympanic membrane a little posterior to its center and running inward to a seat in the oval window of the otic capsule. There is no lateral chamber, but the pars interna of the columella simply occupies the lateral portion of the perilymphatic cistern. The attachment of the columella is complicated by the presence of an ascending process, a portion of which is cut across and indicated at *a* in this figure. This process takes a medial rather than an anteromedial course as in most frogs; it arises at a more posterior level as a broad

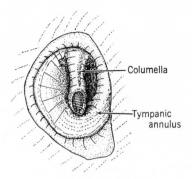

Fig. 4-15. Lateral view of the ear of *Alytes obstetricans* after removal of the tympanic membrane. Scale 12X.

cartilage as shown in Fig. 4-17 and connects to the pars externa of the columella close to the tympanic membrane as shown in the reconstruction drawing of Fig. 4-18. It then runs inward to a firm connection to the anterior wall of the otic capsule.

This connection no doubt has a protective function in limiting the displacements of the tympanic membrane, but also it must impose restraint on the movements of the columella and thus reduce the sensitivity to sounds. That this restraint is not very serious is indicated by sensitivity measurements made by Mohneke and Schneider (1979) whose results are presented in Fig. 4-19. These investigators recorded electrical potentials from neurons of the torus semicircularis (the auditory area of the frog's brain) and obtained the functions indicated in this figure separately for 6 males (broken line) and 6 females (solid line). These data are averaged for two body temperatures, 20° and 28° C, in a range in which these ears operate most effectively, and show a maximum that reaches about −44 db relative to a sound pressure of 1 dyne per sq cm in the region of 2000 Hz. In this region the maximum is nearly the same for both sexes, though elsewhere along the frequency scale the females are generally the more sensitive.

Bombina SPECIES

Three species of *Bombina* were examined: *Bombina variegata, B. orientalis,* and *B. bombina*. These species are alike in that all lack the tympanic membrane and tympanic annulus, and all have an operculum in the oval window. They differ, however, in the presence and form of the columella. The otic cavity is a single space: there is no lateral chamber as found in the advanced frogs. There is also a reduction in the amphibian papilla: it lacks the posterior tectorial body, as was found to be the case in the Ascaphidae, but a sensing membrane is present.

These species are aquatic in habit, spending nearly all their time in the

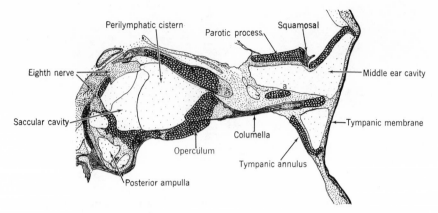

Fig. 4-16. A frontal section through the ear region in *Alytes obstetricans. a* represents the middle portion of the ascending process. Scale 16X.

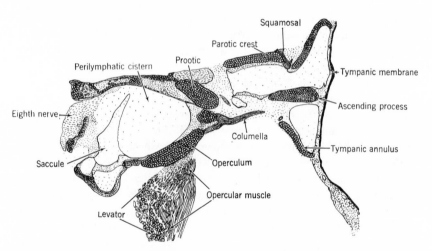

Fig. 4-17. A frontal section in *Alytes obstetricans* ventral to the foregoing showing the lateral portion of the ascending process. Scale 16X.

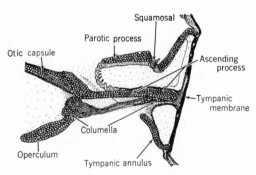

Fig. 4-18. A reconstruction based on a series of sections like the foregoing showing the complete course of the ascending process. Scale 16X.

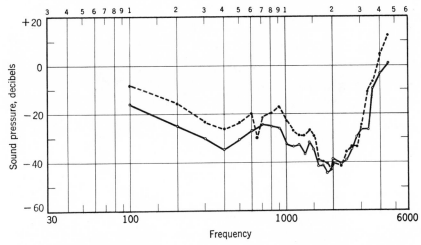

Fig. 4-19. Sensitivity functions for *Alytes obstetricans.* Solid line, mean of 6 females; broken line, mean of 6 males. Data from Mohnecke and Schneider, 1979. Reference level (0 db) = 1 dyne per sq cm.

water. As already noted, the lack of a tympanic membrane is not a serious matter in an aquatic animal, as sound vibrations can pass from the water through peripheral tissues to the inner ear fluids with little reflection, and without the aid of accessory mechanisms.

Bombina variegata.　　This species is found in streams and pools over most of Europe from the coast of France eastward as far as Turkey, including Italy, Greece, and Rumania.

A transverse section through the ear region on the right side is presented in Fig. 4-20. Here the skin and most of the heavy layer of muscles at the side of the head have been removed, but the columella and hyoid process remain lateral to the otic capsule. At this level the columella lies close alongside the capsule wall, and an examination of more anterior sections shows that it is connected to the hyoid process, which here appears ventrolateral to it; this connection is indicated in the figure by broken lines. The columellar cartilage continues posteriorly, and also runs a little farther dorsally along the capsule wall; it ends at a level posterior to the one shown, close to the place where the oval window opens and the operculum appears to occupy it. The end of the columella lies along the edge of the oval window, embedded in the connective tissue there, but makes no firm connection to the otic wall.

Litzelmann (1923) in an embryological study obtained the evidence identifying this cartilage as a remnant of the columella. He found that immediately after metamorphosis this element is present in a form indicated in Fig.

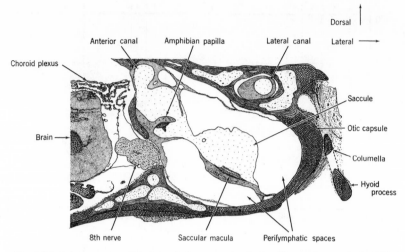

Fig. 4-20. A transverse section through the ear region in *Bombina variegata*. Scale 25X.

4-21, which is a reconstruction drawing from his report. As seen, the columella extends from the operculum to a close contact with the hyoid. Litzelmann reported that early in embryonic development the columella shows an abrupt bend and ends in a small knob adjacent to the closed end of the Eustachian tube, and then later the bent portion of its shaft is resorbed and the downward-directed part remains in contact with the hyoid as shown.

Figure 4-22 presents a section at a far posterior level, where the operculum is cut through near its middle where it occupies a large portion of the otic capsule wall. Although the section presented in Fig. 4-21 has indicated that there is a connection between columella and operculum, this relation appears remote in the region where the oval window has opened and is occupied by the operculum. Thus it seems likely that the columella takes little part in the transmission of sounds inward; rather it appears that sounds pass from the water outside through the skin and muscle layers at the side of the head, then through the operculum to the fluids of the otic capsule, and by this route they finally reach the amphibian and basilar papillae.

No lock mechanism is found in this ear, though on the left side of the specimen represented here a small piece of cartilage appears for a short distance wedged between the operculum and the wall of the otic capsule above and probably represents a remnant of the process of the columellar footplate that in most frogs extends downward along the inner surface of the operculum as the inside prong of the notch in which the upper end of the operculum is retained.

In this ear both opercular and columellar muscles are well developed, and

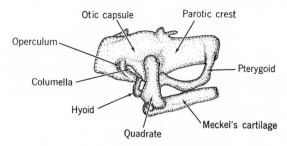

Fig. 4-21. The otic capsule and its relations to adjoining structures in *Bombina variegata* at a stage following metamorphosis. Redrawn from Litzelmann, 1923.

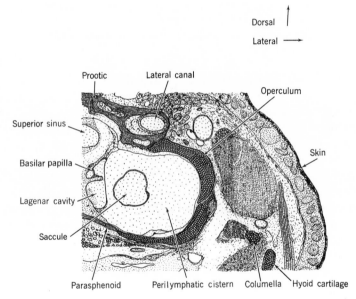

Fig. 4-22. A transverse section in a specimen of *Bombina variegata* (far posterior to the one in Fig. 4-20) where the operculum fills the oval window. Scale 20X.

the opercular muscle inserts on the outer face of the operculum as usual. It is a little surprising that the columellar muscle also makes much the same connections as in other frogs. As shown in Fig. 4-23, it sends its ligament anteriorly along the lateral surface of the operculum and attaches mainly to the columella, though in its passage over the end of the posterolateral wall of the otic capsule it appears to make some slight connection to this capsular cartilage also.

Although the connections of these muscles are little different from nor-

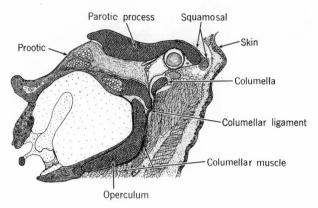

Fig. 4-23. The ear region of *Bombina variegata* at a somewhat posterior level showing the connection of the columellar muscle to the columella. Scale 15X.

mal, their actions cannot be considered as typical. The columella fails to reach the periphery, and can hardly serve for the inward conveyance of vibrations. As this animal is in the water, this is not a serious matter; sound is transmitted through the skin and soft tissues at the side of the head to the operculum directly, without the aid of accessory structures.

Despite the limitations, a degree of control of sound reception appears to be a possibility in this form of ear. A contraction of the columellar muscle will draw the inner end of the columella downward and inward, impinging on the edge of the posterolateral wall of the otic capsule and bringing this wall downward into contact with the upper portion of the operculum, thus decreasing the mobility of the operculum and reducing transmissiom. This system therefore can provide a measure of protection to the ear when the stimulating sounds are excessive. A relaxation of the columellar muscle and a contraction of the opercular muscle will reverse this action and allow the operculum to vibrate freely. The control afforded by this mechanism obviously must be far inferior to that existing in the majority of frogs, but may be useful nonetheless.

The Inner Ear. — The transverse section of Fig. 4-20 has presented a general view of the otic capsule in its anterior region. The amphibian papilla is present, cut across at the level of the anterior tectorial body. In the section represented the operculum is not seen; only at a more posterior level does the lateral wall of the capsule open up to contain this element. A more posterior level was depicted in Fig. 4-22, where also the basilar papilla appears, occupying a recess lying just above the lagenar cavity.

In this specimen there were 349 hair cells in the left amphibian papilla and 361 hair cells in the right one, distributed as indicated in Table 4-I. The

Table 4-I

Hair-Cell Distribution in the Amphibian Papilla of the *Bombina* Species

	To the Anterior Tectorial Body	To the Sensing Membrane	Free-standing Hair Cells	To the Posterior Tectorial Body	Total Hair Cells
		#1 *Bombina variegata*			
Left	73	232	42	*	349
Right	113	224	24	*	361
		#2 *Bombina variegata*			
Left	96	182	28	*	306
Right	74	254	9	*	337
		Bombina orientalis			
Left	205	92	68	*	365
Right	208	99	33	*	340
		Bombina bombina			
Left	208	70	64	*	342
Right	205	87	50	*	342

*Posterior tectorial body absent.

left basilar papilla contained 60 hair cells, and the right one contained 58 hair cells.

A second specimen of this species had the same general arrangement, with a total of 306 hair cells in the left amphibian papilla and 337 hair cells in the right one, distributed as in Table 4-I. In this second specimen the basilar papilla on the left contained 60 hair cells and the one on the right 57 hair cells.

The amphibian papilla in *Bombina variegata* is peculiar in that the sensing membrane appears and connects to the ventral surface of the tectorial body in the anterior region where this body still has a simple unimodal form, as shown in Fig. 4-24. Indeed, there is never a division of the papilla as seen in most frogs, where posteriorly the hair cells form lateral and medial groups, with tectorial connections remaining for cells of the lateral group while the medial cells have free-standing ciliary tufts. In this species the sensing membrane on its first appearance at the anterior end connects to the midportion of the tectorial body, then posteriorly moves to the lateral edge of this body, and then a little farther posteriorly the whole medial part of the papilla disappears, hair cells and supporting cells alike, leaving only a lateral group of hair cells served by the sensing membrane.

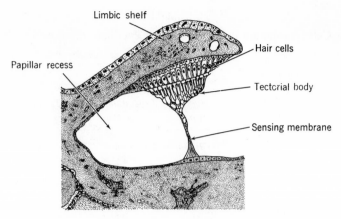

Fig. 4-24. The amphibian papilla in *Bombina variegata* at an anterior level. Scale 125X.

Where farther posteriorly the sensing membrane ends, there is a considerable patch of hair cells remaining without tectorial connections. These continue only a short distance, and no posterior tectorial body is found.

Auditory sensitivity in this species is shown for two specimens in Fig. 4-25. There is fairly good sensitivity in the midtone region, from 290 to 1600 Hz, but with a very rapid decline in the upper frequencies.

Bombina orientalis.   The species *Bombina orientalis* occurs in eastern Siberia, Manchuria, and portions of China. Its ear structure resembles that of *B. variegata* as just described, but the action is a little different.

The Middle Ear Apparatus. — As shown in Fig. 4-26, a frontal section through the ear region, a cartilage is present just outside the posterolateral edge of the otic capsule wall, but without actual contact with it, and runs ventrolaterally through loose tissue posterior to the parotic process for a distance of a little over 0.5 mm, where it ends well below the skin surface. Its outer end lies upon the end of the ascending process of the hyoid coming from below, as the figure indicates. The two processes at their union connect through thick ligamentary strands with the underside of the parotic process, and there are a few fibers also running to the squamosal and covering this process.

This cartilaginous process corresponds closely in form and connections with the one seen in the same location in *B. variegata* and is thus identifiable as a remnant of the columella.

As in the preceding species, there is no lateral chamber and no notch mechanism. The end of the operculum is overlapped on the outside by the posterolateral wall of the otic capsule, and a few elastic fibers connect these

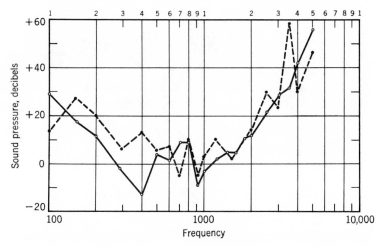

Fig. 4-25. Sensitivity functions for two specimens of *Bombina variegata*. Shown is the sound pressure, in decibels relative to 1 dyne per sq cm, required for an inner ear potential of 0.1 μv.

two cartilages, but otherwise the upper end of the operculum is free. Both opercular and columellar muscles are present, and the opercular muscle attaches to the dorsal surface of the operculum in the ordinary way, as seen in Fig. 4-27. As Fig. 4-26 has shown, its ligament runs along the outer surface of the operculum as usual, but connects not with the columella as in most frogs (including *B. variegata*), but with the lower end of the capsular wall.

Despite this strange insertion of the columellar muscle, it probably functions in the control of sound transmission in this ear, and perhaps quite as effectively as in *B. variegata*. A direct connection to the capsular wall, which is entirely cartilaginous here, will allow the muscle to draw the wall down into contact with the operculum and restrain the motions of that element in response to vibrations conveyed inward through the soft tissues of this region. The reciprocal action of the columellar and opercular muscles will therefore have the usual effect of regulating the acoustic input to the auditory papillae.

The structure of the amphibian papilla in this species resembles that of *B. variegata*: a sensing membrane appears near the middle of the papilla and connects to the undivided tectorial body, then a short distance posterior to this connection the more medial hair cells disappear, leaving only a lateral group that continues in a posterior direction until the sensing membrane ends. Thereafter, over about one-fifth of the length of the papilla, the hair cells have free-standing ciliary tufts; there is no posterior tectorial body.

Table 4-I shows the numbers of hair cells in these three papillar regions.

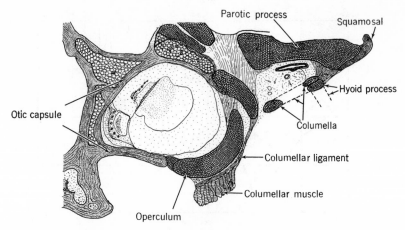

Fig. 4-26. The ear region in a frontal section of *Bombina orientalis*. Scale 15X.

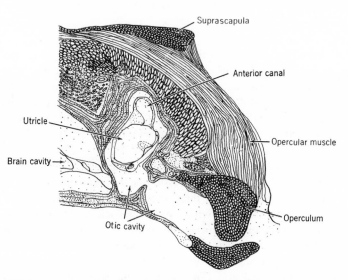

Fig. 4-27. A transverse section through the ear region in *Bombina orientalis*. Scale 20X.

The total number was 365 on the left and 340 on the right. The basilar papilla on the left had 81 hair cells, and the one on the right 78 hair cells.

Sensitivity. — Tests of sensitivity for aerial sounds were carried out on 7 specimens. One of the functions is represented in Fig. 4-28 and shows a fairly good level of sensitivity over a middle range from 200 to 1000 Hz, with the best point reaching −10 db at 500 Hz.

Two other specimens gave responses to aerial sounds as shown in Fig. 4-29. The sensitivity shown here is poorer, but the curves have much the same form, with the best region for one of them between 150 and 800 Hz and for the other between 200 and 1000 Hz. No clear evidence of bimodality is seen in these response functions. An aerial response curve for a fourth specimen, shown by the solid line in Fig. 4-30, is rather convincingly bimodal, with sensitive regions around 200–290 Hz and 600–1200 Hz. A vibratory response for this same specimen, obtained with a vibrating needle applied to the suprascapula, is shown by the broken line in this same figure. There is a sharp peak at 290 Hz and another at 1600 Hz, with these peaks a little higher in frequency than those for aerial stimulation. The sensitivity levels shown are good for both forms of stimulation.

Further tests with vibratory stimulation gave the results of Fig. 4-31 obtained with two different placements of the vibrating needle, on the suprascapula in one series and on the columella in the other. The principal peak for columellar stimulation is at 400 Hz, reaching −28 db, and for suprascapular stimulation is at 290 Hz, reaching −25 db. It seems likely that a vibratory stimulus applied to any firm structure about the head will be effective in actuating the acoustic system.

Bombina bombina. The species *Bombina bombina* occurs in northern Europe, including Denmark and the southern tip of Sweden and eastward through Russia to the Ural mountains. There is overlapping with *B. variegata* over a portion of this range, and hybridization between the two species has been reported.

The Middle Ear. — No trace of a columella is found in this species. An ascending process of the hyoid is present, running anteriorly all the way to the parotic process where it makes a loose contact as seen in Fig. 4-32. It makes no connections relevant to ear structures.

An operculum fills the oval window, as this figure shows. The muscle bundle adjacent to it in the main part of this figure contains large fibers readily recognizable as columellar fibers, but well dorsal to this level (as shown by the inset drawing to the right) there are two sizes of fibers, the more anterior of which are opercular. It was not possible with any certainty to separate the larger fibers here into columellar and levator groups, as ordinarily can be done at least roughly by noting their ligamentary connections; these fibers were seen to connect to delicate ligamentary strands running along the lateral surface of the operculum and mostly ending in the fascia of the joint between operculum and otic capsule, with only a few continuing farther to the parotic floor. Thus the majority of these fibers most likely correspond to the columellar fibers of the preceding species and may serve in a limited way to perform a function like that found in *B. orientalis*: the application of tension to the flexible wall of the otic capsule in such a way as to bring it

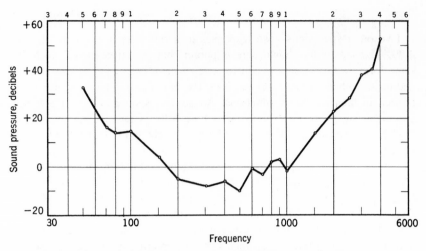

Fig. 4-28. Sensitivity to aerial sounds in a specimen of *Bombina orientalis*. Shown is the sound pressure, in decibels relative to 1 dyne per sq cm, required for an inner ear potential of 0.1 μv.

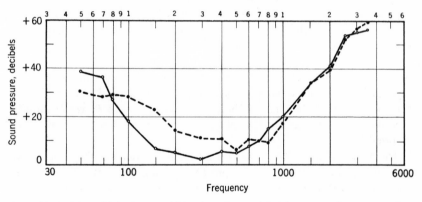

Fig. 4-29. Aerial sensitivity in two additional specimens of *Bombina orientalis*, represented as above.

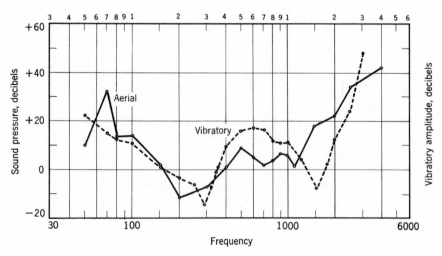

Fig. 4-30. Aerial and vibratory functions for a specimen of *Bombina orientalis*. Aerial sensitivity is represented as above; vibratory sensitivity is shown as the amplitude, in decibels relative to 1 millimicron, required for the standard response. The vibratory stimulus was applied to the suprascapula.

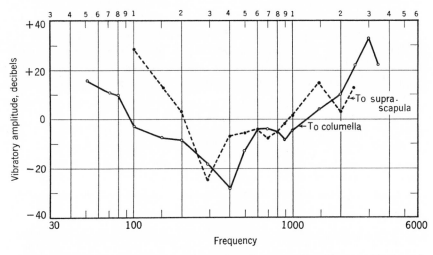

Fig. 4-31. Vibratory sensitivity in a specimen of *Bombina orientalis*, with two applications of the stimulus.

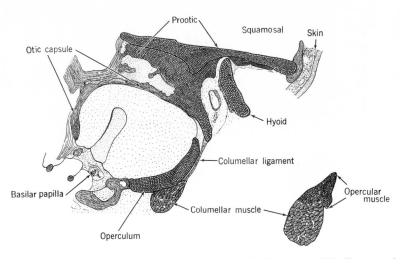

Fig. 4-32. A section through the ear region of *Bombina bombina*. Scale 20X. The opercular muscle is added from a more anterior section.

in contact with the operculum and thereby diminish the vibratory motions of that element. This action must be considered slight because of the limited size of the muscle and the feeble connection to the otic wall. Thus in this species there appears to be minimal control of acoustic input to the auditory receptors.

The Inner Ear. — A specimen of this species had a total of 342 hair cells in the amphibian papilla on each side, distributed as shown in Table 4-I.

Hair cells with free-standing cilia occur in two regions, at the first appearance of the sensing membrane and at the posterior end after this membrane has terminated. As in the two preceding *Bombina* species, there is no posterior tectorial body.

In this species the basilar papilla on the left contained 60 hair cells and the one on the right 51 hair cells.

Sensitivity. — Tests of sensitivity with aerial sounds on two specimens gave the results of Fig. 4-33. One animal showed a fair level of sensitivity in the region of 200-400 Hz, with the maximum at −4 db. The second animal was very poor throughout, only attaining a level of +20 db around 600-700 Hz.

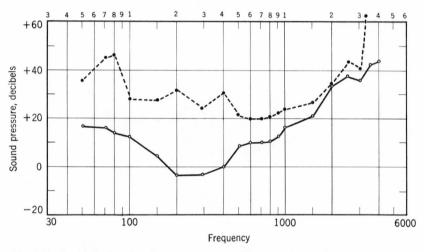

Fig. 4-33. Sensitivity functions for two specimens of *Bombina bombina*. Shown is the sound pressure, in decibels relative to 1 dyne per sq cm, required for a response of 0.1 μv.

5. THE PRIMITIVE FROGS: THE PIPIDAE
AND RHINOPHRYNIDAE

THE PIPIDAE

The consideration of the primitive anurans is continued in the Pipidae. These are frogs with a completely aquatic mode of life, except that they have lungs and breathe air. It is likely that they are derived from frogs with terrestrial habits and that their aquatic adaptation is secondary.

The Pipidae constitute a single family in the suborder Aglossa (the tongueless frogs) and have a disjunct distribution in Africa and South America. Four genera are widely recognized: *Pipa*, containing 5 species in South America, *Xenopus* with 6 species and a few subspecies in South Africa, *Hymenochirus*, with 4 species in west Africa, and *Pseudhymenochirus*, a single species limited to Guinea in northwest Africa. A fifth genus mentioned by Noble (1931), and characterized as less specialized than the others, is *Protopipa* in South America.

The present study included 3 species: *Xenopus borealis, Xenopus laevis*, and *Pipa pipa*.

Xenopus, The Clawed Frog.   The two species examined, *Xenopus borealis* and *Xenopus laevis*, showed closely similar auditory structures except for size. More extensive study was made of the species *X. laevis*, which because of its hardy qualities is often raised in laboratory colonies and is readily available.

The ear of *Xenopus* has undergone some curious modifications, which no doubt has resulted from this animal's adaptations to an aquatic existence. There is no sign of an external ear: the skin continues unmodified over the ear region. The other two divisions of the ear, the middle and inner ear, are well developed.

The Middle Ear Mechanism. — If a patch of skin is removed just behind the eye as in Fig. 5-1 and then a thin layer of fatty tissue is cleared away, an oval, dome-like cartilage is seen in a dorsolateral position. This cartilage, which will be referred to as the tympanic disk, serves as a receiver of aquatic sound vibrations and corresponds to the tympanic membrane of most other frogs.

In a moderately large specimen of *X. laevis* with a body (snout to vent) length of 7.0 cm, this disk measured 7.5 mm along its anteroposterior axis

Tympanic disk

Fig. 5-1. Dorsal view of a specimen of *Xenopus laevis*, with a dissection to expose the tympanic disk on the right side. Scale ½ natural size.

and 6.0 mm transversely; its total area was 25 sq mm. In a female specimen with a body length of 8.5 cm the axes were 5.6 and 4.4 mm, for an area of approximately 31.4 sq mm.

The tympanic disk is shown in further detail in Fig. 5-2. It lies in a shallow funnel-shaped depression formed by the tympanic annulus, which corresponds to the structure that in most frogs supports a thin and flexible tympanic membrane. In this figure only the edges of the annulus are seen, most noticeably at the anterodorsal and posterior ends. The disk is attached to the edges of the annulus over the entire circumference by delicate connective tissue fibers and retains considerable mobility. The transparency of the cartilage in fresh specimens makes it possible to see the deep attachment of the outer end of the columella, which is embedded in the anterior and middle portion of the disk.

Removal of the tympanic disk with care to leave the end of the bony columella intact gives the picture of Fig. 5-3. The tympanic annulus is revealed as a widely flaring funnel-shaped body of cartilage, with the outer portion of the columella occupying the posterior half of its opening. This cartilaginous body is firmly supported by the two arms of the squamosal bone, which are embedded in its undersurface and not visible in this picture.

The relation of the columella to the two portions of the cartilaginous structure is further revealed in Fig. 5-4. In this picture the orientation is the same as in the two preceding figures, but the disk was separated from the annulus except along the lateral edge and then was bent over laterally and held down

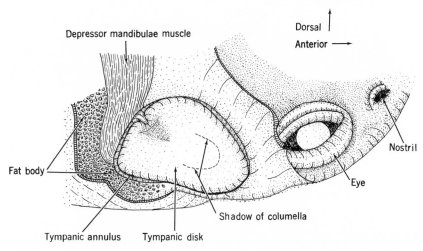

Fig. 5-2. The tympanic disk region viewed laterally and a little posteriorly. Scale 5X.

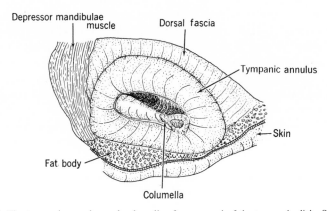

Fig. 5-3. The tympanic annulus and columella after removal of the tympanic disk. Scale 5X.

with a retractor hooked around the columella as shown. The columella is seen extending out of the middle ear cavity and attached along the exposed undersurface of the tympanic disk.

The middle ear cavity, whose outermost portion is seen in the two preceding figures, is a trumpet-shaped air space and extends deeply beneath the parasphenoid bone to the midline, where it meets the cavity from the opposite side of the head. The narrow medial portion of this cavity is often designated as the Eustachian tube, and at the midline this tube communicates with the pharyngeal cavity and thus opens into the mouth.

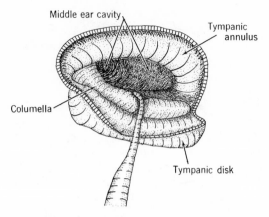

Fig. 5-4. Further details of the tympanic annulus and columella, with exposure of the middle ear cavity. Scale 7.5X.

A general view of the ear mechanism in relation to other structures of the head is given in Fig. 5-5, which is a cross section through a specimen of *Xenopus borealis*. Here may be seen the tympanic disk at the dorsolateral surface with the end of the columella retained near its middle. Above the middle ear cavity is the prootic bone and its large lateral extension as the cartilaginous parotic process, and within this osseous and cartilaginous enclosure lie the several parts of the inner ear and the equilibrial labyrinth.

The space beneath the tympanic disk as well as the entire middle ear cavity is filled with air. This was proved by observations on a specimen that was transferred from its living tank to a plastic water-filled bag, anesthetized, and the tympanic disk exposed and opened, all with the animal kept continuously submerged. On exposure of the tympanic disk the presence of air beneath it was suspected because of its glistening appearance, and when this disk was punctured, a few bubbles of air escaped. Then with pressure on the tissues a small amount of additional air came to the surface, until the whole middle ear space became water-filled.

The mouth seems usually to contain water. When in an intact animal the mouth was forced open under water no air escaped. Evidently the middle ear cavity is kept separate from the mouth, except at times when the air in this cavity is renewed by gulping at the surface.

The deep-lying portion of the columella and its relations to the otic capsule are shown in Fig. 5-6. The inner end of the columella fits in the oval window, a large opening in the dorsolateral wall of the otic capsule. A band of tissue lies over the columella at a point a little beyond its middle. In *Xenopus laevis* this band consists of ligamentary fibers with a covering layer of muscle fibers from the depressor mandibulae. In *X. borealis* the band is formed

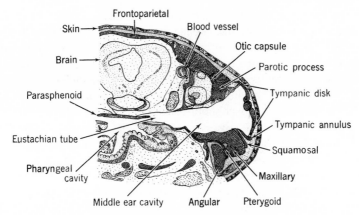

Fig. 5-5. A transverse section through the head of *Xenopus borealis* showing the tympanic region. Scale 10X.

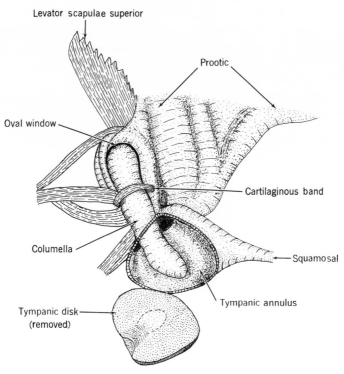

Fig. 5-6. Dorsolateral view of the tympanic region of *Xenopus laevis* after severing the tympanic disk and turning it downward. Scale 3X.

mainly by the union of two cartilages, one extended from each of two portions of the squamosal bone and meeting over the columellar surface as shown in Fig. 5-7. The ends of these cartilages are bound together by ligament, though at one point (shown in Fig. 5-7) the ends appear to be fused. As this figure shows also, the columella is separated from the underlying cartilage (which is the crest of the parotic process) by a bed of connective tissue fibers and thus is free to move over short distances in response to vibratory stimuli.

The further suspension of the columella is indicated in Figs. 5-8 and 5-9. The first of these figures is representative of a single section cut in the frontal plane and passing through the whole extent of the columella. Figure 5-9 is a reconstruction based on frontal sections at several levels and presents a number of additional features.

As these figures show, the external end of the columella, which is osseous with a cartilaginous core, is buried deep in the tympanic disk, runs through the middle of this disk, and exits in a posterior and medial direction. It passes lateral to the outer end of the parotic process, between it and a heavy mass of connective tissue that includes the ligamentary or cartilaginous band already referred to. This ossicle executes a rapid bend ventrally and promptly bends back to its former level, then finally turns to cartilage at its posteromedial end. This end constitutes the columellar footplate in the oval window of the otic capsule.

The cartilaginous end of the columella, indicated as c in Figs. 5-8 and 5-9, forms a footplate of complex form lying obliquely over the end of the

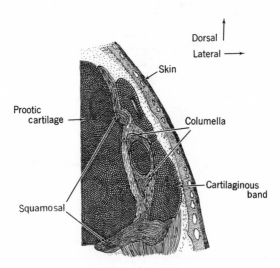

Fig. 5-7. The cartilaginous band covering the columella in *Xenopus borealis*. Scale 40X.

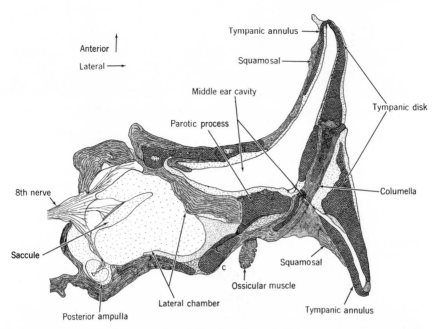

Fig. 5-8. The course and suspension of the columella, and relations to other ear structures in *Xenopus laevis*. Scale 10X.

lateral chamber of the otic capsule. An upper portion of this footplate is split away from the main body by a thick wedge of bone and forms a cartilaginous flap that is suspended by ligament from the ridge along the undersurface of the parotic process (suspension 5 in Fig. 5-9). This detached portion of the footplate in one region extends over almost the whole medial surface of the columella. The posterior portion of the footplate is attached to the outermost edge of a cartilaginous extension of the posterior wall of the otic capsule that most likely is a modified form of the operculum (suspension 6 in Fig. 5-9).

The Operculum and Opercular Muscle. — That the element just mentioned corresponds to the operculum in other frogs is indicated by its location and attachments. Its continuity with the posterior wall of the otic capsule is a characteristic feature, as is also its ligamentary attachment to the end of the columella. Most convincing in this regard is the presence of an opercular muscle, which itself is distinguished by its location and the small size of its fibers. As shown in Fig. 5-9 and in further detail in Fig. 5-10, this muscle lies anterior to a larger muscle mass, the levator scapulae superior, that runs downward from the suprascapula.

These same features are present in *Xenopus borealis*, in which they were

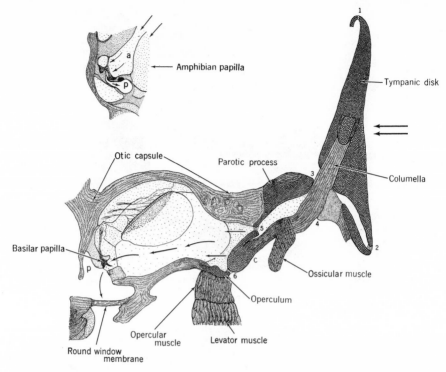

Fig. 5-9. The middle ear structures in *Xenopus laevis*, and the paths of sound conduction for the basilar papilla (main part) and for the amphibian papilla (inset above). *c*, the cartilaginous end of the columella; *p*, the perilymphatic duct. Scale 12.5X.

examined in a series of transverse sections. Here the operculum appears as a cartilaginous extension of the otic capsule in a ventrolateral direction with fibers of the opercular muscle inserted on its dorsolateral surface. These fibers were traced dorsomedially to the suprascapula as the more anterior components of a bundle consisting mainly of fibers attached to the ventrolateral surface of the prootic.

No direct evidence is available concerning the function of the opercular muscle in these two species. The operculum may be flexible enough to undergo displacement downward on contraction of the opercular fibers, but an effect on sound conduction from such action does not seem likely. Probably the opercular muscle in *Xenopus* simply aids in the retraction of the suprascapula.

The Ossicular Muscle. — A second muscle that will be designated as the ossicular muscle arises from the suprascapula and extends into the pocket formed by the abrupt ventral bending of the columella. Many of its fibers

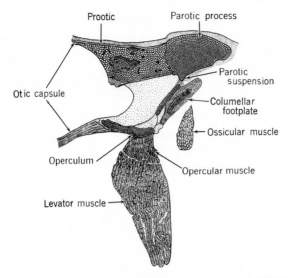

Fig. 5-10. Muscles of the middle ear region in *Xenopus laevis*. Scale 10X.

attach to the upper surface of the columella in this region, though other fibers run to the underside of the parotic process. It seems likely that contraction of this muscle, by pulling on the columellar shaft obliquely with respect to its direction of vibration in response to sounds, will have a damping effect, and thus affords a measure of protection against sounds of excessive magnitude. This protective function is served by the columellar muscle in most frogs, and it is possible that what is here designated as the ossicular muscle can be identified as the same one, but this matter requires further study.

Action of the Middle Ear. — The close suspension of the columellar shaft near its midpoint as shown at points 3 and 4 in Fig. 5-9 suggests at first that here is an axis about which the middle ear mechanism moves in response to acoustic stimuli, and this suggestion seemed to be reinforced by the observation of rocking movements about this point when gross displacements were produced by pressing on the tympanic disk. Further observations, however, have shown that this axis serves only for large-amplitude movements and is not concerned in the minute vibrations produced by sounds. These observations were made on anesthetized specimens of *Xenopus laevis* by driving the middle ear system at various points while recording inner ear potentials.

A needle point driven by a crystal vibrator was used in these experiments. There were two particular objectives: to study the performance of the tympanic disk itself and to ascertain whether the columella operates as a mechanical lever. If the columella had an acoustic axis of rotation, then driving it at different points along its length would reveal this axis: approaching the axis with the driving needle would produce an increase in response, until at the axis itself the action would fail. The results showed clearly that there is no effective axis of rotation for vibrations; the disk and columella together form a unitary system that simply moves in and out in response to acoustic pressures.

Some of the results of these experiments, carried out on *Xenopus laevis*, are shown in Fig. 5-11. In one series of tests the system was driven at nine different points as indicated by the small circles. Seven of these points lay along the longitudinal axis of the columella, of which four were on the disk over the columella, one was on the columella directly, and two were on the disk beyond the attachment of the columella. Two other points were on the free portion of the disk. The amplitude of driving at these different positions was varied so as to produce a constant response of 0.1 μv, and the necessary amplitude is indicated by the numbers in the circles, expressed in decibels relative to an arbitrary level of 1 mμ. Thus the value of -5 db about the middle of the disk over the outer end of the columella signifies that driving at this point required a root-mean-square amplitude of -5 db or 0.56 mμ.

The sensitivity at these various driving positions exhibits a surprising uniformity: over the portion of the disk overlying the columella the values vary only in the range of -5 to -7 db and are within this range also at the two points lateral to the columella where the disk was thick. Less sensitive (-1 and -3 db) were the two points anterolateral to the columella where the disk was thin, and more sensitive (-8 db) was the point on the bony surface of the columella itself. These variations appear to be related to the firmness of the tissues at the driving positions; thus the best point (-8 db) is on the bony surface of the columella, and the poorest (-1 db) is on the relatively thin and soft cartilage of the disk anterolateral to the columella.

These observations were continued in other specimens by driving the columella at numerous points all along its length, and again the amplitudes varied only slightly with the driving position. Thus it is clear that the tympanic disk operates essentially as a rigid plate in response to acoustic pressures, and the columella is transmitting the vibratory motions directly through its footplate to the fluid of the inner ear.

Although a lever action is absent, there is obviously a pressure amplification arising from the areal ratio between the tympanic disk and the columellar footplate: the pressure applied to the large surface of the disk is transmitted to the smaller surface of the footplate. In a specimen of *X. laevis* measuring 7.0 cm in S-V length the area of the tympanic disk was 25 sq mm and the area of the columellar footplate was 1.73 sq mm, giving a ratio of 14.5. The acoustic pressures acting on the disk are thus increased by this factor in their transmission to the inner ear (less a small amount of loss through

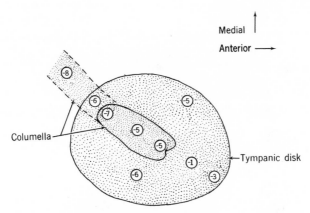

Fig. 5-11. The tympanic disk of *Xenopus laevis* and points of stimulation over its surface.

friction). The areal factor in this species thus improves the ear's sensitivity by an amount of approximately 20 db.

The absence of any lever action in the middle ear mechanism of *Xenopus* is entirely appropriate for an animal whose ear is adapted primarily to the sensing of vibrations in the water. The fluids of the inner ear have an acoustic impedance that is essentially the same as the impedance of the water outside and no transformer action is required (as it is in terrestrial animals). The presence of air in the middle ear cavity, which no doubt is a retention of the condition in ancestral frogs that had ears adapted to receive aerial sounds, is a fortunate circumstance, for it greatly reduces the frictional resistance that the tympanic disk would encounter if this cavity were water-filled. This feature thus adds substantially to the efficiency of the *Xenopus* ear.

The Path of Sound Transmission. — Further to be considered are the locations of the auditory papillae, the character of the round window, and the complete path of sound transmission in *Xenopus*. These features correspond closely to those in other frog species.

The location of the basilar papilla in the posteromedial part of the otic capsule is shown in Fig. 5-9. The amphibian papilla is not shown in the main part of this figure, but appears in the insert above; its proper position in the otic capsule is anterior to the basilar papilla. Its structure is shown in further detail in Fig. 5-12. The location of the round window area was shown in Fig. 5-9 at the posteromedial end of the otic capsule, with the round window membrane almost directly posterior to the basilar papilla. This window leads to the pharyngeal and mouth cavities below.

The path of sound transmission is indicated by the arrows in Fig. 5-9. Sound waves impressed on the surface of the tympanic disk are transmitted

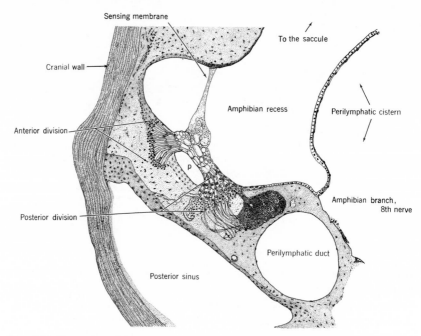

Fig. 5-12. The region of the amphibian recess and papilla in *Xenopus laevis*. *p*, an opening leading to the perilymphatic duct. Scale 100X.

through the bony shaft of the columella to its footplate in the oval window, and there exert alternating pressures on the fluids of the inner ear. These fluids can be set in motion because of the presence of a region of yielding, which extends through the round window membrane into the pharynx. Across this outward path is a thin muscle layer and the mucous membrane lining of the pharynx, but these soft tissues do not represent any significant impediment to the fluid motions. The pathway continues through the mouth to the water outside, and thus the vibratory circuit is completed.

As already described in the general anatomy section, the amphibian papilla in frogs is complex in form and ordinarily consists of two connected portions, shown at one level in Fig. 5-12 as the anterior and posterior divisions. Acoustic vibrations transmitted inward by the columella reach the amphibian recess by way of the perilymphatic cistern and the thin membranous enclosure of the saccule. They then take a somewhat complex path leading into the perilymphatic duct to the round window and pharyngeal cavity, and thus pass through the sensory structures in this pathway. The passage from amphibian recess to perilymphatic duct is not shown in Fig. 5-12, as it lies at a more ventral level; this passage begins in the region indicated in Fig. 5-

12 as *p* and meets a medial extension of the perilymphatic duct that is separated from it by a window of thin connective tissue. The fluid movements involved in this pressure discharge thus include the complex tectorial mass shown in the figure and especially the sensing membrane that lies across the pathway and leads to an archway that connects to both anterior and posterior divisions of the papillar structure.

The hair cells in these two papillar divisions vary in structure, with many of those of the anterior division having an elongate, columnar form, and those of the posterior division showing the more usual flask-shaped form. All are innervated by fibers from a papillar branch of the eighth nerve; some of these fibers supplying the posterior division are indicated in Fig. 5-12. A large bundle of fibers leaves the main nerve mass at a level a little ventral to the one shown here and ramifies to the hair cells of the anterior division. A good many hair cells are found in the amphibian papilla of this ear, with a distribution indicated in Table 5-I.

Sensitivity in *Xenopus.* — The performance of the *Xenopus* ear was investigated by applying aerial sounds through a tube sealed over the tympanic disk area and recording in the usual way with an electrode inserted through a hole drilled in the bone over the anterior semicircular canal. Results for a

Table 5-I

Hair-Cell Distribution in the Amphibian Papilla of the Pipidae
and Rhinophrynidae

	To the Anterior Tectorial Body	To the Sensing Membrane	Free-standing Hair Cells	To the Posterior Tectorial Body	Total Hair Cells
			Xenopus laevis		
Left	267	115	116	*	498
Right	304	156	55	*	415
			Pipa pipa		
Left	107	30	22[a]/17[p]	*	176
Right	115	59	0[a]/16[p]	*	190
			Rhinophrynus dorsalis		
Left	363	153	132[a]/127[p]	*	775
Right	263	196	176[a]/ 97[p]	*	732

*Posterial tectorial body absent.
[a] Anterior hair cells.
[p] Posterior hair cells.

specimen of *Xenopus laevis* are shown in Fig. 5-13. This ear responded best in the region of 1000–2000 Hz, and at the peak at 1200–1500 Hz reached a level of −14 db. A secondary region from 150 to 250 Hz showed sensitivity around 0 db. Another animal gave the results of Fig. 5-14, in which there are likewise two regions of good sensitivity, and in much the same portions of the frequency range. Here the best sensitivity, in the region of 1500–2500 Hz, attained the surprising level of −47 db, and a secondary region of good response appeared around 300–700 Hz, here reaching −40 db at two places. In this low-frequency region the responses were highly variable, and from time to time, as a result of spontaneous activity of the lightly anesthetized animal, or after light tactual stimulation of the skin, the responses fell off by amounts up to 35 db as indicated by the broken line.

Observations on specimens of *Xenopus borealis* showed poorer sensitivity and greater variability, as represented in Fig. 5-15. One of these animals responded best in the low tones, reaching −18 db at 200 Hz, but became generally poorer for the higher tones. Another animal reached −20 db at 800 Hz and was fairly responsive over the whole range from 400 to 2000 Hz, but fell off rapidly for higher and lower tones.

Pipa pipa, the Surinam Toad.   A South American species that has given the name to this family is *Pipa pipa*, sometimes called the Surinam toad. As shown in Fig. 5-16, this animal has a most peculiar form with a squarish body, pointed head, short forelegs with the fingers tipped with little filaments that apparently aid the tactual sensitivity, and powerful hindlegs provided with large webbed feet. This species is perhaps most famous for the curious manner of incubating the eggs: these are placed on the female's back by the male as they are laid and after he has fertilized them, and they hatch out and develop in little pockets in the skin. The young finally emerge as small but well-formed frogs.

These animals are almost wholly aquatic, though they are said to climb out onto the land on occasion, and can survive when the river pools in which they live are much reduced by drought.

Auditory Structures. — In *Pipa pipa* the ear is adapted for the reception of water vibrations just as it is in *Xenopus*, and as in that form there is no external ear and the receptive surface is a disk of cartilage hidden beneath the skin. The dotted circles in Fig. 5-16 indicate the location of the tympanic disks under the skin in a region immediately behind two little skin flaps at the sides of the head.

The skull drawing of Fig. 5-17 shows the location of the middle ear mechanisms, which are well separated at the extreme ends of the occipital region. As shown in further detail in Fig. 5-18, the tympanic disk lies in a shallow depression formed by the tympanic annulus and is connected by a

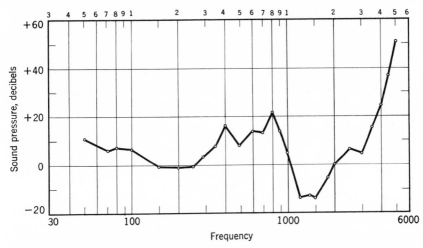

Fig. 5-13. Sensitivity to tones for a specimen of *Xenopus laevis*, shown in decibels relative to a zero level of 1 dyne per sq cm, as required for a response of 0.1 μv.

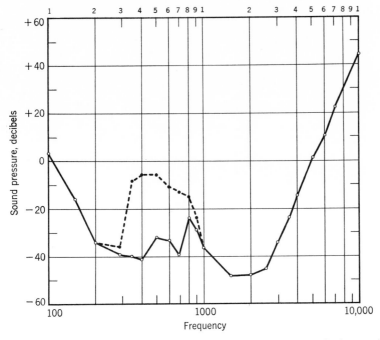

Fig. 5-14. Sensitivity for another specimen of *Xenopus laevis*, represented as in the preceding figure.

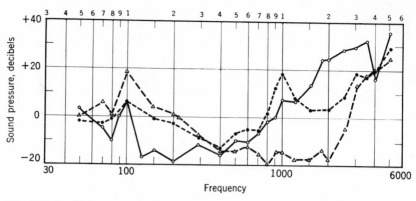

Fig. 5-15. Sensitivity curves for three specimens of *Xenopus borealis*, represented as in the preceding figures.

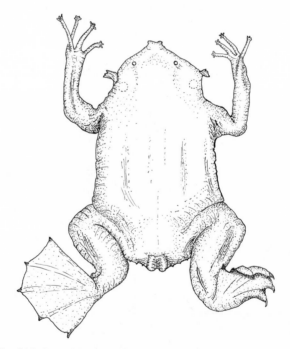

Fig. 5-16. A specimen of *Pipa pipa* in a dorsal view, ½ natural size.

short bony rod, the shaft of the columella, to a footplate in the oval window in the dorsolateral surface of the prootic bone.

On examination of Fig. 5-19, representing a transverse section through the right side of the head, the otic capsule is seen to lie in a thick bony enclosure formed by a fusion of frontoparietals and prootics, in a position just lateral to the cranial cavity, and because of the great width of the head there is a considerable distance between oval window and otic capsule. A long fluid-filled passage connects these two structures. This lateral passage might be regarded as a scala vestibuli, as it leads inward from the vestibular window as this scala commonly does. However, there is a question whether this passage in *Pipa pipa* is strictly homologous with the scala vestibuli in other ears: a membrane separates its fluids from the perilymph of the otic capsule, and these two fluids appear to be different. In all the specimens the fluid of the lateral passage contained a densely staining precipitate that evidently had been formed through the action of the fixatives used in the preparation process, whereas the fluids within the otic capsule, perilymph and endolymph alike, remained entirely clear. It seems likely that this lateral passage is a novel feature with distinctive fluid contents.

As Fig. 5-19 indicates, the anterior edge of the footplate of the columella is attached by ligament to the dorsolateral edge of the prootic, whereas its main portion remains free in the window. It seems likely that this footplate executes a hinge-like motion about its dorsal attachment when acted upon by alternating sound pressures. This motion produces pressure variations in the fluid column of the lateral passage, and these continue in the fluids of the otic capsule and reach both amphibian and basilar papillae.

The relief pathway for vibratory fluid pressures in this species is by way of an opening into the brain cavity that is analogous to a round window. At a far posterior level where the basilar papilla is coming to an end an opening appears in the bony wall separating otic and cranial cavities, as represented in Fig. 5-20. This opening is covered by the arachnoid membrane and readily permits a discharge of pressures into the brain cavity and then through the opposite ear to the outside. As this figure indicates, the window serves directly as an outlet from the basilar papilla, and this same pathway is accessible to the amphibian papilla, which lies farther anteriorly.

The Amphibian Papilla. — This papilla in *Pipa pipa* is represented near its anterior end in Fig. 5-21, and its structure corresponds to the usual anterior portion of this receptor in other frogs. This form continues over a distance of 150 μm as a simple layer of hair cells whose ciliary tufts are covered by a tectorial body as shown, and then this tectorial mass extends downward and attaches to the floor of the papillar cavity to form a sensing membrane as pictured in Fig. 5-22. This structure extends farther posteriorly for a distance of 120 μm, whereupon the papilla ends. A posterior division is not found.

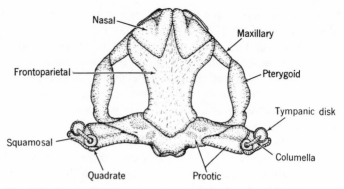

Fig. 5-17. The skull of *Pipa pipa*, redrawn from Parker, 1876. Natural size.

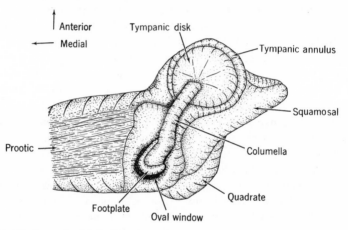

Fig. 5-18. The tympanic disk in *Pipa pipa*, exposed by removal of the skin in the dorsolateral region of the head. Scale 6X.

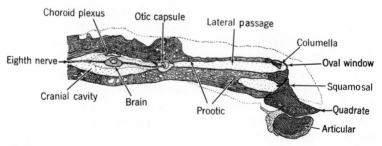

Fig. 5-19. A transverse section through the right side of the head of *Pipa pipa* showing the ear region, with the muscle masses indicated only in outline. Scale 2.5X.

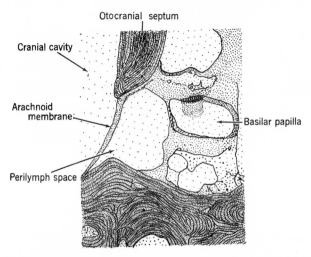

Fig. 5-20. The relief pathway for sounds in *Pipa pipa*: an opening in the bony wall between otic and cranial cavities. Scale 75X.

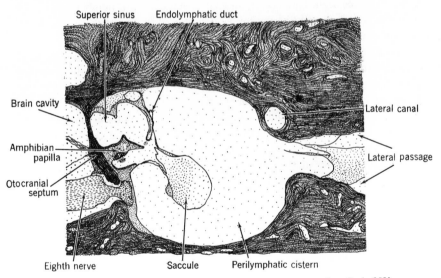

Fig. 5-21. The amphibian papilla in *Pipa pipa* in a transverse section. Scale 30X.

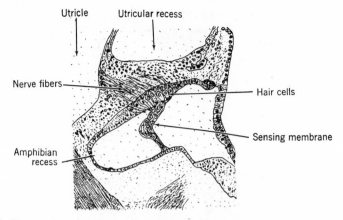

Fig. 5-22. The amphibian papilla in *Pipa pipa* farther posteriorly showing the sensing membrane. Scale 30X.

In one specimen the hair cells of this papilla numbered 301 in the left ear and 300 in the right ear, and in another specimen these numbers were 176 and 190. The hair-cell distribution as determined in one of these specimens is shown in Table 5-I.

The Basilar Papilla. — This second papilla arises at a more posterior level as a diverticulum of the lagena as shown in Fig. 5-23, and a little farther posteriorly it presents about 18 rows of hair cells as indicated in Fig. 5-24. In the first specimen referred to above these hair cells numbered 170 on the left side and 171 on the right, and in a second specimen these numbers were 129 and 179. The averages for these four ears are 242 hair cells for the amphibian papilla and 162 for the basilar papilla, a ratio of about 3:2. This ratio is of the order of Paterson's (1960) estimate of the relative size of the two papillae in *Pipa pipa* (1.6 to 1) and does not support the often repeated assertion that the basilar papilla is the chief acoustic receptor in amphibians.

Sensitivity. — The sensitivity of the ear was measured in terms of the inner ear potentials recorded with an electrode inserted in a small hole drilled in the middle part of the lateral passage. The results in response to aerial sounds presented through a tube sealed over the tympanic disk are represented for one of the specimens in Fig. 5-25. These results show rather poor sensitivity, with the best region around 2000 Hz.

The poor performance shown in these tests may be explained by two conditions, the location of the electrode and the use of aerial sounds. The electrode was somewhat remote from the sensory cells of the two papillae, and in the small number of specimens available it was not possible to work out a more favorable location; the thick bone surrounding the otic capsule makes

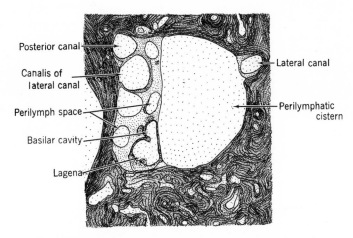

Posterior canal

Canalis of lateral canal

Perilymph space

Basilar cavity

Lagena

Lateral canal

Perilymphatic cistern

Fig. 5-23. A transverse section through the inner ear region of *Pipa pipa* showing an extension from the lagena forming the basilar cavity. Scale 30X.

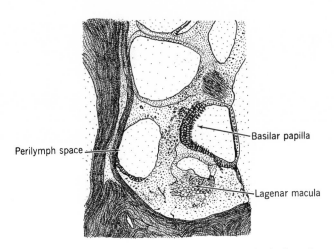

Perilymph space

Basilar papilla

Lagenar macula

Fig. 5-24. The basilar cavity and papilla in the same specimen as in the foregoing, at a more posterior position. Scale 75X.

difficult a close approach. The use of aerial sounds is of course inappropriate for an ear adapted to aquatic vibrations, and the actual sensitivity must be regarded as greater than that shown by about 30 db.

These tests were continued in the same animal by the use of vibratory stimuli applied with a needle to the exposed surface of the tympanic disk, with results shown in Fig. 5-26. Again the best sensitivity is indicated for the high tones, in the region of 2000–3000 Hz.

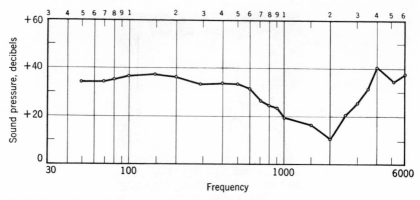

Fig. 5-25. Aerial sensitivity in a specimen of *Pipa pipa*. Zero sound pressure = 1 dyne per sq cm.

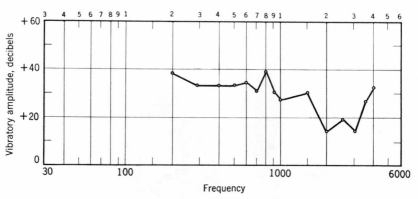

Fig. 5-26. Vibratory sensitivity in the same specimen of *Pipa pipa*. Here is shown the vibratory amplitude, in decibels relative to 1 millimicron, required for a response of 0.1 microvolts.

From these results it appears that the ear in *Pipa pipa* is a serviceable receptor for a moderate range of tones in the medium high frequencies.

THE RHINOPHRYNIDAE

The Rhinophrynidae, a family from Central America represented by a single species, *Rhinophrynus dorsalis*, have a highly uncertain position in respect to the other primitive frogs. Duellman (1975) placed this family together with the Pipidae and the extinct Paleobatrachidae in the superfamily Pipoidea, though these frogs differ from the pipids in the presence of a tongue, in the absence of ribs, and in having a vertical pupil rather than a round one. Walker

(1938) remarked that *Rhinophrynus dorsalis* occupies a peculiarly isolated position, primitive in many respects but well advanced over the leiopelmids.

Rhinophrynus dorsalis. This frog is of medium size, 5–6 cm in body length when adult, with a noticeably pointed head. The tongue is attached to the floor of the mouth at its back margin and can be extended for licking up termites, this frog's main article of diet. The male has vocal sacs and produces calls during the breeding season that are described as bird-like.

There is no eardrum or columella, but a large operculum fills the oval window, and the inner ear is well developed, including amphibian and basilar papillae as in other frogs. A drawing through the ear region showing otic capsule and operculum is presented in Fig. 5-27.

As may be seen in this figure, the anterior wall of the otic capsule is extended laterally in a prominent parotic process whose cartilaginous end lies close beneath the skin and a thin muscle layer. The operculum covers the

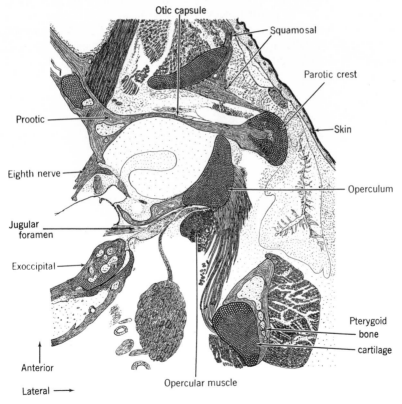

Fig. 5-27. A frontal section through the ear region of *Rhinophrynus dorsalis*. Scale 10X.

whole lateral face of the otic cavity, and at the level shown is firmly attached posteriorly to the bone of the capsule but connected only by sparse elastic fibers to the capsular roof at its anterolateral end. A peculiar feature is the massive ligament connecting the outer surface of the operculum with a muscle below that runs mainly in a dorsal direction and also sends long fibers to the inner surface of the pterygoid cartilage. The muscle just below the operculum is a part of the levator scapulae superior complex, and the other muscle lateral to it is evidently the pterygoideus.

The Amphibian Papilla. — In terms of hair-cell numbers this papilla is well developed. In one of the specimens studied there were 7 rows of hair cells at its anterior end, and ventrally this number increased progressively to reach 31 rows a little beyond the middle of the papilla where the sensing membrane begins. This point marks the end of the anterior division of this papilla and the beginning of what corresponds to the middle division in the advanced frogs. (There is no portion corresponding to the posterior division of the advanced frogs.)

The tectorial body, with the sensing membrane spanning the area between it and the floor of the papillar recess, sends forth two arms of tectorial tissue that make contact with the hair cells as shown in Fig. 5-28. A small conical arm goes to the cells of the medial part of the papilla and another arm, much broader and elaborated into a series of canals at its end, is applied to the

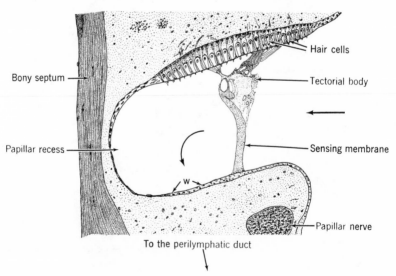

Fig. 5-28. A transverse section through the anterior portion of the amphibian papilla. w represents an area that about 100 μm farther posteriorly will open into the perilymphatic duct. Scale 100X.

cells of the lateral part. The conical arm is present only over a short distance, so that few of the medial hair cells are served. The lateral arm continues farther, and many of the cells are closely approached or covered by the terminal canals while others, especially the most lateral ones, are connected by fine filaments that extend from the main part of the canal structure. Some of these filaments are very long, and it is likely that they have been drawn out artificially in these preparations by shrinkage of the tectorial tissue. In the living condition these canals probably approach the cell ends more closely and their filamentary connections are shorter.

A little farther posteriorly, as seen in Fig. 5-29, the epithelial layer exhibits a degree of bimodality: here the more medial portion of the structure grows a little larger, while the lateral portion to which the lateral sensing arm has its connections maintains its size and form; between these two areas the hair cells are few and scattered. Then farther posteriorly there are rapid changes: the medial portion of the structure diminishes and finally disappears, and the lateral portion also diminishes over a short distance and then grows larger, until it contains about two-thirds as many rows of hair cells as the anterior division had at its maximum. The tectorial structure then extends broadly as an extremely delicate fibrous network over the whole array of hair cells.

In one ear of the specimen studied in detail the anterior portion of this

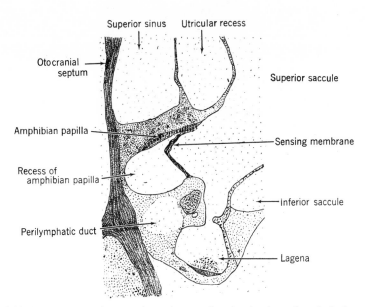

Fig. 5-29. A transverse section posterior to the preceding showing the perilymphatic duct. Scale 40X.

papilla, before the first signs of a division, included 540 hair cells, and the remainder of the papilla, made up of the continuing and expanding second division, contained 280 hair cells.

Although this second division contains only about a third of the total number of hair cells in this papilla, there is reason to expect that it provides the keenest sensitivity: though direct evidence is lacking, the action of the sensing membrane ought to assist the responses.

Thus this papilla presents a feature found to be characteristic of the amphibian papilla in the advanced forms: there are two types of responsive mechanisms working together. The direct mechanism consisting of webs of tectorial tissue suspended in the vibratory pathway and picking up vibratory energy immediately from the fluid is supplemented by a sensing membrane that lies athwart the vibratory pathway and adds its pressure effects. Hair-cell distributions are shown in Table 5-I.

The Vibratory Pathway. — In the higher anurans, as shown for *Rana* in Chapter 3, fluid pulsations introduced into the lateral opening of the amphibian recess continue through this recess itself and have their relief opening far posteriorly, finally reaching the round window that faces the pharyn-

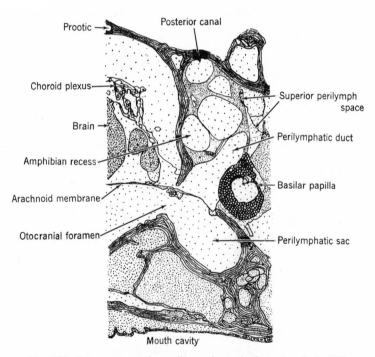

Fig. 5-30. A transverse section well posterior to the foregoing. Scale 20X.

geal cavity. Species lacking the posterior division of the papilla lack this extension of the papillar recess, but employ a route of vibratory outflow that is closely similar. This pathway follows a perilymphatic duct that runs posteriorly between the otocranial septum and the lagena until it reaches the otocranial foramen, into which it opens as shown in Fig. 5-30. A little farther posteriorly the perilymphatic sac leads to the round window whose membrane bounds the pharyngeal cavity. Figure 5-30 includes the basilar papilla, which utilizes the same relief pathway for vibratory movements.

Sensitivity. — Response curves for two specimens of *Rhinophrynus dorsalis* are shown in Fig. 5-31. One of these animals showed a fair level of response in the midfrequency region, between 200 and 1200 Hz, with peaks at 400 and 900 Hz. The other animal was less sensitive, but with the best responses in the same general region, peaking at 700 Hz.

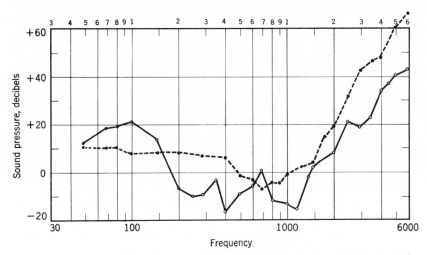

Fig. 5-31. Sensitivity functions for two specimens of *Rhinophrynus dorsalis*. Shown are the sound pressures, in decibels relative to 1 dyne per sq cm, required for the standard response of 0.1 μv.

6. THE INTERMEDIATE FROGS:

THE PELOBATIDAE

THE FAMILY PELOBATIDAE

The pelobatids include three principal genera, *Scaphiopus*, *Pelobates*, and *Megophrys*, along with a few others, widely distributed over the northern hemisphere. *Scaphiopus* includes a number of species in the southern regions of the United States and Mexico that are commonly known as spadefoot toads because of the presence of a sharp tubercle on the foot that is used in digging. *Pelobates* includes a few species of Europe that are similarly equipped. *Megophrys* is a genus of many species without the digging instrument but otherwise similar to the spadefoot group occurring in the Orient in the region of the South China Sea.

Noble (1931) early noted the similarities of these widely separated species and also observed that they take an intermediate position between the most primitive frogs, the leiopelmids and discoglossids, and the remaining frogs that are advanced in many respects. This intermediate position for the pelobatids has been generally accepted. Lynch (1973) made a special category of transitional families in which the Pelobatidae and the closely similar Pelodytidae were included. Duellman (1975) recognized a superfamily Pelobatoidea that bracketed these two families, and gave this superfamily the most advanced position in the suborder Archaeobatrachia—the primitive frogs. In his classification the oriental megophryines were included in the Pelobatidae.

The *Scaphiopus* species inhabit arid regions and live in burrows dug deep in the sandy or friable soil where sufficient moisture is present for their survival. They may come to the surface in the relative cool of night and appear in great numbers in the breeding season after heavy rains probably have flooded them out of their burrows and fortuitously have produced temporary pools suitable for breeding purposes. They then make their presence known by calling, and after mating the eggs are attached in long strings to water weeds in the shallow pools, where they hatch out quickly, the whole process of development to fully formed frogs taking about a month's time. There is often a fateful race between the maturation process and evaporation of the temporary pool.

Two subgenera of *Scaphiopus* have been recognized (McAlister, 1959, following Blair, 1955, 1956), the subgenus *Scaphiopus* including three species (*Scaphiopus holbrooki*, *S. hurteri*, and *S. couchi*) and the subgenus *Spea* containing another three (*Scaphiopus hammondi*, *S. bombifrons*, and *S. intermontanus*). All except the last-named are included in the present study. These two subgenera differ significantly in the character of the call notes of the males; as characterized by McAlister the species of the *Scaphiopus* group produce groaning sounds or harsh "meows," whereas those of the *Spea* group make sounds like the plucking of a rubber band or striking the teeth of a comb. He found differences in the structure of the vocal mechanism that he thought might account for this distinction in call quality.

A division of the vocal pouch into two chambers by a membrane in the *Spea* species may explain the observation that the frequency of the trills—the pulses produced by the vocal cords—is not in agreement with the fundamental frequency of the call; this agreement is reported as characteristic of frogs of the *Scaphiopus* subgenus.

In the general form of the auditory structures all these species are closely similar and are well advanced over the species already considered as members of the Archaeobatrachia. A tympanic membrane is present, though often it is small and only slightly differentiated in color and texture from the surrounding skin. Other parts of the middle ear are well developed, with an ascending process, a columellar footplate of complex form, and a locking mechanism involving the operculum. Basilar and amphibian papillae are present but the latter organ has only the reduced form seen in the primitive frogs: the posterior division occurring in the advanced frogs is absent. These structures have been examined in close detail in some of the *Scaphiopus* specimens.

Scaphiopus hurteri: Hurter's Spadefoot Toad.   This species from eastern Texas reaches an adult size around 40–80 mm in body length and has a short head about as wide as it is long. The tympanic membranes are easily seen, but are small (2.7 mm in diameter in a specimen with a body length of 51 mm). Figure 6-1 shows a section through the ear region in this animal. Dorsal to the bony columella and closely investing its outer end is a cartilaginous ascending process that runs inward and attaches to the lower surface of the parotic crest. The columella runs inward below this process and expands in a large cartilaginous footplate that forms a deep notch for insertion of the upper end of the operculum. A portion of the opercular muscle is seen along the ventral surface of the operculum, and the columellar muscle attaches by a moderately short ligament to the ventrolateral arm of the footplate. Here is a well-developed lock mechanism for the regulation of acoustic transmission to the inner ear.

The Amphibian Papilla. — A series of transverse sections through the

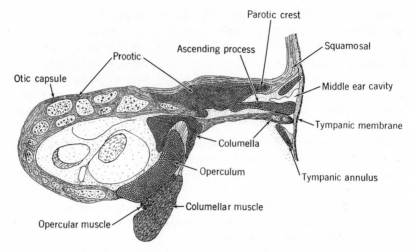

Fig. 6-1. The ear region of *Scaphiopus hurteri*. Scale 10X.

amphibian papilla showing its changing form from anterior to posterior ends is presented in Figs. 6-2 to 6-8.

Figure 6-2 shows this papilla near its anterior end, where there are 14 rows of hair cells and a tectorial structure of relatively simple form. Only a portion of the limbic shelf is included in this figure; it extends medially to its attachment to the bony septum between cranial and otic cavities. As usual the hair cells are suspended between columnar processes from the supporting cells whose nuclei form a continuous array along the limbic surface above, and these hair cells send their ciliary tufts downward to make contact with the tectorial body. This body consists of a central mass made up principally of extremely small vesicles, with fine fibers running upward and spreading inward and outward to the hair cells.

Figure 6-3 shows this papilla a little farther posteriorly, where the number of rows of hair cells has increased to 24, and the tectorial structure has changed in form to include large vesicles in its central body and distinct canals extending to the hair cells. Many of these canals have thick walls, especially those on the medial side, but some of the others have relatively thin walls. There is reason to believe that each hair cell sends its kinocilium into a canal and makes contact with its inner wall, though this connection can be seen only in exceptional instances. Many nerve fibers are found in the limbic plate above, and occasionally one can be followed into the hair-cell layer. The actual connection with the hair cells is usually not seen, as this relation is near the limit of the light microscope.

The Sensing Arm. — Figure 6-4 shows this papilla at a still more posterior position, where there are 29 rows of hair cells. The main portion of the

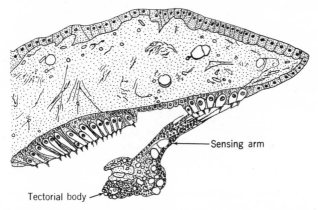

Sensing arm

Tectorial body

Fig. 6-2. The amphibian papilla of *Scaphiopus hurteri* in a transverse section at an anterior level, right side. Scale 200X.

tectorial body remains much as before, with a central mass of large and small vesicles and an array of thick-walled canals extending over the inner and middle hair cells, but a new feature appears in the lateral portion of the structure. A part of the tectorial tissue now forms thin ribbons and strands that constitute a sensing arm extending outward to the most lateral hair cells.

This sensing arm becomes more obvious when examined a little farther posteriorly as in Fig. 6-5. Here the connections between the tectorial body and the more medial hair cells have been lost.

As the figure shows, a number of hair cells remain in this medial region of the papilla and appear to have free-standing ciliary tufts—tufts that seemingly extend into the surrounding fluid without tectorial connections. It is possible, however, that many of these hair cells have indirect connections with the tectorial body, through bridges formed by filaments that attach to the tufts of neighboring cells. This possibility is suggested by the frequent observation of short fibers or wisps of tectorial tissue clinging to the ciliary tufts of these otherwise isolated cells. Such indirect connections would greatly increase the excitability of these cells.

The sensing arm forms little canals at its terminal end and makes connection with only a small group of hair cells at the lateral side of the papilla.

In Fig. 6-6 the medial hair cells have disappeared, and only the lateral (and posterior) group remains. The sensing membrane is progressively extended ventrally and is here approaching the limbic floor of the papillar recess. It attaches to this floor a short distance farther posteriorly as seen in Fig. 6-7. Here it still connects with a small array of lateral hair cells, seen in 7 rows.

Still farther posteriorly, as Fig. 6-8 shows, the number of hair cells increases rapidly, spreading into the medial region that for a space in this pos-

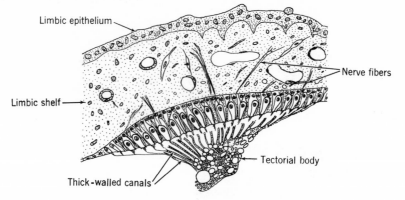

Fig. 6-3. The amphibian papilla of *Scaphiopus hurteri* a little farther posteriorly than in the preceding figure. Scale 200X.

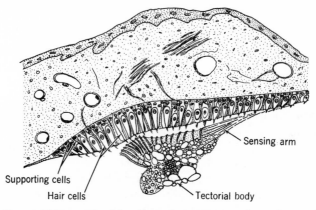

Fig. 6-4. The amphibian papilla of *Scaphiopus hurteri* just posterior to the preceding, where the sensing arm of the tectorial structure appears. Scale 200X.

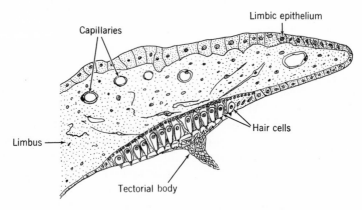

Fig. 6-5. The amphibian papilla of *Scaphiopus hurteri* just posterior to the preceding, where the sensing arm is detached from the inner group of hair cells. Scale 200X.

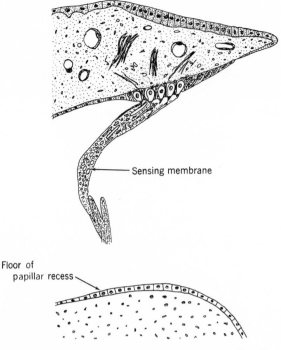

Fig. 6-6. The amphibian papilla of *Scaphiopus hurteri* farther posterior where the innermost hair cells have ended. Scale 200X.

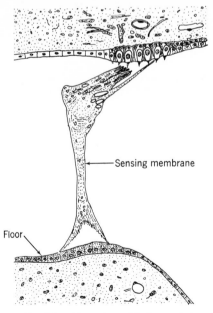

Fig. 6-7. The amphibian papilla of *Scaphiopus hurteri* farther posterior where the sensing membrane attaches to the floor of the papillar recess. Scale 200X.

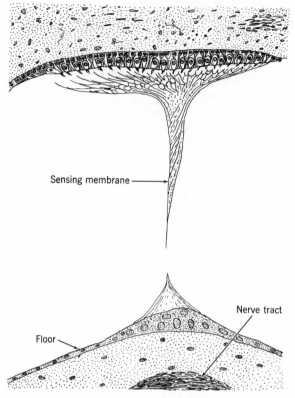

Fig. 6-8. The amphibian papilla of *Scaphiopus hurteri* where the sensing membrane is ending. Scale 200X.

terior portion of the structure has been free of them; the number of rows reaches 23 at the level illustrated.

The sensing membrane, which has formed a curtain of delicate tectorial tissue over the area just examined, now shows a break in its connection to the floor of the papillar recess, and in the next section farther posteriorly it comes to an end. This sensing membrane thus is not a complete barrier across the path of fluid vibrations from oval window to the relief path through the perilymphatic window and duct to the round window, but forms a suspended curtain, attached at upper and lower edges, and responsive to the vibratory current. Along its upper edge this curtain makes direct contact with only a portion of the total complement of hair cells—in this ear to about 30 percent of the total number. This connection would seem to add substantially to the sensitivity of this special group of hair cells. Thus in certain of the hair cells the usual response to vibrations by movements of the tectorial networks lying in the sound path is augmented by the sensing membrane. This membrane should be effective in transmitting the vibratory energy to the hair cells in contact with it because it summates the action of the fluid particles over its surface, converting their somewhat random motions to a resultant motion in phase. This action may be compared with that of a pressure receptor like the basilar membrane of higher vertebrates.

A clearer conception of the structure of the amphibian papilla in *Scaphiopus* can be gained from Fig. 6-9, which is a reconstruction based on four consecutive sections cut in the frontal plane. The view is from the dorsal side, looking into the structure as though it were partially transparent. Swinging around the central mass, which is the tectorial body, are 5–7 rows of hair cells (represented by their nuclei as black dots) in a rather orderly arrangement, forming an incomplete circle. The tectorial mass is made up of many vesicles in the midregion, outside of which appears a series of radiating septa, which are the walls of tectorial canals cut across. The sensing membrane is seen edgewise extending from one region of the mass and running beneath (i.e., ventral to) a portion of the hair cells to which it attaches along its upper border.

The papillar nerve is present, but only a few of its nerve fibers near their terminations can be seen. The trunk of this nerve runs dorsally (toward the observer), spreads out in the limbic roof of the structure, and courses to the hair cells from above.

A further examination of the amphibian papilla in a specimen of *Scaphiopus hurteri* showed additional details of the distribution of its hair cells; the results are presented in Table 6-I for right and left sides. As will be seen, about half of the total number of hair cells were in contact with the principal

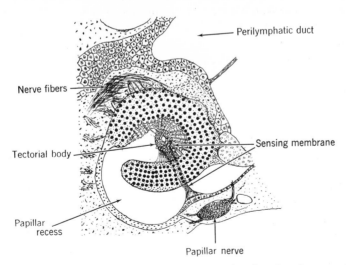

Fig. 6-9. The amphibian papilla of *Scaphiopus hurteri* seen in a frontal section, reconstructed from four successive sections. Anterior is above and lateral to the right. Scale 80X.

Table 6-I

Hair-Cell Distribution in the Amphibian Papilla of *Scaphiopus hurteri*

	To the Tectorial Body	To the Sensing Membrane	Free-standing	Relation Uncertain	Total Hair Cells
Left	199	140	17	27	383
Right	192	125	54	29	400

mass of the tectorial body, about a third were attached to the sensing membrane, and the remaining hair cells were either free-standing or perhaps made indirect connections with the sensing membrane. Thus the sensing membrane plays a significant role in the stimulation process.

The Basilar Papilla. — In these ears the basilar papilla shows no unusual features. In the specimen considered above, this papilla on the left side contained 51 hair cells and the one on the right contained 73 hair cells.

Hair cells were counted in a second specimen of this species that was sectioned frontally, though this plane (which gives a view like Fig. 6-9) presents difficulties in separating the cells into the different groups. The results showed a total of 383 hair cells for the amphibian papilla on the left and 421 hair cells for the one on the right. For the basilar papilla the numbers were 42 for the left side and 60 for the right.

Sensitivity. — Tests were made by stimulating through a sound tube sealed over the tympanic membrane for the application of aerial sounds and recording with an electrode inserted in a minute hole drilled in the region of the anterior semicircular canal so as to make contact with the perilymph. Results of these tests on two specimens of *S. hurteri* are presented in Fig. 6-10. These curves are in agreement in showing good sensitivity over a range from 100 to 1500 Hz, with fairly well-defined peaks in two regions, at 150 and 400 Hz, and for higher tones well-sustained sensitivity up to 900 Hz in one animal and to 1500 Hz in the other. A third specimen, represented in Fig. 6-11, showed similar peak regions, at 150 and 800 Hz, but the general level of sensitivity was considerably lower.

Scaphiopus couchi: Couch's Spadefoot Toad. Couch's spadefoot is a relatively large species found in a broad area that includes most of Texas, adjoining parts of Utah and Arizona, and northern Mexico. It is toadlike in appearance, with a tympanic membrane that is difficult to see because its color and texture are the same as the surrounding skin. This membrane lies in a slight depression in front of the swollen parotoid region, outlined by a

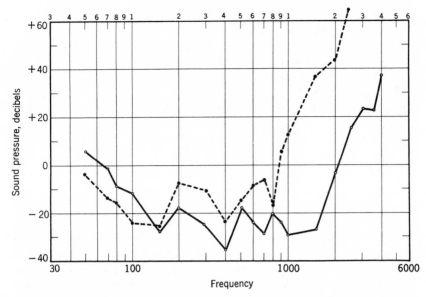

Fig. 6-10. Sensitivity curves for two specimens of *Scaphiopus hurteri*, with aerial stimulation. Shown are the sound pressures, in db relative to 1 dyne per sq cm, required for the standard response of 0.1 μv.

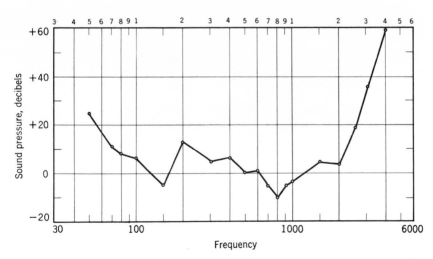

Fig. 6-11. Aerial sensitivity for a third specimen of *Scaphiopus hurteri*, expressed as in the foregoing.

barely perceptible ridge. In a male of 66 mm body length it measured 3 × 4 mm, with the longer axis nearly vertical.

In the breeding season the male of this species produces a call that is said to resemble the bleat of a lamb; the female also is able to vocalize, producing a series of short grunts.

The middle-ear mechanism, represented in Fig. 6-12, in most respects takes the standard form: the outer end of the columella is connected to the middle of the tympanic membrane by a boss of cartilage from which a little rod, the ascending process, swings upward and attaches to the undersurface of the parotic process. The inner end of the columella has a complex form, with a notch for the end of the operculum and a descending bony process on which the ligament from the columellar muscle is inserted. There is a point of departure from the usual form of this connection, however: the free portion of the columellar ligament is greatly reduced. The columellar muscle extends much farther anteriorly than usual so that only a short ligament suffices to make the connection to the footplate. A sheet of ligamentary tissue runs along the anteromedial surface of the muscle, adjacent to the operculum, and summates the forces exerted by the more remote parts of the muscle as usual. This sheet, however, does not continue as a long ligament to the footplate in the customary fashion. It transmits the muscle action in two ways: by the short, thick ligament at its upper end as shown and by a more or less con-

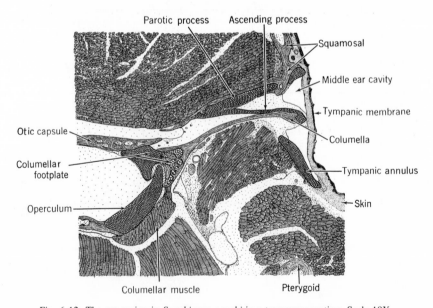

Fig. 6-12. The ear region in *Scaphiopus couchi* in a transverse section. Scale 10X.

tinuous connection with a cartilaginous process of the footplate that extends downward ventral to the body of the muscle. This feature has not been noted in other spadefoot toads; in these others the usual extended columellar ligament was present. Couch's spadefoot thus appears to be exceptional in this respect, though functionally the point seems of little importance.

The amphibian papilla in *S. couchi* has a form closely resembling that pictured and described for *S. hurteri:* the anterior and middle divisions are well developed and a posterior division is lacking. In one specimen studied there were 415 hair cells on the left side, of which 100 had connections to the sensing membrane, and on the right side there were 372 hair cells, with 119 connected to the sensing membrane. The proportion of hair cells served by the sensing membrane is thus 24 and 32 percent on the two sides, comparable to the relation found in *S. hurteri.*

The basilar papilla in this specimen of *S. couchi* showed 46 hair cells on the left side and 44 on the right, but the counting was difficult and these figures must be taken as approximate.

Sensitivity. — Tests in terms of the inner ear potentials were carried out on a number of specimens of *S. couchi.* In Fig. 6-13 are given curves for two of these specimens, one with good sensitivity over the range from 150 to 1500 Hz, and another that is somewhat poor over a similar range. The better of these curves reaches − 18 db at 200 Hz and shows further peaks of good sensitivity at 700 and 1500 Hz. The other animal gives the best response near the lower end of the range, at 80 Hz, and grows poorer as the frequency is raised, with a steep decline beyond 1500 Hz. Figure 6-14 gives results for two additional specimens that agree in showing a peak in the low frequencies, around 180 Hz, and for one of these there is a further peak in the region of 1500 Hz. The general performance for these two ears is poorer than for those represented in the preceding figure, yet the general picture is similar: a considerable degree of variation through the frequency range, but a fair level of sensitivity within the middle range from 150 to 1500 Hz. A fifth specimen whose results are shown in Fig. 6-15 presents a striking picture of bimodal sensitivity: a broad region of good response between 150 and 500 Hz with a maximum at 200–300 Hz, and a secondary maximum between 800 and 1500 Hz. A fourth function for this species, shown in Fig. 6-16, is unusual for its degree of uniformity: the curve is nearly flat through the low and medium frequencies and then rises rapidly above 900 Hz.

Clearly the ear in this species is subject to wide variations, but a preference for the low and middle frequencies is evident, and on the whole two regions are favored, one in the low range around 200 Hz and another in the middle range about 800–1500 Hz. These observations are in good agreement with the results of auditory nerve recording in this species made by Capranica and Moffat (1975). These investigators found two populations of auditory nerve fibers, one with best frequencies distributed over the range

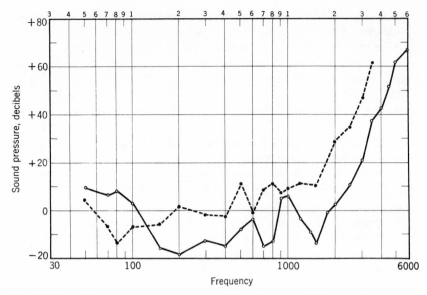

Fig. 6-13. Sensitivity curves for two specimens of *Scaphiopus couchi*, for aerial stimulation. Shown are 0.1 μv curves for sound pressures relative to a zero level of 1 dyne per sq cm.

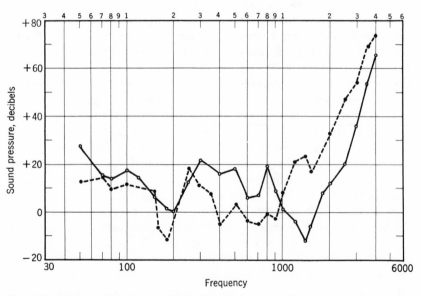

Fig. 6-14. Aerial sensitivity in two additional specimens of *Scaphiopus couchi*, represented as above.

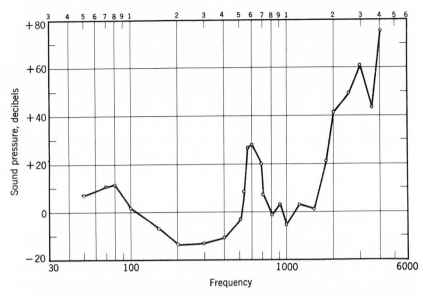

Fig. 6-15. Aerial sensitivity in a further specimen of *Scaphiopus couchi* with a bimodal form of function, represented as in the above figures.

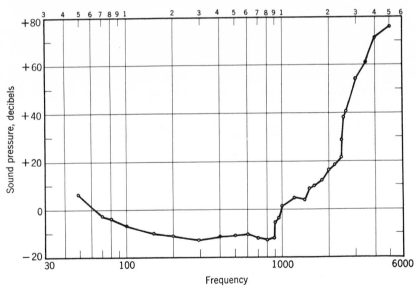

Fig. 6-16. Aerial sensitivity in another specimen of *Scaphiopus couchi* with a highly regular form of function, represented as above.

from 100 to 700 Hz and another population with the range 900 to 1500 Hz. These investigators concluded that the low-frequency responses are principally due to the action of the amphibian papilla and the high-frequency responses reflect the action of the basilar papilla.

Scaphiopus bombifrons: the Plains Spadefoot Toad.   The Plains spadefoot, *Scaphiopus bombifrons*, assigned to the *Spea* subspecies, occupies an area running obliquely through the middle of the United States from Montana to Oklahoma. This form is rather small, of the order of 40–60 mm in body length. The call produced by the male in the breeding season is described as a kind of squawk. The tympanic membrane is inconspicuous and somewhat variable in form; sometimes it is nearly round, but more often it is oval with the longer axis vertical. In one specimen with a body length of 50 mm this membrane was 2.3 mm wide and 2.6 mm high; in another of about the same body length it was 2.3 mm wide and 3.5 mm high.

The inner ear has the form already indicated for this genus, with anterior and middle divisions of the amphibian papilla but lacking the posterior division possessed by the advanced frogs. In one specimen the amphibian papilla contained 469 hair cells on the left side, of which 191 were connected to the sensing membrane, and 451 hair cells on the right, with 163 connected to the sensing membrane. The basilar papilla in this specimen contained 53 hair cells on the left side and 87 hair cells on the right.

A second specimen of this species gave corresponding figures, with a total of 342 hair cells in the amphibian papilla on the left, of which 128 were connected to the sensing membrane, and a total of 394 on the right, with 133 connected to the sensing membrane. In this second specimen the basilar membrane on the left contained 74 hair cells, and the one on the right contained 92 hair cells.

The sensitivity of the first of these two specimens is represented by the solid-line curve of Fig. 6-17. This level is poor, with the more favored region around 400–800 Hz, where the sensitivity reaches +16 to +18 db. A second specimen gave the results shown by the broken curve, which are even poorer in the lower frequencies but elsewhere are in good agreement.

Scaphiopus hammondi: Hammond's Spadefoot Toad.   A second species belonging to the *Spea* subgenus is *Scaphiopus hammondi*, spread along a belt from middle and lower California through New Mexico and Arizona to the edges of Texas and Mexico. It is of small to medium size, around 37 to 60 mm in body length. The tympanic membrane is inconspicuous and in the single specimen examined, whose body length was 45 mm, it was an oval 2.3 mm wide and 2.7 mm high. This specimen had a total of 358 hair cells in the left amphibian papilla and 361 hair cells in the right papilla, distrib-

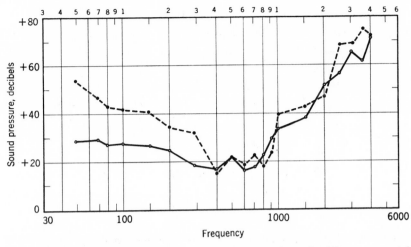

Fig. 6-17. Aerial sensitivity functions for two specimens of *Scaphiopus bombifrons*, represented as above.

uted as shown in Table 6-II. The basilar papilla on each side contained 39 hair cells.

A sensitivity curve for this specimen is shown in Fig. 6-18. This curve is rather uniform in the region of +14 to +17 db in the low frequencies with only a suggestion of a maximum at 200 Hz, and thereafter, beyond 500 Hz, shows a generally progressive loss. This level of sensitivity is perhaps consistent with the small size of the tympanic membrane and the limited number of hair cells.

Megophrys monticola nasuta. A second subfamily of the Pelobatidae is the Megophryinae, containing about 40 species in southeast Asia, mostly in areas bordering the East China Sea; some of these are on the mainland in Thailand and Cambodia, but more are scattered over the islands from Sumatra and Java to the Philippines. The species available for study was *Megophrys monticola nasuta*, a medium-sized form of striking appearance because of the presence of triangular flaps of skin extending over the eyes and a nose process in the form of a cone, as shown in Fig. 6-19.

The Middle Ear. — In this species there is a delicate tympanic membrane covered with skin at the side of the head just behind the eye. Here a triangular area may be seen, outlined by ridges formed by arms of the squamosal bone above and below and the edge of the tympanic membrane as shown in Fig. 6-20. In this drawing only the edges of the membrane are seen; the middle portion, which lies closely over the endplate, had been removed to reveal the deeper structures. Shown are the endplate and the outer end of the

Table 6-II

Hair-Cell Distribution in the Amphibian Papilla of *Scaphiopus hammondi*

	To the Anterior Tectorial Body	To the Sensing Membrane	Free-standing Hair Cells	To the Posterior Tectorial Body	Total Hair Cells
Left	183	147	28	*	358
Right	173	173	15	*	361

*Posterior tectorial body absent.

columellar shaft running anterodorsally through the middle ear cavity to the inner ear region. A strong ligament attaches one edge of the endplate to an arm of the squamosal.

The tympanic membrane in a specimen 75 mm in body length had an area of 7.6 sq mm, and the columellar endplate measured 2.0 sq mm.

Megophrys presents the standard elements of the middle ear, though with certain complications. As Fig. 6-21 shows, there is a long tapering columella with an expanded outer end, which as usual is cartilaginous and attaches dorsally to the tympanic membrane about midway between a squamosal process and the ventral part of the tympanic annulus. The shaft portion of the columella is a thin tube of bone with a core of cartilage. It extends

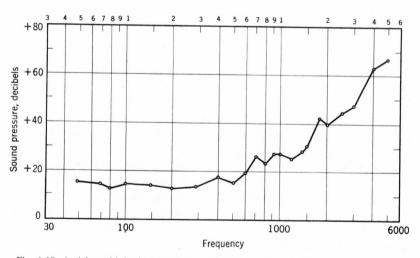

Fig. 6-18. Aerial sensitivity in a specimen of *Scaphiopus hammondi*, represented as above.

Fig. 6-19. *Megophrys monticola nasuta*. Drawing by Anne Cox.

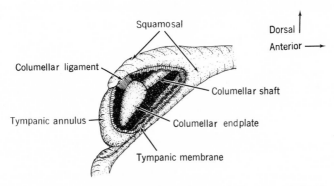

Fig. 6-20. The tympanic region in *Megophrys monticola nasuta*. Scale 6X.

inwardly to the region of the otic capsule, where it becomes fully osseous and thereafter turns to cartilage again as it expands at the oval window and enters the capsular cavity.

The osseous portion of the columella sends a branch process posteroventrally and then fuses to a cartilaginous process that extends from the parotic process of the capsule (indicated as the descending process in part *b* of this figure). The bony extension of the columella is firmly attached to the descending process over its whole lower portion.

Between the ventrally directed branch process and the cartilaginous terminal expansion of the footplate is a deep notch into which the dorsolateral end of the operculum is inserted. The operculum is held in this notch by numerous elastic fibers. At the level shown in part *a* of Fig. 6-21 the operculum is seen outside the otic capsule, but at a level a little posterior to this one the oval window appears, and the operculum (as well as the columella)

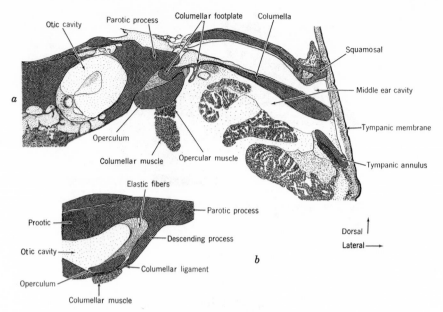

Fig. 6-21. The ear region in *Megophrys monticola nasuta*, from frontal sections, *a*, at a level through the columella, and *b*, at a more ventral level through the descending process. Scale 10X.

enters the capsular cavity; it is seen in this inner position in part *b*. The operculum is attached by a prominent bundle of elastic fibers to the lower concave surface of the parotic region.

Two muscles are identified in this region: an opercular muscle that attaches to the lower surface of the operculum, and a columellar muscle that sends a short ligament dorsolaterally to the end of the descending process. As is common, these two muscles show further marks of distinction: the opercular muscle contains small fibers that were deeply stained in the procedure used, and the columellar muscle is made up of larger fibers that stained more lightly.

This middle ear mechanism presents a peculiar picture: most elements are present that in other frogs operate as a mechanism for the control of sound input to the inner ear, but there is the disturbing presence of the descending process and its attachment to an extension of the footplate. These ears still function, however, even though somewhat poorly as the sensitivity results soon to be presented will show.

The opercular muscle is connected to the operculum in the usual way, and its contraction should pull the end of the operculum out of the notch in the footplate to give to the columella whatever freedom of vibratory motion it retains in the presence of the attachment to the descending process. A contraction of the columellar muscle, however, can be expected to have a much reduced effect, in that it is now pulling on the descending process, and relatively little of its tension can be transferred to the columella. Indeed, there is hardly any need for the protective effect of this muscle, because of the considerable degree of restraint already present.

It may be noted that the columellar muscle in this species is smaller than in most frogs; possibly this is an example of atrophy from disuse.

The Auditory Papillae. — The amphibian papilla in *Megophrys* has the reduced form, with anterior and middle divisions, containing a sensing membrane but lacking a posterior division. In one specimen there were 555 hair cells on the left side, of which 294 were connected to the tectorial body, 156 were connected to the sensing membrane, and 105 were either free-standing or had doubtful connections. On the right side the total was 556, of which 294 were connected to the tectorial body, 184 to the sensing membrane, and 78 were free-standing or doubtful. Thus the percentage of hair cells served by the sensing membrane was 28 percent on one side and 33 percent on the other. This species shows a considerable complement of hair cells in the amphibian papilla, with a large number in the favorable position of being connected to the sensing membrane.

The basilar papilla in this same specimen showed 63 hair cells on the left side and 65 on the right. These numbers are somewhat larger than observed in this papilla in other pelobatids.

Sensitivity. — Tests of sensitivity in terms of the inner ear potentials were carried out in 11 specimens. This ear shows only a fair degree of sensitivity, though there are large individual variations. Curves for two animals that represent about the average of the group are presented in Fig. 6-22. One of these, shown by the solid line, had a level of sensitivity around +3 db over a range of 250–350 Hz, and there is a second region around 700 Hz that contains a peak at +7 db. The other animal, represented by the broken curve, likewise exhibits two maximum regions, one at 100–250 Hz running about +4 db and another around 700 Hz that reached −4 db. About the same order of sensitivity was found for two other animals whose curves are given in Fig. 6-23. One of these, represented by the solid line, shows best sensitivity at 200–300 Hz where it runs close to 0 db, and then shows some irregular peaks at 800 and 1500 Hz. The other animal, shown by the broken curve, has a sensitive point at 100 Hz and a rather broad region of good sensitivity at 700–1800 Hz within which a level of 0 db is reached.

The specimen exhibiting the greatest sensitivity of all those tested in this group is represented by the solid curve of Fig. 6-24. Here a somewhat irregular region of fairly good sensitivity is indicated around 100–500 Hz, where the level reaches −8 db, and a very sensitive band appears at 600–800 Hz where a level of −20 db is attained. This same animal was kept under anes-

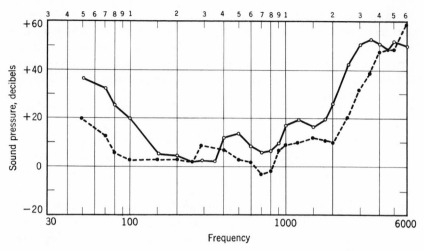

Fig. 6-22. Sensitivity functions for two specimens of *Megophrys monticola nasuta*. Represented is the sound pressure in db relative to 1 dyne per sq cm required for a response of 0.1 μv.

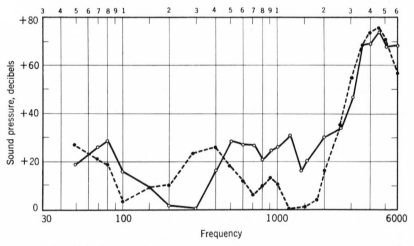

Fig. 6-23. Sensitivity functions for two additional specimens of *Megophrys monticola nasuta*, represented as in the foregoing.

thesia and tested again on the following day, with results shown by the broken curve. This second curve has much the same form as before, but with a considerable loss in the level of sensitivity. The reason for this loss is not altogether clear; it may have been due to an accumulation of fluid in the otic tissues.

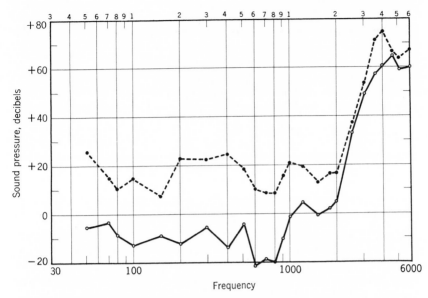

Fig. 6-24. Sensitivity functions for a specimen of *Megophrys monticola nasuta*, recorded on the first day (solid line) and one day later (broken line). Represented as above.

It appears from these observations that *Megophrys*, though suffering the presence of a particularly firm connection between columella and otic capsule that would seem to have a damping effect on sound transmission, nevertheless exhibits a fair degree of auditory sensitivity. Perhaps the rather large population of hair cells in the auditory papillae has partially compensated for this structural feature.

The characters observed in the ears of this family of Pelobatidae give good support to the intermediate status accorded to this group on other, more general grounds. The anterior tectorial body is well developed, and a sensing membrane is present, connected to a good proportion of the hair cells. Absent, however, is a posterior division of the amphibian papilla, to be seen as a prominent feature in all the advanced frogs.

7. THE ADVANCED FROGS:

THE LEPTODACTYLIDAE

AND BUFONIDAE

The advanced frogs were given the formal designation of Neobatrachia by Reig (1958) and are primarily distinguished by their vertebral structure; with a few exceptions all have holochordal centra: the original cartilage of the notochord is largely replaced by bone. In Dowling and Duellman's classification (see Chapter 1, p. 13) this group is divided into three superfamilies, the Bufonoidea, Microhyloidea, and Ranoidea. The first of these, the Bufonoidea, includes a large number of toad-like forms in nine families as shown in Table 7-I. The second superfamily is the Microhyloidea containing the single family of Microhylidae, though sometimes the *Phrynomerus* species are split away as a separate group. The third of these superfamilies is the Ranoidea, the common frogs of ponds and streams in many areas, and including four families, the Sooglossidae, Ranidae, Hyperoliidae, and Rhacophoridae. Now to be considered is the large group of Bufonoidea, beginning with the family Leptodactylidae.

THE LEPTODACTYLIDAE

The Leptodactylidae constitute one of the largest amphibian families. Lynch (1971) recognized 650 species in 57 genera, arranged in 7 subfamilies. These subfamilies are of wide distribution: two are in Australia and adjacent areas, one occupies the southern tip of Africa, and the remaining four are in South America, covering almost the entire continent; they extend northwest through Central America and Mexico into the United States as far as Texas, and are also scattered along the islands of the Caribbean Sea and in Florida.

Breeding practices among the Leptodactylidae vary greatly with species, and for many are not completely known. Most species are said to lay their eggs in water or on land near by. The *Eleutherodactylus* species are reported to lay their eggs on land, even away from water, usually in a cavity in moist soil. Often the eggs undergo early growth and differentiation and reach a stage where a well-formed froglet emerges (a process known as direct development, in which a tadpole stage is lacking).

Only a small sampling of this large group was possible, with representa-

Table 7-1

The Advanced Frogs

Superfamily Bufonoidea
  Family Myobatrachidae*
  Family Leptodactylidae
  Family Bufonidae
  Family Brachycephalidae
  Family Rhinodermatidae*
  Family Dendrobatidae
  Family Pseudidae*
  Family Hylidae
  Family Centrolenidae
Superfamily Microhyloidea
  Family Microhylidae
Superfamily Ranoidea
  Family Sooglossidae*
  Family Ranidae
  Family Hyperoliidae
  Family Rhacophoridae

*No specimens available for study.

tives of four genera collected in the Panama Canal Zone. These included 5 species of *Leptodactylus*, 3 of *Eleutherodactylus*, and one each of *Pleurodema* and *Physalaemus*. In ear structure these 10 species present a high degree of differentiation, much as has already been seen in *Rana*; all give a picture of advanced development of the amphibian papilla, with a fairly large complement of hair cells, and a moderate development of the basilar papilla. The amphibian papilla in all these species presents three divisions (as described earlier in *Rana*), with the anterior division most prominent, then giving way to the middle division with its sensing membrane, and the structure finally ending with the long, slender posterior division. This posterior division presents an extended ribbon of hair cells contained in a relatively narrow tube, through which fluid surges are set up by vibrations of the columella, with a relief path at the far posterior end of the tube leading to the round window.

In most species this ribbon of cells as followed from anterior to posterior ends gradually dwindles and then disappears with only a barely noticeable increase in hair-cell numbers as the papilla terminates in a sharp bend. However, in certain species of Leptodactylidae, as will be described, there is a marked departure from this typical form: the posterior end of the structure presents a notable flare in hair-cell numbers. Details of these and other features will be presented for a few of the species examined.

Leptodactylus labialis. This is a lowland frog occurring mainly in Mexico but extending into southern Texas, as well as southward into middle America to northern Colombia and Venezuela. It lives in such places as roadside ditches, drains, and wet meadows.

A cross-sectional view of the ear region is shown in Fig. 7-1, where a relatively large, thin tympanic membrane is presented, with operculum and columella filling the oval window.

The general form of the amphibian papilla is much like that described earlier in *Rana*. This papilla, as shown in Fig. 7-2 in a transverse section through its anterior end, presents 10 rows of hair cells from which depends a tectorial mass made up of extremely fine strands that mostly run dorsoventrally and are only tenuously connected to the hair cells at their dorsal ends; these strands are themselves closely interwoven along their ventral border. Farther posteriorly the inner and dorsal strands become greatly thickened and form little canals, each of which receives at its upper end the ciliary tuft of a hair cell. This form continues posteriorly with the number of thick-walled canals increasing in number until all the hair cells of the row are thus served. The number of rows of these cells is greatly increased at this end, reaching 23 in the specimen examined. Then abruptly—in the very next section 20 μm away, as shown in Fig. 7-3—this papilla divides, and the hair cells form two groups of about 8 cells each, with a space between where only supporting cells remain. At the same time the tectorial structure also has greatly altered; it takes the form of a "Y," with an arm going to each group of hair cells and the stem of the "Y" expanded to a foot that attaches to a limbic plate serving as the floor of the papillar recess. This remaining ribbon of tectorial tissue is the sensing membrane and stands as a barrier in the path of the fluid movements set up by vibrations of the columellar footplate under the action of sound. The width of this sensing membrane was measured as approximately 80 μm in the specimen examined, and its effective area was a little over 140 sq μm. This membrane thus serves as a delicate transmitter of vibration to the small group of hair cells to which it makes connection.

A little farther posteriorly the medial arm of the "Y" disappears, leaving only the lateral one, which together with the stem portion constitutes the sensing membrane.

Still farther posteriorly as seen in Fig. 7-4 the more medial hair cells disappear and only the lateral ones are left with their connections to the sensing membrane. These connections continue, and at the same time the dorsal part of the sensing membrane becomes enlarged and comes to contain a series of parallel, thick-walled canals that fit over the ciliary tufts of the hair cells, as seen in Fig. 7-5. The sensing membrane then ends, and the enlarged

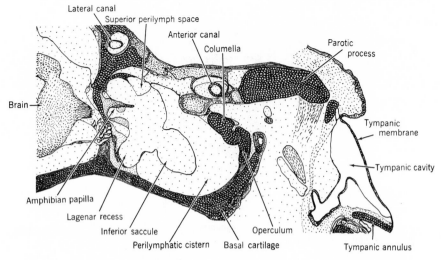

Fig. 7-1. A transverse section through the ear region of *Leptodactylus labialis*. Scale 20X.

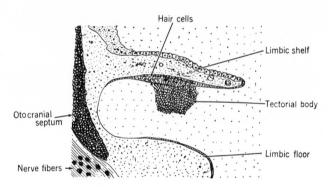

Fig. 7-2. The anterior region of the amphibian papilla in *Leptodactylus labialis*. Scale 100X.

canaliculate tectorial structure remains over the extended ends of the hair cells, as shown in Fig. 7-6. This is the posterior division of the amphibian papilla and it continues for a short distance after the sensing membrane has ended, with a steady reduction in the size and number of hair cells covered. Finally, near the end, there is in this species a flaring and bending of the tube that encloses this portion of the papilla and a noticeable increase in the number of hair cells. Figure 7-7 shows the beginning of this increase, along with a prominent extension of the tectorial body. The following section (Fig. 7-8) indicates a great proliferation of hair cells, after which the papilla comes to an end.

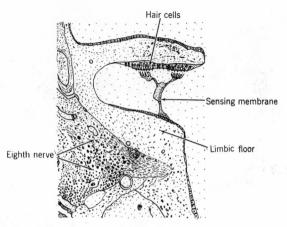

Fig. 7-3. The midregion of the amphibian papilla in *Leptodactylus labialis*, where the sensory structure is duplex. Scale 80X.

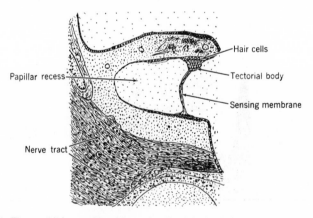

Fig. 7-4. The amphibian papilla of *Leptodactylus labialis* farther posteriorly. Scale 80X.

Counts were made of the hair cells in the different regions of the papilla in both left and right ears with results shown in Table 7-II. The number of hair cells contained in the posterior tectorial body is impressive, exceeding the number in the anterior division of this structure.

By comparison the basilar papilla is poorly developed; there were only 33 hair cells in this papilla on the left side and 26 hair cells in the one on the right.

Leptodactylus bolivianus. A single specimen of *Leptodactylus bolivianus* was examined. This is a medium-sized species inhabiting the northcentral

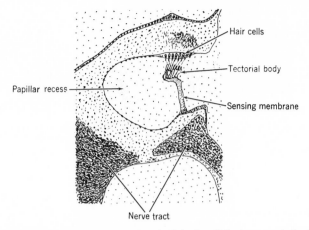

Fig. 7-5. The amphibian papilla of *Leptodactylus labialis* still farther posteriorly. Scale 80X.

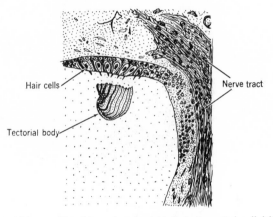

Fig. 7-6. The amphibian papilla of *Leptodactylus labialis* in its posterior division. The separation of the tectorial body from its hair cells is an artifact. Scale 200X.

area of South America and extending northwestward into Panama. The specimen studied measured 7.9 cm in body length and had a nearly round eardrum 5 mm in diameter.

As shown in Fig. 7-9, the columella in this species presents a relatively large cartilaginous pars externa attached to the inner surface of the tympanic membrane a little posterior to its middle and runs dorsally to the parotic region where it sends off two branches, one a dorsolateral process that expands to a knob lying beneath the skin and another (the ascending process), more ventrally located, that fuses with the cartilage and bone of the parotic

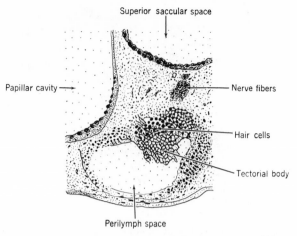

Fig. 7-7. The amphibian papilla of *Leptodactylus labialis* near the beginning of its posterior flare. Scale 125X.

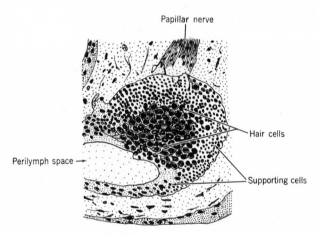

Fig. 7-8. The posterior division of the amphibian papilla in *Leptodactylus labialis* at its maximum expansion. Scale 200X.

region. The columella in its course inward from the lateral knob in the parotic region, as shown in Fig. 7-10, first runs ventromedially, where its bony pars media makes a close contact with the lateral end of the parasphenoid cartilage, with an attachment formed by a small group of connective tissue fibers, and then continues medially as the cartilaginous pars interna. This innermost part of the columella now fills the dorsolateral opening of the otic

Table 7-II

Hair-Cell Distribution in the Amphibian Papilla of the Leptodactylidae

	To the Anterior Tectorial Body	To the Sensing Membrane	Free-standing Hair Cells	To the Posterior Tectorial Body	Total Hair Cells
		Leptodactylus labialis			
Left	235	60	27	247	569
Right	224	62	16	262	564
		Leptodactylus bolivianus			
Left	140	20	†	211	371
Right	110	35	†	225	370
		Pleurodema brachiops			
Left	135	66	31	176	408
Right	146	36	23	124	329
		Physalaemus pustolosus			
Left	191	68	6	106	371
Right	146	73	25	118	362
		Eleutherodactylus gaigei			
Left	163	37	0	377 ††	577
Right	153	37	6	369 ††	565
		Eleutherodactylus longirostris			
Left	183	55	†	162	400
Right	198	70	†	145	413

† Not determined.

†† Sum for left ear includes total for posterior tectorial body and accessory papilla; for right ear includes 160 for posterior tectorial body and 209 for accessory papilla.

capsule—the oval window. Seen at this level is the saccular macula lying along the floor of the saccule, and suspended from a shelf near the middle of the otocranial septum is the amphibian papilla.

The Amphibian Papilla. — This papilla begins with 7 hair cells in a row that, when the papilla is viewed in a single section, increase posteriorly to around 18 or 20, whereupon the papilla divides, with a somewhat greater number of hair cells appearing in the medial portion of the structure. Then both parts diminish, and the medial one disappears. The number of hair cells in the lateral part of the papilla varies irregularly for a short distance and then rises slowly to about 11, whereupon almost suddenly there is a great

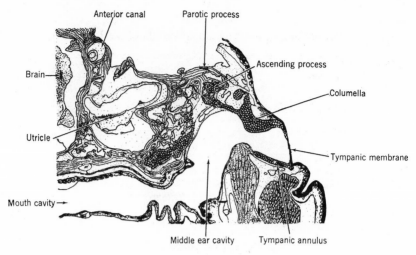

Fig. 7-9. A transverse section through the ear region of *Leptodactylus bolivianus*. Scale 25X.

expansion that reaches 80 or more, after which the papilla quickly comes to an end. The hair-cell distribution is shown in Table 7-II.

At the anterior end of the papilla there is the usual lateral opening to the papillar cavity, as Fig. 7-10 has shown, through which sound vibrations have ready access to the hair-cell layer. This opening continues posteriorly over about two-thirds of the extent of the papilla and then is closed by a thick lateral wall that serves as a path through which the papillar nerve runs from the anterior portion of the eighth nerve below to a position within the limbic roof of the papillar structure, from which the fibers find their way ventrally to the hair cells. This level is shown in Fig. 7-11. More posteriorly this thickening of the lateral wall disappears, and the wall thereafter consists only of an exceedingly thin membrane. There is a corresponding thinning of the inner wall also, and finally this whole enclosure is reduced and disappears as the papilla comes to an end.

As this description has indicated, the papilla passes through a series of changes in its course from anterior to posterior ends. It begins at its anterior end as a simple array of hair cells suspended from the ventral surface of the limbic shelf along the otocranial septum, with a tectorial mass below that embeds the ciliated ends of the hair cells.

About two-fifths of the distance toward the posterior end of this papilla a degree of duplexity enters and soon is displayed as a complete separation of the papilla into two parts, a small medial portion and a larger lateral one. Then a little farther posteriorly the medial portion ends, and only the lateral one remains, with the ciliary tufts of the hair cells connected to a tectorial

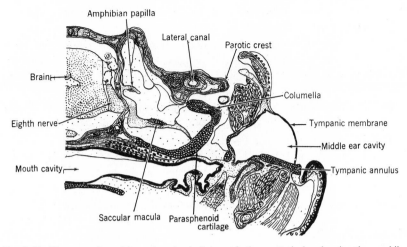

Fig. 7-10. The ear region of *Leptodactylus bolivianus* farther posteriorly, showing the amphibian papilla. Scale 20X.

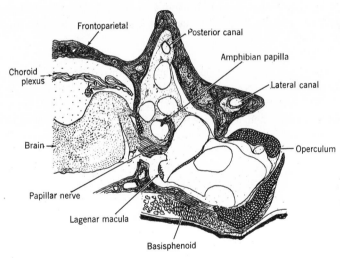

Fig. 7-11. The ear region of *Leptodactylus bolivianus* far posteriorly, passing through the posterior division of the amphibian papilla. Scale 30X.

ribbon whose ventral edge is attached to the floor of the papillar cavity, thus constituting a sensing membrane. Still farther posteriorly a tectorial body appears that now makes the connections to the hair cells and is attached ventrally to the sensing membrane. After this point the sensing membrane ends, and the tectorial body alone remains suspended below the hair-cell layer, with the ciliary tufts of the hair cells inserted into the upper ends of its parallel-running canals.

Near the posterior end of the papilla as the hair cells begin their great proliferation the tectorial body loses its cap-like form and then disappears. As shown in Table 7-II the number of hair cells in the posterior division considerably exceeds that in the anterior division.

The Basilar Papilla. — In this specimen there were 45 hair cells in the basilar papilla on the right side, and 38 hair cells in the one on the left.

The Stimulation Pathways. — The principal mode of entrance of sounds in this species, as in others, is from the inner end of the columella in the oval window, across the perilymphatic cistern to the lateral opening of the papilla, and then through the tectorial structures to the hair cells. As the sensing membrane appears near the middle of this papilla an improvement in sensitivity can be expected since this membrane serves as a receptive curtain across the vibratory path.

After the lateral wall of the papilla has closed, the action of sounds is no longer direct, but must involve a continuation of the vibratory flow posteriorly through the narrow papillar duct, passing out of this duct through its thin side walls at its lowermost end and then entering first the perilymphatic duct and finally the perilymphatic sac that leads outward through the round window.

Pleurodema brachiops. Ten species are recognized in the genus *Pleurodema* (Lynch, 1971), most of them restricted to the southern part of South America. The species *brachiops*, collected in Panama, is small in size; the specimen examined had a body length of 3.7 cm and an eardrum of oval form, inclined forward, that was 1.8 mm wide and 2.2 mm high. The amphibian papilla is well developed, containing the usual three divisions, and in the anterior division there were 23 rows of hair cells as the papilla began to divide into medial and lateral portions. Here the medial portion is the more prominent, but a little farther posteriorly it dwindles and disappears, leaving the lateral portion to continue its connection to the sensing membrane until, still farther posteriorly, this membrane ends and a small cap-like tectorial body appears to make the attachments to the ciliary tufts of the hair cells. In this posterior region the ventral wall of the papillar cavity thins out to a simple membrane consisting of squamous cells in a layer 1 or 2 cells thick, bordering a perilymph space that ventrally leads to the round window. At its posterior end this papilla shows a moderate expansion, which on one side reaches 39 hair cells as seen in a single section. The hair-cell distributions as determined in this specimen are shown in Table 7-II.

Physalaemus pustulosus. *Physalaemus* is one of the larger genera of leptodactylids; Lynch (1971) recognized 34 species distributed in open lowland areas from southern Mexico to Argentina. Five specimens of the species *pustulosus* were studied. These are moderately small frogs, with body weights varying from 1 to 2 grams, and body lengths around 3 cm.

In these species a tympanic membrane is not immediately obvious, but careful probing in the region behind the eye reveals a yielding area, and on removal of a thin skin layer a disk of cartilage comes to view, held in the cartilaginous frame of the tympanic annulus. In one specimen this disk measured 1.1 by 1.3 mm, and connected to its center was the end of the pars externa of the columella. This element runs inward and connects through the pars media to a relatively large footplate in the oval window.

The papillar structures in this species follow generally the forms already described for leptodactylids. In one specimen the amphibian papilla on the left side contained 371 hair cells and the one on the right 362 hair cells, distributed as in Table 7-II. The posterior portion of the amphibian papilla is relatively short, and shows only a moderate increase in hair-cell numbers near its termination posteriorly.

ELEUTHERODACTYLUS SPECIES

A further genus belonging to the Leptodactylidae is *Eleutherodactylus*, with more than 340 species in Mexico, Central America, the West Indies, and South America. Three species were available for study, all collected in Panama and tentatively identified as *gaigei*, *planirostris*, and *longirostris*. These frogs produce a few relatively large eggs that are deposited in terrestrial locations, and these exhibit direct development: a small froglet is hatched out, without a tadpole stage.

Eleutherodactylus gaigei. The specimens of these small frogs had body lengths of about 4 cm and weighed 4 grams. An orientation view of the right side of the head passing through the ear region at a somewhat anterior level is presented in Fig. 7-12. Here is seen a large, thin tympanic membrane covering a fairly spacious middle ear cavity, with the bony pars externa of the columella attached near the membrane's dorsal edge. This bony process runs inward to join the upper end of the otic capsule wall and then continues as the cartilaginous pars interna, which fuses to a plate of cartilage on the surface of a posterior extension of the prootic bone just below the lateral canal. Farther anteriorly a cartilaginous element (the ascending process) had run from an attachment on the tympanic membrane just ventral to the columellar attachment shown here and had extended obliquely upward to fuse with the parotic crest; the end of this connecting process is still to be seen just inward of the squamosal.

Farther posteriorly, as shown in Fig. 7-13, the operculum comes into view running within the curve of the otic capsule wall and medially attaching to the expanded middle portion of the prootic. A portion of the pars interna

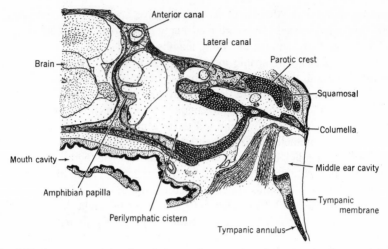

Fig. 7-12. A transverse section through the anterior ear region of *Eleutherodactylus gaigei*. Scale 15X.

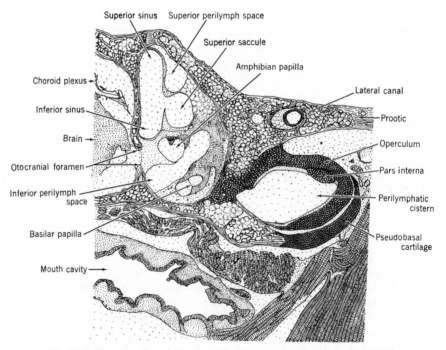

Fig. 7-13. The ear region of *Eleutherodactylus gaigei* far posteriorly. Scale 25X.

remains, now enclosed by the operculum. Farther posteriorly the pars interna disappears, and the operculum is seen as a complete ring of cartilage, and then farther still, as its posterior end is cut through, the operculum appears as an ovoid disk, now only half enclosed by the capsule wall. At a still more posterior location the capsular wall ends, and the operculum appears smaller as its posterior end is approached. Thereafter the opercular muscle comes into view, with its fibers directed dorsally.

The Amphibian Papilla. — The amphibian papilla has been seen near its anterior end in Fig. 7-12, where there are 7 rows of hair cells, and the tectorial body is only beginning. In the next section the number of hair-cell rows increases to 14, and the tectorial body shows a corresponding number of canals. Posteriorly this tectorial structure takes its usual course, dividing into two portions, with the more lateral one favored by a connection with the sensing membrane, and then continuing alone as the other portion disappears.

In Fig. 7-13 the amphibian papilla is seen well toward its posterior end, where the papillar nerve tract has run around the ventral end of the papilla and is approaching the hair cells from the dorsolateral side. Below the row of hair cells is a tectorial body containing 10 canals corresponding to the hair-cell numbers. The ventrolateral wall of the papillar cavity here is only a thin membrane, which adjoins a perilymph space. The basilar papilla appears in this same section.

An unusual feature is shown in Fig. 7-14, representing a section a little

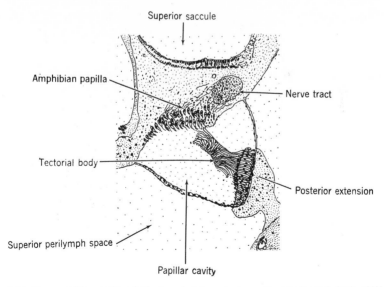

Fig. 7-14. The amphibian papilla of *Eleutherodactylus gaigei* near its posterior end. Scale 100X.

posterior to the level just pictured: a layer of tall columnar hair cells lies on the ventrolateral wall of the papillar cavity and the tectorial body bridges this cavity, connecting to the ciliary tufts of these hair cells as well as to those of the main body above. Then in the following section the gap between these two arrays of hair cells is filled in, and the papilla forms a semicircle of about 40 rows of hair cells, as represented in Fig. 7-15. A little farther posteriorly the perilymph space shown here connects through the round window membrane with the pharyngeal cavity, which is air-filled, and thus provides a region of vibratory outflow.

Table 7-II gives the number of hair cells in this papilla for the different regions of each ear, except that on the left side there was no clear division between the posterior tectorial body and its extension as the accessory papilla and thus the figure given represents the combined number in these two regions. The total number of hair cells in these ears is impressive.

The special arrangement toward the posterior end of the amphibian papilla appears to be highly favorable for sound reception; a large population of hair cells lies in the path of vibratory flow.

Eleutherodactylus longirostris. In this species the middle ear takes a curious form: the tympanic membrane is of a type often referred to as "concealed": it lies deep, well below the skin surface, and is covered by a thin

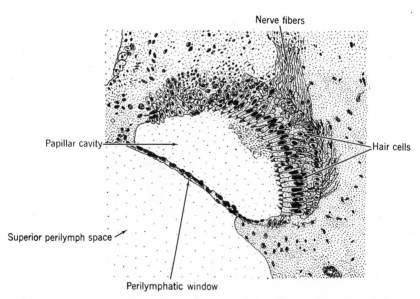

Fig. 7-15. The posterior division of the amphibian papilla of *Eleutherodactylus gaigei* at its greatest expansion. Scale 200X.

membrane that bears a plate of cartilage on its outer surface. This cartilaginous plate has a bilobed form and sends outward through a thick, loose subdermal layer a stalk of cartilage that expands dorsally as a lozenge-shaped body in close contact with the skin. This cartilage thus serves as a sensor of the vibrations reaching the skin surface through the action of a sound. The form of the structure at one level is shown in Fig. 7-16.

As may be seen, the sensing cartilage is embedded in the superficial skin layer that covers the large middle ear air cavity, which gives the superficial structures considerable mobility. This plate, with its extending knob, is a part of the columella and undergoes complex changes in form as it extends posteriorly. It finally covers a wide opening in the lateral surface of the otic capsule, which is the oval window. The columella thus conveys its vibrations in response to sounds to the fluids of the perilymphatic cistern, and thus to the two auditory papillae.

The distribution of hair cells in the amphibian papilla is indicated for one of the specimens in Table 7-II.

The course of vibratory fluid flow could readily be followed through the series of sections representing this ear and closely resembled that already described for other species. The amphibian papilla ends in a simple manner without the expansion of hair cells at the posterior end as just described for *Eleutherodactylus gaigei.*

THE BUFONIDAE

The second family of advanced frogs to be considered is the Bufonidae, whose members are commonly identified as toads and characterized by their com-

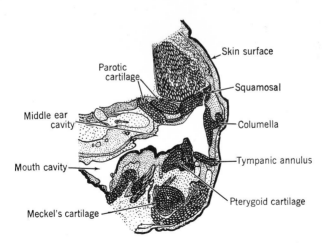

Fig. 7-16. The middle ear region of *Eleutherodactylus longirostris* showing the sensing cartilage of the columella. Scale 25X.

pact body and short legs, and their habit of walking or hopping rather than leaping as frogs do. They are properly designated as cosmopolitan, inhabiting all of the major land masses except Australia, Antarctica, and the island of Madagascar. Their general success is attributed to a high degree of adaptability to a variety of habitats; they are mainly terrestrial, though a few are aquatic or arboreal, and they make fewer demands in the way of water sources than most frogs do. They require only a moderate amount of moisture throughout the year, and for most species a shallow pool or its equivalent in the form of a bromeliad or a water pocket in a hollow tree serves adequately for use during the breeding season.

A peculiarity of the Bufonidae is the presence in the male between the gonad and the fat body on each side of the midline of a small structure known as the organ of Bidder, which appears to be a mass of immature egg cells. If the gonads are removed these structures in the course of several months develop into functional ovaries.

Eight species of the genus *Bufo* were studied and will be briefly described.

Bufo alvarius, the Colorado River Toad. *Bufo alvarius* has a range in the southwestern part of North America centering in the Sonoran Desert area, but extending widely over the regions covered by the Colorado and Gila rivers, occurring nearly always in close proximity to permanent streams and pools. This toad grows to a large size, up to 15 cm in body length in the males and 17 cm in females. The specimens studied were smaller, varying from 7 to 10 cm in length.

The Middle Ear. — The tympanic membranes in the specimens examined were of oval form, with the longer axis nearly vertical, and measured from 4.0×5.0 mm to 6.0×7.0 mm. The columella attaches to the middle of this membrane, with a rather small pars externa and a relatively large ascending process, and then extends inward as a straight bony shaft whose expansion at the otic capsule is continued by a particularly large mass of cartilage that fills the outer portion of the lateral chamber. This large footplate contains a deep notch into which the anterolateral end of the operculum is inserted, and there is also, just medial to this notch, a posteriorly directed process of the footplate that fits into a depression in the upper middle portion of this end of the operculum, as may be seen in Fig. 7-17.

For connection to the columellar ligament the posterior process of the footplate is extended by a cartilaginous process ("link process") that appears to be secondary and is partly formed within the ligament. As Fig. 7-17 shows, this cartilage is attached at its broad anterior end to the posterior process of the footplate and then at its extreme posterior end is embedded in the ligamentary band extending from the columellar muscle.

The columellar muscle is well developed, but in proportion to the body size is smaller than in most frogs. The opercular muscle is relatively large,

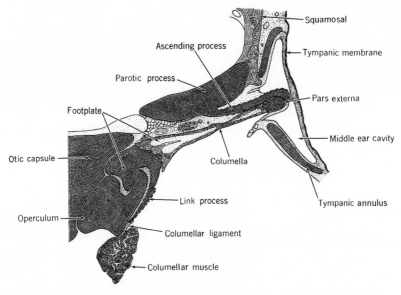

Fig. 7-17. The right ear region of *Bufo alvarius* in a transverse section. Scale 12.5X.

having about half the bulk of the columellar muscle; it lies far anterior to the level shown in Fig. 7-17.

The Inner Ear. — The amphibian papilla of *Bufo alvarius* begins with the usual tectorial body, and then about a third of the way along the papilla the sensing membrane appears, making contact with the extended tectorial body toward its lateral end. Almost immediately this anterior tectorial body comes to an end, and the sensing membrane is left to serve the more lateral hair cells. The remaining hair cells at this level have only free-standing ciliary tufts. Then, only a little farther along the papilla, the posterior tectorial body appears with the sensing membrane attached to it, and with both structures serving the same group of hair cells on the lateral side of the papilla. Supporting cells continue for a short distance on the medial side, but no hair cells are clearly discernible among them. The sensing membrane then ends at a point a little over halfway along the structure, leaving the posterior tectorial body by itself to cover the remainder of the papilla. The numbers of hair cells contacted by these different tectorial structures are indicated in Table 7-III.

The Basilar Papilla. — The basilar papilla in this species is relatively well developed. The one on the left side contained 110 hair cells, and the one on the right 98 hair cells.

Sensitivity. — Measurements of sensitivity were made in three specimens with an electrode inserted into the anterior semicircular canal; the results are

Table 7-III

Hair-Cell Distribution in the Amphibian Papilla of the Bufonidae

	To the Anterior Tectorial Body	To the Sensing Membrane	Free-standing Hair Cells	To the Posterior Tectorial Body	Total Hair Cells
		Bufo alvarius			
Left	206	89	87	69	469
Right	222	62	131	57	472
		Bufo americanus			
Left	163	83	104	19	369
Right	154	88	99	17	358
		Bufo boreas boreas			
Left	221	83	103	105	512
Right	236	73	97	58	464
		Bufo cognatus			
Left	163	106	22	113	404
Right	192	81	24	153	450
		Bufo granulosus			
Left	109	62	16	97	284
Right	131	50	15	99	295
		Bufo marinus			
Left	244	146	5	230	625
Right	274	161	6	220	661
		Bufo valliceps			
Left	220	119	11	128	478
Right	248	100	9	148	505
		Bufo viridis			
Left	263	80	66	59	468
Right	273	51	18	111	453

indicated in the two following figures. Figure 7-18 presents a curve for one of the animals in which a fair degree of sensitivity is indicated in the region of 500–2000 Hz, with a clear maximum at 1000 Hz where the curve reached −16 db. There is only a suggestion of a secondary maximum in the low frequencies around 100–200 Hz.

A second animal whose curve is the broken line of Fig. 7-19 shows only slightly better sensitivity in general, reaching a maximum of −24 db at 1500 Hz, and presenting a fairly good level over the range of 800–2000 Hz.

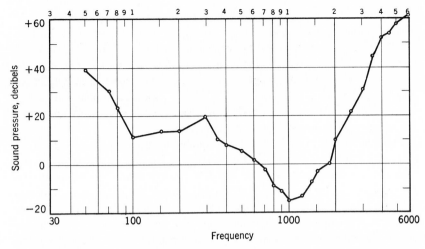

Fig. 7-18. Sensitivity in a specimen of *Bufo alvarius*. Shown is the sound pressure in decibels relative to a zero level of 1 dyne per sq cm required for an inner ear response of 0.1 μv. over the frequency range.

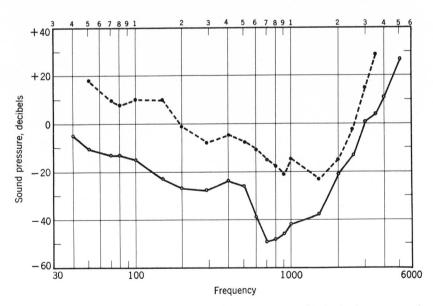

Fig. 7-19. Sound pressure functions for two additional specimens of *Bufo alvarius*, represented as in the preceding figure.

The third animal, whose function is shown by the solid line in this figure, exhibits a much greater degree of sensitivity. This curve is best in the region of 600–1500 Hz and reaches the excellent level of −48 db at 700 Hz. A secondary maximum is indicated around 200–500 Hz, but is not prominent. These three functions are in agreement in showing the best sensitivity in the medium high frequencies around 600–2000 Hz, and also in indicating a secondary region in the low tones where the gain in response is only moderate, but agrees well in its location in all three ears.

The mating call of this species was studied by Blair and Pettus (1954) in a large breeding aggregation in a stock tank in the Sonoran desert that contained at least three other anuran species at the same time. A significant feature reported for the male *Bufo alvarius* specimens was the extreme faintness of their calls. Also an examination of the vocal pouches in these animals indicated very small cavities or sometimes no functional cavity at all. These authors therefore concluded that the vocal apparatus in this species is vestigial and probably nonfunctional in the breeding process. In their view the close crowding of these animals in small and scarce desert pools makes a mating call unnecessary.

The observations of Blair and Pettus on the character of the mating call in this species, and an analysis made with the sonograph, showed a fundamental frequency of 1096 Hz, with progressively weaker harmonics up to 4384 Hz. An examination of Figs. 7-18 and 7-19 shows close agreement between this fundamental frequency and the region of the ear's maximum sensitivity. This relation suggests that the mating call, even though of low intensity, may have a useful function after all: it may serve to sort out suitable mates among the various species present in the same pool.

Bufo americanus. The American toad occurs in many areas of the northeastern United States and extends as far north as Canada and westward to the edge of Kansas. It flourishes in many habitats from grassy stretches in towns and cities to mountain slopes where there are pools and moist areas suitable for breeding purposes. These frogs are of small to medium size, usually of the order of 5–9 cm in body length, though larger ones are occasionally found.

A tympanic membrane is present, though it is sometimes difficult to see. It is located just below the junction of the eye and the prominent parotoid gland and is of oval form with the longer axis vertical. In a male with a body length of 4 cm this membrane measured 1 mm wide and 2 mm high; in a female with a body length of 4.8 cm it measured 1.6 by 3.0 mm.

The columella has the typical form: a cartilaginous pars externa is attached to the middle and posterior portions of the tympanic membrane, and a bony shaft extends posteromedially to a deeply notched footplate that is mostly cartilaginous, with a particularly long and slender posterior process to which the columellar tendon is attached. The medial end of the footplate nearly fills the lateral chamber of the otic capsule. The acoustic control system is well formed, with the end of the operculum extending deeply into the notch in the footplate, and well-defined opercular and columellar muscles are present to regulate the locking mechanism.

A cross-sectional view of the ear structure is given by Fig. 7-20 at a level where the columella is present alongside the operculum in the oval window, and these occupy the anterolateral opening of the lateral chamber. At this level the amphibian papilla is just beginning to appear.

The Amphibian Papilla. — When this papilla is followed in transverse sections from anterior to posterior ends, as the series of drawings (Figs. 7-21 to 7-27) will illustrate, this papilla is first seen to increase in complexity until as many as 25–27 rows of hair cells are present, all with their ciliary tufts entering the canals of the tectorial body as in Fig. 7-21. Then almost abruptly the character of this structure changes: the main portion of the tectorial mass disappears, and a sensing membrane remains to make contact with only the more lateral hair cells (Fig. 7-22). The medial hair cells at this point have free-standing ciliary tufts. Then farther posteriorly these medial hair cells disappear (Fig. 7-23), leaving only the lateral ones to make contact with the sensing membrane. This abrupt change contrasts with the condition found in the pelobatids, in which the papilla first becomes duplex, and then the medial portion dwindles and terminates.

The sensing membrane in *Bufo* is much extended and serves a good many hair cells—about half as many as are contacted by the anterior tectorial body. Then at a point about two-thirds of the distance along the papilla the upper end of the sensing membrane, the end that splits up to form the terminal canals into which the ciliary tufts extend, undergoes a progressive expansion

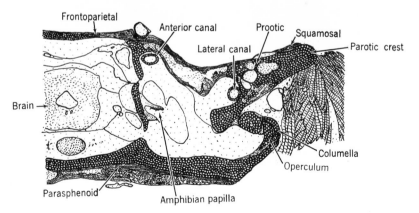

Fig. 7-20. The right ear region in *Bufo americanus*, seen in a transverse section. Scale 10X.

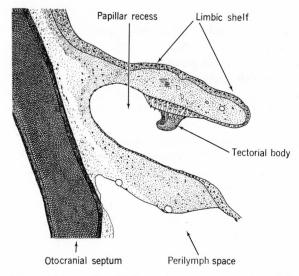

Fig. 7-21. The amphibian papilla of *Bufo americanus* near its anterior end. Scale 87X.

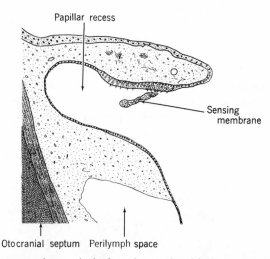

Fig. 7-22. The same specimen as in the foregoing, sectioned farther posteriorly where the sensing membrane is present. Scale 87X.

as the rows of hair cells themselves increase and comes to embrace the whole width of the papilla as Fig. 7-23 shows. The sensing membrane then appears as a dense column along the lateral edge of the tectorial mass (Fig. 7-24). Finally the sensing membrane ends altogether (Fig. 7-25), leaving a tectorial

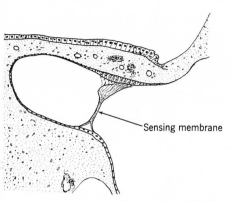

Fig. 7-23. The above specimen sectioned farther posteriorly. Scale 87X.

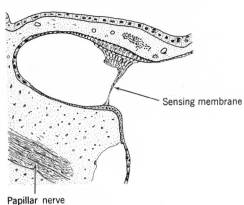

Fig. 7-24. The above specimen sectioned still farther posteriorly. Scale 87X.

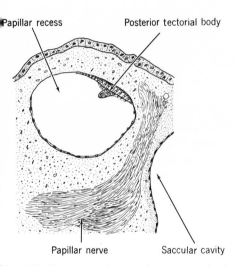

Fig. 7-25. The same specimen as above, cut at a level posterior to the sensing membrane. Scale 87X.

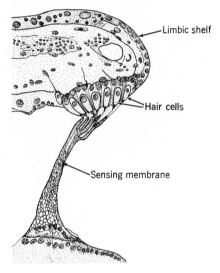

Fig. 7-26. Another specimen of *Bufo americanus*, with a transverse section passing through the sensing membrane. Scale 250X.

body that continues along the posterior portion of the papilla. An enlarged picture of the sensing membrane and its connections to the lateral group of hair cells was obtained from another specimen and is shown in Fig. 7-26.

Near the posterior end of the papilla, as shown in Fig. 7-27, a window opens into a broad chamber that is an expansion of the perilymphatic duct.

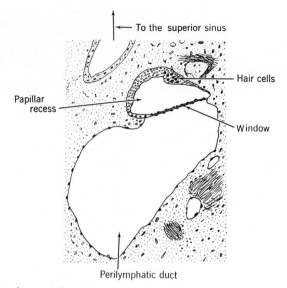

Fig. 7-27. The above specimen in a transverse section near the posterior end of the papilla. Scale 87X.

This window is covered by an extremely thin layer of connective tissue, and in one specimen was measured as only 3.3 μm thick; the membrane is so thin that the cell nuclei appear as bulges over its surface.

Three groups of hair cells can thus be distinguished in this papilla: an anterior group attached to the principal (anterior) tectorial body, a middle group served by the sensing membrane, and a posterior group with attachments to the posterior tectorial body. For a short distance (a distance of 40 μm in the specimen studied in detail) the second and third groups overlap: some of their hair cells are served by both the sensing membrane and the posterior tectorial body. In Table 7-III showing the distribution of hair cells these doubly connected hair cells are included with those served by the sensing membrane alone since this membrane appears to operate more significantly in the determination of sensitivity.

These observations show that a substantial number of hair cells, nearly a fourth of the total, are served by the sensing membrane. No doubt it is this portion of the hair-cell population that mainly determines the frog's ability to detect the faintest sounds. The remaining hair cells extend the dynamic range of this receptor and aid in loudness discrimination.

The Basilar Papilla. — The basilar papilla in this species presents the usual structure. The specimen described above had 81 hair cells in both right and left papillae.

Sensitivity. — Tests of sensitivity with an electrode located in a hole drilled through the otic capsule adjacent to the round window showed poor responsiveness to sounds. As seen in Fig. 7-28 the sensitivity function runs generally between +10 and +40 db, with its best points in the low frequencies where at 200 Hz a level of +8 db is reached. This poor sensitivity is in part to be attributed to the small size of the tympanic membrane; in the specimen represented in Fig. 7-28 this membrane was only 2.5 mm wide and 2.8 mm high.

Bufo boreas boreas, the Northwestern Toad.   This species inhabits a large area over the northwestern part of the United States and extends north into Canada and along the coastline into Alaska. It is highly tolerant of cold, often observed as mating when the temperature is low and even when the ground is covered with snow. It has a preference for high altitudes up to 11,000 feet and mostly is found in the vicinity of large streams. These toads reach an adult length of 100 mm in males and 125 mm in females.

The call of the male is described as a high-pitched tremolo, reinforced by a vocal sac. The structure of the amphibian and basilar papillae follows much the form already described for *Bufo* species.

The Amphibian Papilla. — The amphibian papilla begins with its hair cells connected to a rather large tectorial body, and the structure grows rapidly in size as it extends ventrally until it contains 25 rows of hair cells about a third of the way to the ventral end. In this course the tectorial structure develops thick-walled canals along its medial side and more of a webbed or light spongy tissue elsewhere; then it passes through rapid changes as it continues in the

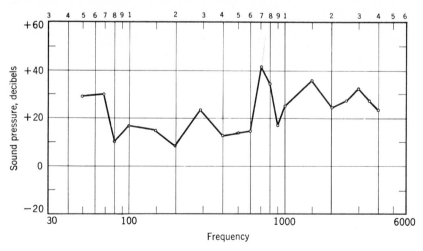

Fig. 7-28. A sensitivity function for a specimen of *Bufo americanus*. Shown is the sound pressure, in decibels relative to 1 dyne per sq cm, required for an inner ear potential of 0.1 μv.

ventral direction. The canal walls everywhere thicken, and then suddenly a sensing membrane arises along the posterolateral side of the structure, with a dense-appearing arm extending from the lower portion of the mass and separating into distinct canals as it reaches the lateral hair cells. Here the outermost hair cells are connected to the canal walls of the sensing arm, the innermost hair cells remain connected by delicate fibers to the medial tectorial mass, and the middle hair cells are left without connections. Then in the next adjacent region the sensing membrane separates off and only a fragment of the main tectorial tissue remains connected to its posterior end. This sensing membrane continues its connections to the most lateral hair cells, but no other tectorial attachments are present here. The most medial hair cells have only free-standing ciliary tufts, and a space is left between lateral and medial groups in which the hair cells are lacking. In the next section all these medial hair cells along with their whole supporting-cell structure have disappeared, and all that remains on the limbic surface is a thin layer of squamous epithelial cells. At this point the sensing membrane has made a firm anchorage along its posterior edge to the floor of the papillar recess.

This condition continues over about a third of the length of the papilla until the sensing membrane loses its limbic anchorage and its anterior portion transforms into an ovoid mass that is the posterior tectorial body. This body contains numerous canals whose walls are thick medially and become progressively thin laterally where they form openings to receive the ciliary tufts.

The loss of the limbic anchorage by the sensing membrane at this posterior end of the papilla does not prevent its exertion of restraint on the ciliary tufts throughout its course, though because of the elasticity of the tectorial filaments this force may be reduced in some degree. Toward the end of the papilla a small web of tectorial tissue remains and when cut across in the sections appears as a perforated membrane lying close over the ends of the hair cells, with clear connections both with the ciliary tufts and with the expanded feet of the columnar processes that run between the hair cells to support their outer ends.

These relations were especially clear in this specimen and were given close study. The picture presented is no doubt the typical one for the amphibian papilla of anurans.

In the specimen examined, as shown in Table 7-III, the anterior tectorial body served about half the hair cells in the right amphibian papilla and somewhat fewer in the left papilla, the sensing membrane connected to only about 16 percent of the cells, and the posterior tectorial body served one-fifth or fewer, leaving a good many hair cells without tectorial connections. As indicated, the total number of hair cells in this papilla is considerable.

The Basilar Papilla. — The basilar papilla in this specimen contained 75 hair cells on the left side and 84 hair cells on the right.

Sensitivity. — A sensitivity function for this specimen is presented in Fig.

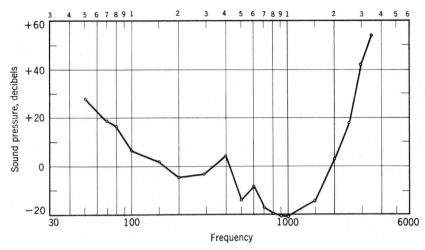

Fig. 7-29. A sensitivity function for a specimen of *Bufo boreas*, represented as in the preceding figure.

7-29. This curve shows a primary region of fairly good sensitivity in the higher frequencies from 500–1500 Hz, with the best level of −22 db at 900–1000 Hz. A secondary maximum appears in the region of 200–290 Hz where the sensitivity reaches −9 db.

Bufo cognatus, the Great Plains Toad. *Bufo cognatus* is a large-bodied toad that inhabits the Great Plains of the western United States, extending from Montana and North Dakota in the north and sweeping southward to Arizona and Texas and also including some adjacent portions of Canada and Mexico. This species flourishes along streams and ditches, and in breeding makes use of temporary pools after rains. The voice is described as harsh, generally of low pitch, and explosive in character; it is reinforced by a particularly large vocal sac. The sound produced by large numbers calling in concert is said to be of nearly deafening intensity.

The tympanic membrane is a broad oval, with its longer axis nearly vertical. In a medium-sized specimen with a body length of 65 mm this membrane measured 4.5 mm wide and 5.3 mm high.

As represented in Table 7-III, the distribution of hair cells in the amphibian papilla is much as shown in other *Bufo* species.

The basilar papilla in this specimen contained 104 hair cells on the left side and 94 on the right.

Sensitivity. — Measurements of auditory sensitivity were carried out in 3 specimens, with results as shown in Figs. 7-30 and 7-31. These curves indicate only a poor level of acuity, best in the middle range where in one

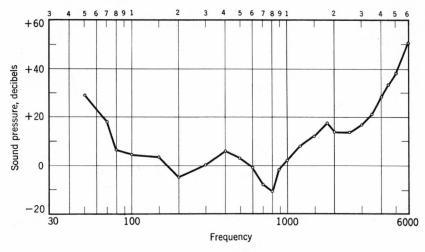

Fig. 7-30. A sensitivity function for a specimen of *Bufo cognatus*, represented as above.

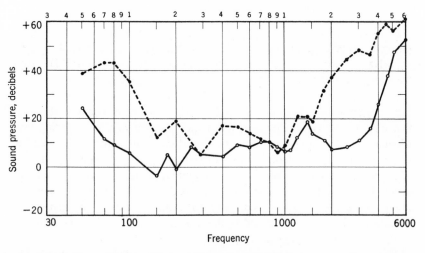

Fig. 7-31. Sensitivity functions for two additional specimens of *Bufo cognatus*, represented as above. The solid line is for a female, and the broken line for a male.

animal it reaches −11 db at 800 Hz. There is perhaps a secondary region of better sensitivity in the low frequencies, at 150 Hz in one animal, 200 Hz in another, and 290 Hz in a third, but the functions are irregular and nowhere does the acuity reach impressive levels.

It is suggested that the high intensity of the mating call in chorus may be essential in attracting females with this poor level of auditory function.

Bufo granulosus. *Bufo granulosus* belongs to the broad-skulled toads of South America. A drawing of the ear region is shown in Fig. 7-32. The columellar muscle is highly developed, and connects to the columella by a strong ligament.

The amphibian papilla in one of the specimens contained 284 hair calls on the left side and 295 hair cells on the right, distributed as shown in Table 7-III. The basilar papilla in this specimen contained 14 hair cells on the left side and 29 on the right. Thus the complement of hair cells is small, especially for the basilar papilla.

Sensitivity curves for two specimens are shown in Fig. 7-33. One of these ears exhibits fairly good sensitivity in the region of 130–200 Hz, where it reaches a level of −14 db. The other ear presents great irregularities with generally poor hearing but there are peaks at 250, 900, and 3000 Hz.

Bufo marinus, the Giant Toad.   This large species occurs in the northern part of South America from Colombia to Guiana and extends northward through Central America and southern Mexico to the border of Texas. The frogs occur in a variety of habitats, with preferences for places containing litter under which they hide by day and from where they come forth to feed

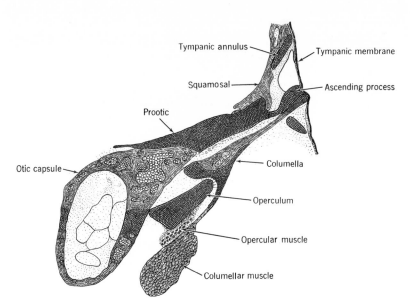

Fig. 7-32. The ear region in a specimen of *Bufo granulosus* in a transverse section. Scale 15X.

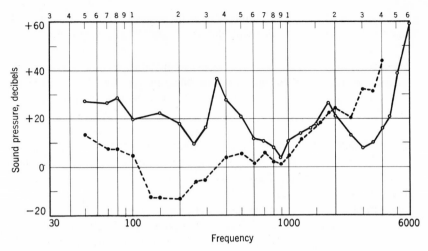

Fig. 7-33. Sensitivity functions for two specimens of *Bufo granulosus*. Shown is the sound pressure, in decibels relative to a zero level of 1 dyne per sq cm, required for an inner ear potential of 0.1 μv.

mainly at night. They are often found in or near brackish pools; one experimenter found the eggs to hatch more quickly in a weak saline solution than in fresh water. The species has been introduced into a number of regions, such as Hawaii and Australia, where it has flourished, often to the detriment of native amphibian populations. It is distinguished by a large size, 132 mm in body length on the average in males and 145 mm in females, though specimens as large as 220 mm have been reported (Wright and Wright, 1949). The paratoid glands are enormous, and the venom is particularly toxic.

The Middle Ear. — The tympanic membrane is prominent, and a little taller than wide. In a specimen with a body length of 123 mm this membrane measured 6.5 × 7.0 mm. The lock mechanism is well developed, with a double notch as shown in Fig. 7-34. The columellar muscle and its ligament are particularly large.

The Inner Ear. — The amphibian papilla is well developed. One specimen showed 625 hair cells on the left side and 661 on the right, distributed as in Table 7-III. The basilar papilla in this animal had 189 hair cells on the left and 221 cells on the right, which is well above the average population in anurans.

Sensitivity. — Inner ear potential measurements were carried out in 13 specimens. The responses to aerial sounds for one of these are shown in Fig. 7-35. There are two prominent maximum regions, one in the low tones with its lowest point at 400 Hz, where it reaches −24 db, and another in the medium high frequencies with its lowest point of −19 db at 1200 Hz. In a

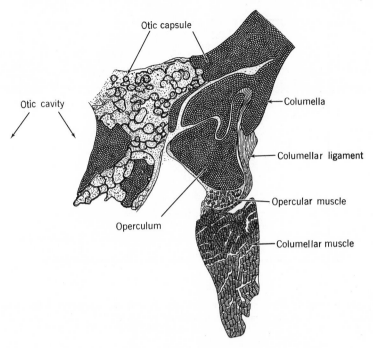

Fig. 7-34. The middle ear region of *Bufo marinus*. Scale 10X.

second specimen both right and left ears were examined, with results shown in Fig. 7-36. The two curves agree in showing two maximum regions, one in the low tones around 200 Hz and the other for the medium high tones between 800 and 1800 Hz. There are considerable variations in sensitivity to particular tones, but the general form of the function is similar for the two ears.

Vibratory stimulation was used in a few animals, with results as indicated in the next two figures. In one instance, represented by the dashed-line function of Fig. 7-37, the vibrating needle was applied to the middle of the tympanic membrane, over the end of the pars externa. Measurements were made also with aerial sounds, giving the solid-line curve of this figure. The two functions are generally similar. Both show a maximum in the low frequencies, though the one for vibratory stimulation is shifted downward in the frequency scale as compared with the other. In the higher frequencies, around 700–1500 Hz, a contrary shift is seen, but it is not very marked. The presence of two maximums for these two methods of stimulation strongly supports the idea that two different sensory mechanisms, the amphibian and basilar papillae, are operating.

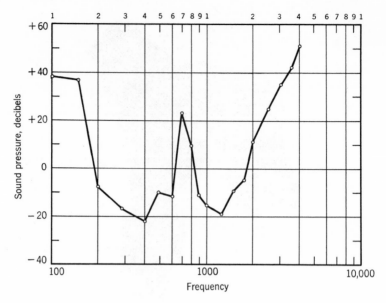

Fig. 7-35. A sensitivity function in a specimen of *Bufo marinus*, showing the sound pressure in db relative to 1 dyne per sq cm giving an inner ear potential of 0.1 μv.

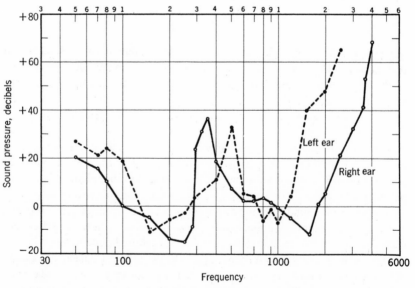

Fig. 7-36. Sensitivity functions in right and left ears of a specimen of *Bufo marinus*. Shown is the sound pressure, in decibels relative to 1 dyne per sq cm, required for an inner ear potential of 0.1 μv.

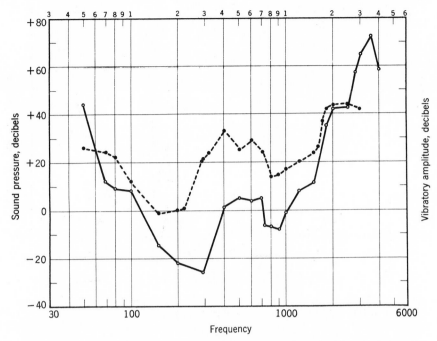

Fig. 7-37. Sensitivity in a further specimen of *Bufo marinus* as determined for stimulation with aerial (solid line) and vibratory (broken line) stimulation. Zero level for vibratory stimulation = 1 μm.

Further results with these two methods of stimulation are presented in Fig. 7-38. Here the vibratory stimulus was applied to the top of the head, about the middle of the frontoparietal bone. Again the two types of stimulation give similar functions, with two maximum regions, though there are considerable variations in detail.

These results for *Bufo marinus* indicate a fairly good level of sensitivity, around −20 db in the maximum regions.

The consistency of sensitivity measurement over a period of time was tested by making a series of measurements on one day and then maintaining the animal under deep anesthesia for similar tests on the following day. As the two curves of Fig. 7-39 indicate, there are variations but the general form of the sensitivity function is maintained. This result can be expected if the anesthesia is deep enough to depress the activity of the middle ear muscles.

Bufo valliceps, the Gulf Coast Toad.   This is a large species, pictured in Fig. 7-40, occurring along the Gulf of Mexico, in the southern parts of Louisiana and Texas and the eastern region of Mexico, and extending along the

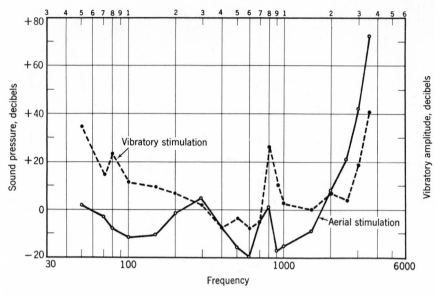

Fig. 7-38. Results on an additional specimen of *Bufo marinus* with both aerial and vibratory stimulation. The vibrator was applied to the exposed skull.

coast as far as Costa Rica. It is reported also as crossing the isthmus of Tehuantepec to the Pacific (Blair, 1972). The body length in males is from 53 to 98 mm and may be greater, up to 125 mm, in females (Wright and Wright, 1948). The habitat is varied, including coastal forest and prairie as well as lowland areas containing pools and ditches. Baldauf (1958) made a thorough study of the cranial morphology in this species.

The Middle Ear. — There is a prominent tympanic membrane, which in a specimen with a body length of 63 mm measured 4.7 mm wide and 5.0 mm high, with the upper edge slightly overhung by the squamosal ridge. The columellar mechanism is of standard form as represented in Fig. 7-41. The columellar shaft is particularly slender, expanding medially to a large footplate that has a deep notch for the end of the operculum.

The Inner Ear. — The amphibian papilla in one of the specimens contained 478 hair cells on the left side and 505 hair cells on the right, with tectorial connections as shown in Table 7-III. The basilar papilla in this specimen contained 94 hair cells on the left side, and 77 hair cells on the right.

Sensitivity. — Tests with inner ear potentials were carried out in five specimens, and results for two of these are shown in Fig. 7-42. The sensitivity appears to be very poor, with the lowest portion of the curves running along the +40 db line over the range from 150 to 700 Hz. The mating call

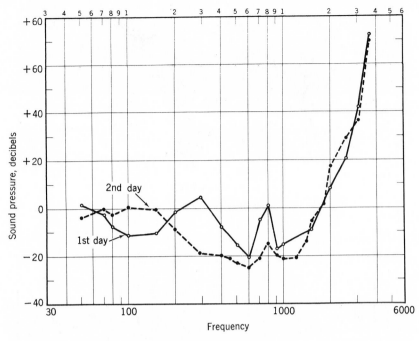

Fig. 7-39. Sensitivity curves for aerial stimulation on two successive days for a specimen of *Bufo marinus*.

Fig. 7-40. A specimen of *Bufo valliceps* with the vocal sac distended. Drawing by Anne Cox.

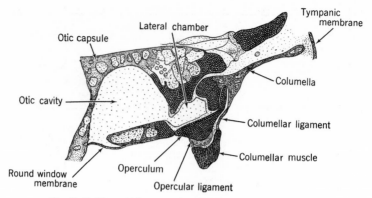

Fig. 7-41. The middle ear region in *Bufo valliceps*. Scale 10X.

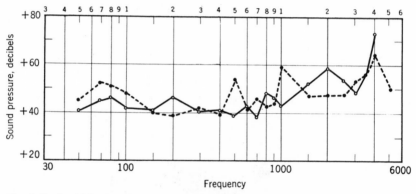

Fig. 7-42. Sensitivity functions in two specimens of *Bufo valliceps*, for aerial sounds. Zero level = 1 dyne per sq cm.

is described as a trill with a fundamental frequency that varies regionally, from about 1400 Hz in the north to 2100 Hz in the south (Porter, 1970).

Bufo viridis. This species has a wide distribution over Eurasia, extending from Germany and Italy eastward as far as Siberia, Mongolia, and the northern parts of China. The specimens examined came from a pool in Jerusalem into which they had been introduced.

This is a medium-sized species, with a prominent tympanic membrane, usually slightly oval in form with the vertical axis the larger. A specimen measuring 74 mm in body length had a tympanic membrane 3.8 mm wide and 4.0 mm high. The middle ear mechanism has the standard structure, but with a particularly close and complex interdigitation of operculum and foot-

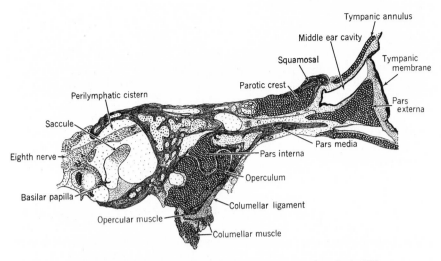

Fig. 7-43. The ear region in *Bufo viridis*, in a transverse section. Scale 10X.

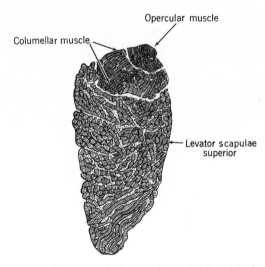

Fig. 7-44. The middle ear muscles in a specimen of *Bufo viridis*. Scale 15X.

plate, as shown in Fig. 7-43. The opercular and columellar muscles, along with the very large levator scapulae superior, are represented in Fig. 7-44.

The Inner Ear. — The amphibian papilla in one of the specimens contained 468 hair cells on the left side and 453 hair cells on the right, connecting to the tectorial structures as shown in Table 7-III. In this same specimen there were 53 hair cells in the left basilar papilla and 60 hair cells in the right one.

Sensitivity. — Inner ear potential tests of sensitivity were made on 9 specimens and in general showed best sensitivity in two regions of frequency in the middle range. Figure 7-45 shows the function obtained for one of these animals, in which a peak appears at 400 Hz and another, somewhat broader, at 1500 Hz. The maximum of this second peak is at −1 db, representing only a fair level of sensitivity. A second specimen gave the results of Fig. 7-46, in which one peak appears at 500 Hz and another at 1600–1800 Hz, with best sensitivity a little above that of the preceding animal, reaching −6 and −3 db in the two regions.

The hearing of this species has been studied also by the use of electrodermal responses, a procedure in which changes in the electrical potential of the skin are observed while stimulating with brief tones. Results obtained by Werner and Strother are presented in Figs. 7-47 and 7-48. The first of these figures shows two regions of good sensitivity, one around 200–290 Hz and another around 1000–2200 Hz with a peak at 1800 Hz where the maximum level reached is −17 db. The second animal (Fig. 7-48) likewise showed two sensitive regions, one at 290–500 Hz and the other at 1000–2000 Hz, with maximums in these regions at −23 and −24 db.

The two procedures, employing inner ear potentials and skin potentials,

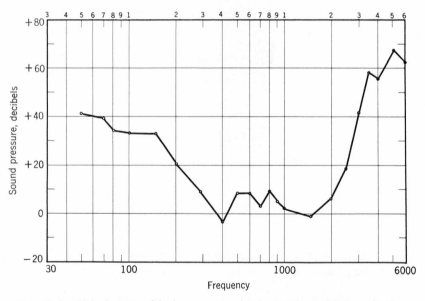

Fig. 7-45. Sensitivity in terms of the inner ear potentials in a specimen of *Bufo viridis*, shown as the sound level, in decibels relative to 1 dyne per sq cm required for a response of 0.1 μv.

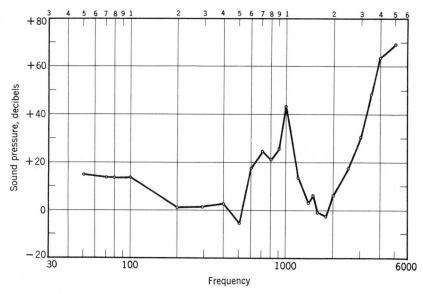

Fig. 7-46. Sensitivity in a second specimen of *Bufo viridis*, represented as above.

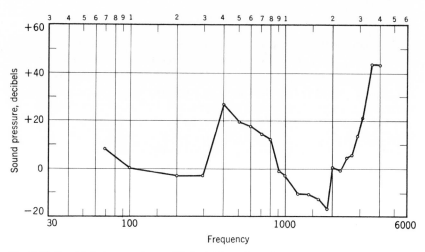

Fig. 7-47. Sensitivity in a specimen of *Bufo viridis* as determined with electrodermal responses by Werner and Strother.

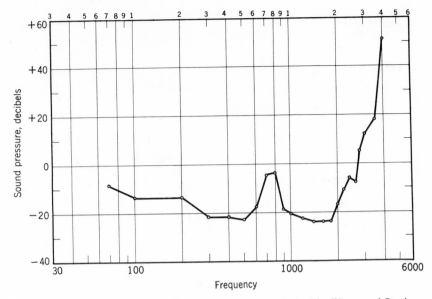

Fig. 7-48. Results for a second specimen of *Bufo viridis* as obtained by Werner and Strother. Shown are the minimum levels of sound, in decibels relative to 1 dyne per sq cm, required for a detectable response.

are in good agreement in revealing two regions along the frequency scale at which this ear performs relatively well, though the level of performance is hardly outstanding. The mating call of this species as observed by Andrén and Nilson (1979) falls in the range of the uppermost of these two regions of sensitivity; according to their measurements the fundamental frequency of this call varies between 1200 and 1400 Hz, with the mean of one series at 1308 Hz.

8. THE ADVANCED FROGS:

BRACHYCEPHALIDAE, RHINODERMATIDAE,

DENDROBATIDAE, HYLIDAE,

AND CENTROLENIDAE

A number of frog families generally considered to be derived from the Bufonidae, or at least closely related to these, will now be examined.

THE BRACHYCEPHALIDAE

A group of frogs to which Noble in 1931 had accorded the status of a subfamily has more recently been regarded as a full family, the Brachycephalidae, including four genera, *Atelopus*, *Brachycephalus*, *Dendrophryniscus*, and *Oreophyrnella*, with a number of species distributed over South America from Guiana to Argentina. Of these genera *Atelopus* has by far the widest representation in terms of species, numbering about 34, whereas *Brachycephalus* has only one; this condition may justify the designation of the family as the Atelopodidae, as some have preferred (Griffiths, 1963).

Four *Atelopus* species were available for study. These are small frogs of slender body form, all lacking a tympanic membrane and columella.

Atelopus varius. This is a particularly slender species, with long hindlegs; one specimen measured 31 mm in body length, and the hindlegs were 37 mm long. As Fig. 8-1 shows, there is no sign of a tympanic membrane or columella, and the operculum alone occupies the oval window. Between the lateral body wall and the operculum are large muscle masses that no doubt are involved in conducting sounds inward, and perhaps the relatively massive squamosal, which covers the parotic process, serves as a further path, especially for aquatic vibrations.

Atelopus senex. This species is of about the same size as the foregoing, but the body is more compact and the legs especially thin. The skin continues over the ear region without interruption, as in all atelopids. As seen in Fig. 8-2, the operculum is particularly large and presents the unusual feature of complete encasement in bone. Also its cartilaginous core is strongly calcified.

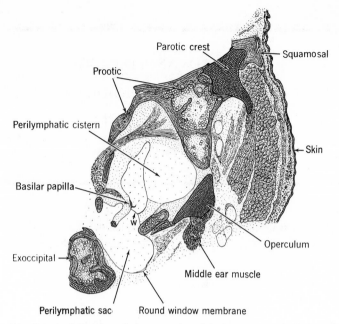

Fig. 8-1. A section through the ear region of *Atelopus varius*. Scale 20X.

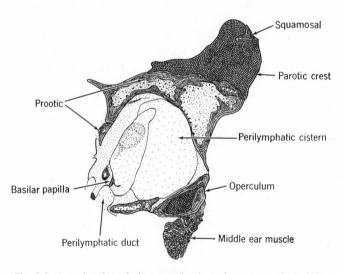

Fig. 8-2. A section through the ear region in *Atelopus senex*. Scale 20X.

A moderately large muscle lies ventral to the operculum and connects by a heavy ligament to its shoulder on the ventrolateral side. This muscle appears to be single; there is only a suggestion of a distinction of large and small fibers as commonly found in this area; all the fibers are comparatively large. In most frogs there are two muscles in this region, the opercular and columellar muscles, readily distinguished by their courses and fiber characteristics: a bundle made up of fine fibers ends on the operculum and another of larger fibers attaches through a ligament to a process on the columellar footplate. In the *Atelopus* specimens examined there was sometimes a suggestion of duplexity in the muscle mass, but only a single insertion on the operculum. In an atelopid specimen unidentified as to species the series of sections was extensive enough to permit a following of the muscle bundle anterodorsally all the way to its origin on the suprascapula; here one portion consisting of small fibers came from the anterior end of the suprascapula and the other, of larger fibers, came from the inner surface of the suprascapula farther posteriorly. The two portions of the bundle are distinct over much of its ventral course, but near the attachment of the bundle to the operculum the fibers were mostly of the larger size. A suggestion is that in the course of the degenerative processes that led to the loss of the columella, or subsequent to these, the fibers of the columellar muscle transferred their insertions to the operculum, so that the usual distinction of opercular and columellar muscles was lost.

The action of this muscle would seem to be a pulling of the anterolateral end of the operculum outward and downward, opening a gap between the operculum and the lower end of the prootic. Such an action would reduce the mobility of the operculum and at the same time provide a bypass to vibrations coming from the outside into the otic capsule through the mass of prootic and squamosal. This action would reduce the sound pressure discharge through the basilar papilla to the round window, and probably would reduce the discharge through the more anteriorly located amphibian papilla as well. If this is the case, then the muscle serves to protect the ear against overstimulation as in the majority of frogs.

Sensitivity measurements were made in two specimens of *Atelopus senex*, and the results are presented in Fig. 8-3. One of these animals, whose curve is shown by the solid line, presents a fair degree of sensitivity in the region of 290 to 1500 Hz, with a maximum of −14 db at 600 Hz. The other animal is considerably less sensitive, though the region of the maximum agrees rather well. The level of sensitivity is surprisingly high for animals lacking both tympanic membrane and columella.

THE RHINODERMATIDAE

This family, considered to have been derived from the Leptodactylidae, probably during the Cretaceous period, is represented by a few species in

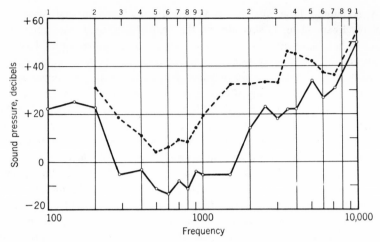

Fig. 8-3. Sensitivity functions in two specimens of *Atelopus senex*. Shown is the sound pressure, in decibels relative to 1 dyne per sq cm required for a response of 0.1 μv.

South America. One of these is *Rhinoderma darwini*, a frog distinguished by the practice of the male of carrying the eggs and then the hatched-out larvae in his vocal pouch. Another species of note is *Sminthillus limbatus*, considered to be the world's smallest frog, with a body length of 1 cm. Species in this family are further characterized by the absence of Bidder's organ, and also the absence of a columella.

No specimens belonging to this family were available for study.

THE DENDROBATIDAE

The family Dendrobatidae includes a small number of species from Central and South America often referred to as "poison arrow frogs" because of the potency of the venom produced by their cutaneous glands and the use of this venom by the South American Indians to tip their arrowheads.

Dendrobates auratus. One of the species studied, *Dendrobates auratus*, is a small, slender frog from the region of Panama. A small eardrum is present, and in a specimen with a body length of 29 mm it was nearly round and 1.4 mm in diameter. The middle ear mechanism is of the usual form, with the pars externa of the columella extending inward from the edge of the tympanic membrane as shown in Fig. 8-4 and connecting to a greatly expanded inner portion containing a deep notch for the end of the operculum. At the level shown the tympanic membrane resembles the surrounding skin, but more anteriorly this area may be seen to be thinned out and more differentiated.

Figure 8-5 presents a transverse section in which the tympanic membrane is more clearly seen, and an ascending process runs from a cartilaginous plate on the inner surface of the membrane to extend upward to the parotic crest. At this level there is a wide connection between the middle ear cavity and the pharynx.

Hair-cell populations were studied in a specimen of *Dendrobates auratus* with results indicated in Table 8-I. The total number of hair cells in the amphibian papilla was 289 on the left side and 266 on the right. Of these a little more than half were connected to the anterior tectorial body. Then more posteriorly the sensing membrane appeared and for a short distance formed the sole tectorial connections for the hair cells, which here were few in number. This sensing membrane was then joined by the posterior tectorial body, and these together served a few additional hair cells. The sensing membrane ended early in the posterior region, leaving only the posterior tectorial body to serve the remaining hair cells, which comprised about 29 percent of the population.

The basilar papilla in this same specimen had 23 hair cells on the left side and 30 hair cells on the right.

The hair-cell populations for both amphibian and basilar papillae are small relative to those of most frog species, and especially scanty are the hair cells served by the sensing membrane of the amphibian papilla.

Sensitivity. — Measurements were made on three specimens of *Dendrobates auratus* with results indicated in Figs. 8-6 and 8-7. One of the animals in Fig. 8-6 shows its best region around 800–1000 Hz, but at the most favorable point at 1000 Hz only reaches +34 db. The other curve shows even poorer sensitivity over most of the range, though it is only slightly worse at 800–900 Hz. A third specimen, represented in Fig. 8-7, shows a little better sensitivity, reaching +28 and +30 db at 1500 and 1800 Hz. There is a suggestion of bimodality in Fig. 8-7, with one relatively favored region at 500–900 Hz and another at 1500–2000 Hz, but this bimodal feature is not very clearly evident in Fig. 8-6. The results indicate extremely poor hearing in this species.

Colostethus inguinalis.   A second member of the Dendrobatidae is *Colostethus inguinalis*, occurring in the New World tropics. This small frog has a pear-shaped tympanic membrane with the smaller end extending dorsally; the lower portion of this membrane was measured as 1.0 mm wide, and the total height was 2.1 mm. The ridge over the anterior semicircular canal was prominent, and the bone in this region was sufficiently transparent to show the location of the anterior crista.

The amphibian papilla is well developed; as Table 8-I shows, the number of hair cells in this specimen reached 1114 in the left ear and 1196 in the right ear, with the largest number contained in the posterior tectorial body.

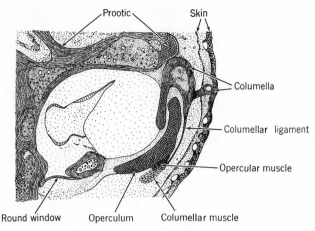

Fig. 8-4. A frontal section through the ear region in *Dendrobates auratus*. Scale 25X.

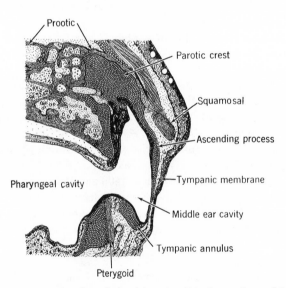

Fig. 8-5. A transverse section through the right ear region of a specimen of *Dendrobates auratus*. Scale 20X.

Table 8-I

Hair-Cell Distribution in the Amphibian Papilla of the Dendrobatidae

	To the Anterior Tectorial Body	To the Sensing Membrane	Free-standing Hair Cells	To the Posterior Tectorial Body	Total Hair Cells
		Dendrobates auratus			
Left	157	19	11	83[t]/19[s]	289
Right	157	23	0	77[t]/ 9[s]	266
		Colostethus inguinalis			
Left	197	76	24	817	1114
Right	216	78	9	893	1196

[t]To the posterior tectorial body.
[s]To both the posterior tectorial body and the sensing membrane.

Thus there is a resemblance to *Eleutherodactylus* species in which a great proliferation of hair cells was found in the posterior portion of this papilla. There is an important difference, however, in the manner of stimulation of these posterior hair cells.

It will be recalled that in *Eleutherodactylus* the most posterior hair cells are in a region of effective vibratory fluid flow, made possible by the presence of a window in the papillar wall leading into a perilymph space at the extreme posterior end of the organ. In *Colostethus*, however, such a window is absent, and a different (and possibly less effective) mode of stimulation of these posterior hair cells is utilized. The main portion of this papilla operates in the usual manner with a path of vibratory outflow as shown in Fig. 8-8, where at this level the papillar cavity over more than half its extent is bounded by a thin membrane permitting a discharge of vibratory pressures into a wide perilymphatic duct that leads ultimately to the round window. At the level shown the great proliferation of the hair cells begins more posteriorly: there is a wall of limbic tissue around the hair cells that increases in thickness posteriorly so as to constitute a solid plate embedding this end of the papilla, as seen in Fig. 8-9. Consequently the stimulation of the hair cells in this posterior region is only indirect: by way of the tectorial body extending from the anterior portion. This body is set in motion by fluid vibrations passing through the main region and transmits these movements posteriorly, making connections with the ciliary tufts of these deep-lying hair cells.

A sensitivity function for a specimen of *Colostethus inguinalis* is shown

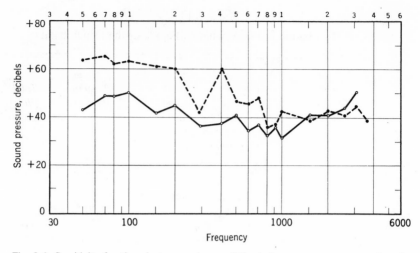

Fig. 8-6. Sensitivity functions in two specimens of *Dendrobates auratus*, represented as the sound pressure, in decibels relative to 1 dyne per sq cm, required for an inner ear potential of 0.1 μv.

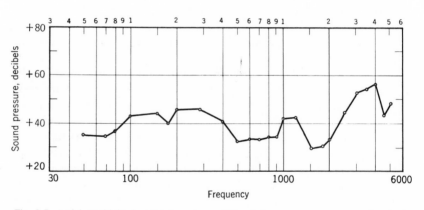

Fig. 8-7. Aerial sensitivity in a third specimen of *Dendrobates auratus*, represented as in the foregoing figure.

in Fig. 8-10. This ear performs well in the upper frequencies, with a sharp maximum at 1500–2000 Hz.

THE HYLIDAE

The family Hylidae contains a varied assembly of about 30 genera made up of more than 500 species commonly known as tree frogs. Among these gen-

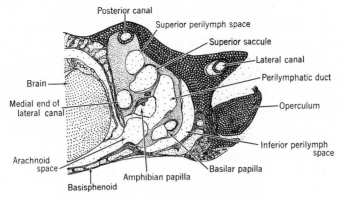

Fig. 8-8. The amphibian papilla of *Colestethus inguinalis* at a posterior level showing a broad area of pressure relief. Scale 25X.

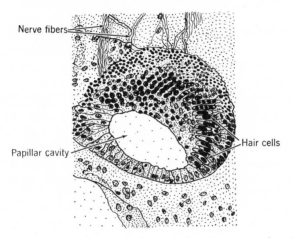

Fig. 8-9. The same specimen at a more posterior level where the escape path to the perilymph becomes closed. Scale 137X.

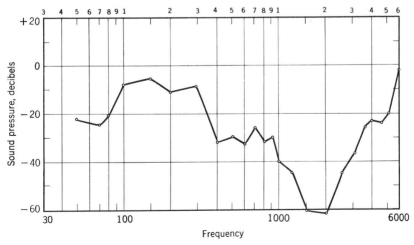

Fig. 8-10. A sensitivity function for a specimen of *Colestethus inguinalis*, shown as the sound pressure, in decibels relative to 1 dyne per sq cm, required for an inner ear response of 0.1 μv.

era all but two occur in the New World: the genus *Hyla* is cosmopolitan, having spread over most land areas around the world except India, southern Asia, and most of Africa and Arabia. The genus *Nyctimystes* is restricted to New Guinea.

The hylid frogs have undergone modifications of the fingers and toes that enable them to grasp twigs and branches and thus to be effective in climbing. These modifications include the addition of an extra element in the digits, a cartilage between the terminal phalanx and the one usually next to it, which adds length and flexibility and thus assists in grasping, and also the development of an adhesive disk beneath the tips of the digits that helps these frogs in clinging to leaves and branches. Consequently this group has made extensive use of the arboreal environment. Nevertheless a few hylid frogs are found elsewhere, living in thickets, on grasslands, or even in burrows in the ground.

Duellman (1970) separated the hylids into four subfamilies, of which two, the Hylinae (with 20 genera) and the Phyllomedusinae (with 3 genera), are represented in the present study. Among the Hylinae, specimens were obtained in 5 genera, *Acris*, *Hyla*, *Pseudacris*, *Smilisca*, and *Triprion*, to be examined in that order. The Phyllomedusinae are represented only by *Agalychnis*.

ACRIS, THE CRICKET FROGS

The genus *Acris* contains two species living in eastern and central North America, each with two subspecies; both subspecies of one of these were available for study.

Acris crepitans crepitans, the Northern Cricket Frog.   This small frog, named for its click-like call, occurs in the eastern and southern states of North America from New Jersey to the edge of Texas, except for the southeastern seaboard areas. Its adult body size varies from 16 to 35 mm. The tympanic membrane is small and inconspicuous, though bounded above by a little thickening of skin. In a specimen of 22 mm body length this membrane was nearly round, with a diameter of 1 mm.

The locking mechanism of the middle ear is well developed, as pictured in Fig. 8-11. Extending ventrally alongside the lateral surface of the operculum is a tapered spike of bone, a part of the pars interna of the columella, to which is connected a long, thin ligament from the columellar muscle. As shown in the figure, this ligament runs dorsolaterally along the surface of the operculum and connects by a number of fine fibers to a deep notch in the underside of the columella. The columellar footplate is somewhat complex and fills the lateral portion of the otic capsule.

The amphibian papilla of *Acris crepitans* has the typical structure, with the hair cells distributed as in Table 8-II. About 40 percent of these cells are

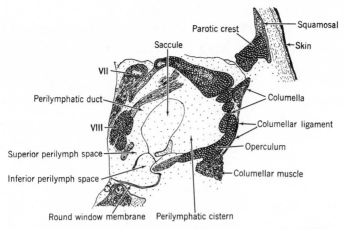

Fig. 8-11. The ear region in a specimen of *Acris crepitans*. Scale 20X.

connected to the anterior tectorial body, about 25 percent go to the sensing membrane, and the remainder are connected to the posterior tectorial body.

In the specimen examined, the basilar papilla contained 68 hair cells on the left side and 66 hair cells on the right.

Sensitivity. — Measurements of sensitivity were made on 6 specimans of *Acris c. crepitans*, and results for two of these are shown in Figs. 8-12 and 8-13. These curves indicate best sensitivity around 400–1000 Hz, which in one animal reaches −12 db and in the other reaches +2 db.

Duellman (1970) carried out measurements on the mating call of *Acris crepitans* and found this click-like signal to contain a fundamental frequency of 175 Hz, with energy distributed widely over the spectrum and a dominant frequency at 3150 Hz. From the form of the sensitivity function as shown in Figs. 8-12 and 8-13 it appears that components in the range of 400–1000 Hz are chiefly utilized.

Acris crepitans blanchardi. In a specimen of the subspecies *Acris crepitans blanchardi*, as Table 8-II shows, the numbers and distribution of the hair cells in the amphibian papilla are much the same as in the foregoing. The total number of hair cells was 373 on the left side and 358 on the right, and of these about 42 percent were connected to the anterior tectorial body, about one-fourth to the sensing membrane, and the remaining third to the posterior tectorial body.

The basilar papilla in this specimen contained 69 hair cells on the left side and 58 hair cells on the right.

A low level of sensitivity was found in this species, which is probably explained, at least in part, by the small size of the tympanic membrane and

Table 8-II
Hair-Cell Distribution in the Amphibian Papilla of the Hylidae

	To the Anterior Tectorial Body	To the Sensing Membrane	Free-standing Hair Cells	To the Posterior Tectorial Body	Total Hair Cells
	Acris crepitans crepitans				
Left	166	96	†	132	394
Right	141	94	†	120	355
	Acris crepitans blanchardi				
Left	155	92	2	124[s]	373
Right	155	92	0	111[s]	358
	Hyla boulengeri				
Left	186	51	7	197	441
Right	152	22	†	141	315
	Hyla cineria cinerea				
Left	191	69	12	392	664
Right	196	77	22	331	626
	Hyla ebraccata				
Left	190	89	6	146	431
Right	224	74	8	158	464
	Hyla microcephala				
Left	99	44	19	107	269
Right	115	36	8	142	301
	Hyla phlebotes				
Left	128	48	7	148	331
Right	133	39	3	133	308
	Hyla septentrionalis				
Left	158	76	†	210	444
Right	190	97	†	210	497
	#1 *Hyla versicolor*				
Left	139	78	5	80	302
Right	147	85	0	81	313
	#2 *Hyla versicolor*				
Left	123	100	18	111	352
Right	146	82	5	122	355
	Pseudacris clarkii				
Left	205	65[s]	6	85	361
Right	216	65[s]	0	77	358

	To the Anterior Tectorial Body	To the Sensing Membrane	Free-standing Hair Cells	To the Posterior Tectorial Body	Total Hair Cells
		Pseudacris ornata			
Left	193	90[ss]	†	117	400
Right	199	86[ss]	†	109	394
		Pseudacris streckeri			
Left	157	62	9	164	392
Right	164	56	5	159	384
		Smilisca phaeota			
Left	196	65	6	269	536
*Right***	0	77	188	171	436
		Triprion spatulatus spatulatus			
Left	152	127	7	206	492
Right	225	60	5	116	406
		Agalychnis callidryas			
Left	164	76	0	302	542
Right	157	86	0	263	506
		Pachymedusa dacnicolor			
Left	174	69	0	299	542
Right	184	83	6	297	570

† Not determined.
** Defective ear.
[s] Hair cells in sensing membrane and posterior tectorial body.
[ss] The sensing membrane and its expansion.

the limited population of hair cells. It seems likely that this low level is adequate for mating purposes, and a higher level may not be needed to aid in the escape from predators by a frog equipped with unusually effective venom glands.

HYLA SPECIES

Among the numerous tree frogs in the genus *Hyla*, specimens of seven species were obtained for this investigation.

Hyla boulengeri. This is a medium-sized species inhabiting the lowland forests of Central America from Nicaragua through Panama to South America. According to Duellman, males reach maximum body lengths of 48.7

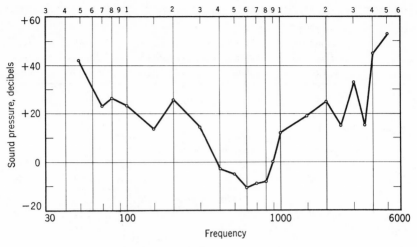

Fig. 8-12. Aerial sensitivity in a specimen of *Acris crepitans*, shown as the sound pressure, in decibels relative to 1 dyne per sq cm giving an inner ear potential of 0.1 μv.

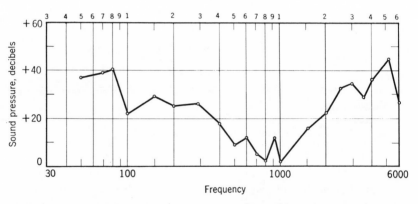

Fig. 8-13. Aerial sensitivity in a second specimen of *Acris crepitans*, represented as in the foregoing.

mm and females 52.8 mm. The specimen examined, which was collected in Panama, was a male 37 mm long. The tympanic membrane was nearly round, and 2.3 mm in diameter.

The amphibian papilla in this specimen was of normal appearance on the left side, but on the right was malformed, the entire structure seeming to be displaced a little posteriorly from its usual position and rotated counterclockwise about 30°. A general reduction of about 30 percent in the hair-cell population, as compared with the left side, was found also, as Table 8-II shows.

The basilar papilla in this specimen contained 40 hair cells on the right side and 59 hair cells on the left.

Sensitivity. — A sensitivity function for this animal is shown in Fig. 8-14. This curve presents a sharp maximum at 1200 Hz, within a general region of fairly good sensitivity from 1000 to 2500 Hz. The call note of this species is reported as a low-pitched growl, with two equally prominent harmonics at 1600 and 2800 Hz (Duellman, 1970).

Hyla c. cinerea, the Green Treefrog. *Hyla c. cinerea* is a medium-sized hylid occurring in coastal areas from Virginia through all of Florida to eastern Texas and extending northward to the southeastern portion of Arkansas. It is often hidden in vegetation or can be seen (usually with difficulty) on the stems or leaves of plants, mostly above water or on the banks of ponds and streams. The mating call is described as very loud and with the quality of a cowbell. The tympanic membrane is round, but its upper edge is partially obscured by a thick ridge of skin. In a specimen with a body length of 50 mm its diameter was 3.0 mm.

The amphibian papilla is well developed, and the tectorial structures show abrupt changes in form along the anteroposterior axis. The posterior tectorial body extends over half the length of the papilla, and makes contact with a corresponding number of hair cells, as Table 8-II indicates. The total complement of hair cells in this papilla is strikingly large. The basilar papilla, on the other hand, was found to contain only a modest number of hair cells: 42 on the left side and 33 on the right.

Sensitivity. — A sensitivity function for the specimen described above is presented as the solid curve of Fig. 8-15. The response is poor and falls off in an irregular manner as the frequency rises. In another specimen with the recording electrode on the round window membrane a curve of similar form

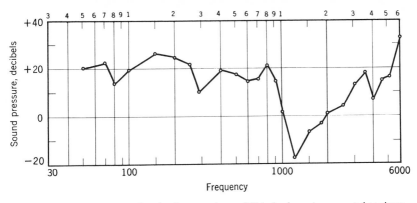

Fig. 8-14. An aerial sensitivity function for a specimen of *Hyla boulengeri*, represented as above.

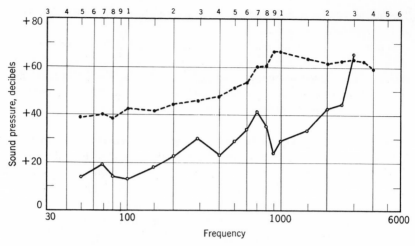

Fig. 8-15. Aerial sensitivity in two specimens of *Hyla c. cinerea*, determined as the sound pressure, in decibels relative to 1 dyne per sq cm required for a response of 0.1 μv.

was obtained (the broken curve of this figure), but its level is about 20 db poorer except at the upper end.

Weiss and Strother (1965) examined the auditory sensitivity in this species by the use of galvanic skin potentials arising on the presentation of a tone and recorded from electrodes that penetrated the skin at two positions along the anteroposterior line of the back, one at a point between the tympanic membranes and the other far posteriorly, a little ahead of the vent. The threshold intensities required to produce detectable responses were observed in 7 specimens and showed wide variations for tones along the frequency scale and from one animal to another; they were in agreement in exhibiting two marked peaks of sensitivity, one at 600 and the other at 2000 Hz, as indicated for one animal in Fig. 8-16.

There is no very meaningful correspondence between these measurements of galvanic skin potentials and results obtained by the use of cochlear potentials; further investigation is needed.

Hyla ebraccata. This small hylid is found in two lowland tropical forest areas in Central America, one area running through lower Mexico to the edges of British Honduras and Guatemala and the other sweeping through Nicaragua and Panama. Duellman (1970) reported males of this species as attaining a maximum body length of 27.8 mm and females a maximum of 36.5 mm. There are geographic variations in size, and individuals from Panama (which was the source of the six males of the present study) tend to be larger than those from Guatemala and Costa Rica. These specimens varied from

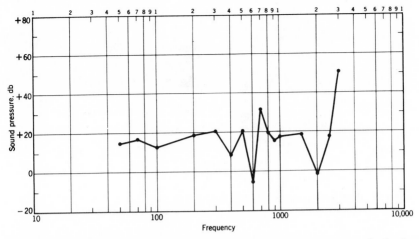

Fig. 8-16. Aerial sensitivity in a specimen of *Hyla c. cinerea*, from observation of galvanic skin potentials (after Weiss and Strother, 1965).

26.0 to 26.5 mm in body length, and their tympanic membranes were nearly round, or very slightly higher than wide, and measured around 1.1 mm.

The middle ear mechanism has the typical anuran form, as pictured in Fig. 8-17. The columellar footplate sends off long, slender processes of cartilage to which the tendon of the columellar muscle is attached. Particularly sturdy is the ascending process, which assists in the attachment of the outer end of the columella to the tympanic membrane and arches upward to the parotic process.

The tectorial connections of the hair cells of the amphibian papilla were examined in one of the specimens, with results shown in Table 8-II. Somewhat less than half these connections are made to the anterior tectorial body and about a third to the posterior tectorial body.

The basilar papilla in this specimen contained a good many hair cells in relation to other species: there were 75 on the left side and 81 on the right.

Sensitivity. — Functions representing auditory sensitivity are shown for two animals in Fig. 8-18. In general this sensitivity is poor. One of the curves has a fairly broad maximum in the high frequencies at 900–2500 Hz, with the peak at 1500 Hz where the level reached is +16 db. The other curve has its maximum region higher along the scale, at 2500–4500 Hz, where the best point is +28 db. Duellman (1970) reported the mating call of this species as an insect-like creaking sound with a dominant frequency around 2300–2650, averaging 2504 Hz. The maximum sensitivity in one of these animals corresponds roughly to the range of the call note according to Duellman's observations, but in the other animal is below this range.

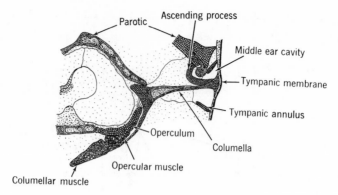

Fig. 8-17. The ear region in a specimen of *Hyla ebraccata*. Scale 20X.

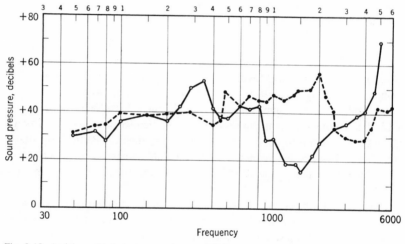

Fig. 8-18. Aerial sensitivity in two specimens of *Hyla ebraccata*, shown as the sound pressure, in decibels relative to 1 dyne per sq cm, required for an inner ear potential of 0.1 μv.

Hyla m. microcephala. This small, slender hylid is found in open fields or cut-over forest areas of southeastern Costa Rica and lowland coastal areas of Panama. According to Duellman (1970), males attain a body length of 24.5 mm, with a mean of 22.4, and females reach 30.9 mm with a mean of 27.9. Three specimens, all males, were studied; two of these measured 22.0 mm in body length and the third measured 24.0 mm. The tympanic membrane is a broad oval; in one specimen it measured 0.8 mm wide and 1.2 mm high.

The middle ear mechanism is of standard form, with a large columellar footplate filling a lateral portion of the otic capsule. There is no septum dividing off a lateral chamber, however. The lock mechanism shows a double notch, including a deep cleft in the footplate between pars interna and pars media into which the end of the operculum inserts, and a broad depression in the operculum is occupied by a posterior lobe of the footplate.

The amphibian papilla in one of the specimens contained 269 hair cells on the left side and 301 hair cells on the right, with tectorial connections as in Table 8-II.

The basilar papilla in this same specimen had 33 hair cells on the left side and 25 hair cells on the right.

Sensitivity measurements on two specimens showed rather poor response, with the maximum in the high tones around 2000 Hz as indicated in Fig. 8-19.

Hyla phlebotes. This is a small hylid inhabiting lowland humid forest areas in Central America, ranging from southeast Nicaragua to Panama. The single specimen examined came from the Canal Zone and was a male 23 mm in body length. Duellman reported the mean length in males as 22 mm and that of females as 26.8 mm, making this the smallest of the *microcephala* group of hylids.

A tympanic membrane is present, though small and inconspicuous; in this specimen it measured 1.0×1.1 mm. As in most small frogs the anterior canal was marked by a prominent ridge and made simple the placement of the recording electrode.

The amphibian papilla was studied in serial sections and showed a total of 331 hair cells on the left side and 308 hair cells on the right. As Table 8-II shows, the anterior and posterior tectorial bodies serve about equal

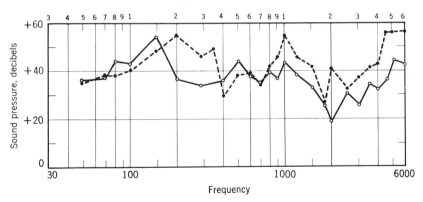

Fig. 8-19. Aerial sensitivity in two specimens of *Hyla m. microcephala*, expressed as in the foregoing figure.

numbers of hair cells, with a rather small portion connected to the sensing membrane.

The basilar papilla in this specimen contained 73 hair cells on the left side and 49 hair cells on the right.

A sensitivity curve for this animal is shown in Fig. 8-20. The performance in general is poor, with the curve running around +40 db in the low frequencies and reaching its best point of +26 db at 1000 Hz, with another small swing downward around 2000–2500 Hz. According to Duellman (1970) the mating call has its dominant frequency between 3220 and 4067 Hz, with a mean of 3578 Hz, in a region where the sensitivity as shown here is very poor.

Hyla septentrionalis, the Giant Treefrog.   This large hylid is found in Cuba and the Bahamas and also on the southern tip of Florida and the Keys. It is often abundant around buildings and in cisterns and wells, as well as amongst trees and shrubs. The adult body length varies greatly, ranging from 38–90 mm in males and 51–130 mm in females (Conant, 1958). The skin of the head is often partly co-ossified with the skull, and the fingers and toes bear especially large adhesive disks. The tympanic membrane is nearly round, and in a female whose body length was 78 mm it had a diameter of 4.7 mm.

The middle ear of this species was examined in particular detail. The footplate of the columella nearly fills the rather small lateral chamber. With the operculum this footplate shows a double notch arrangement, as shown in Fig. 8-21; in addition to the usual notch in the footplate into which the antero-lateral end of the operculum is inserted, there is a deep notch in the body of the operculum itself that receives a ventral process of the footplate.

At the level shown in this figure the columellar ligament does not appear;

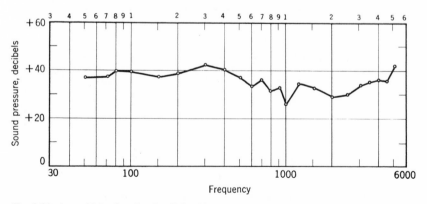

Fig. 8-20. A sensitivity function for *Hyla phlebotes,* expressed as in the foregoing.

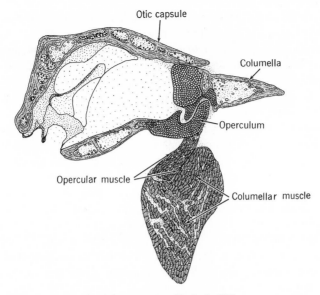

Fig. 8-21. The ear structures in *Hyla septentrionalis*. Scale 15X.

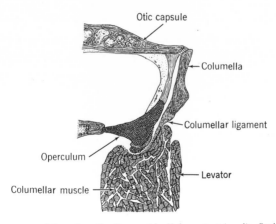

Fig. 8-22. The course of the columellar ligament in *Hyla septentrionalis*. Scale 15X.

it is at a more ventral level as shown in Fig. 8-22. This ligament runs from the upper end of the muscle bundle to the lower edge of the bony footplate. Only a few fibers of the opercular muscle are present at this level; the bulk of this muscle is much more dorsal. Also seen in this figure is the strong bundle of elastic fibers running from the end of the operculum to an attachment on the otic capsule; these fibers tend to hold the operculum in the notch of the footplate when the opercular muscle is relaxed, even in the absence of an activation of the columellar muscle.

One specimen of *Hyla septentrionalis* was found to have 444 hair cells in the amphibian papilla on the left side and 497 hair cells in the one on the right, with tectorial connections as shown in Table 8-II.

In this same specimen the basilar papilla on the left contained 38 hair cells and the one on the right contained 37 hair cells.

Sensitivity. — Auditory sensitivity tests were made by the use of inner ear potentials in four specimens, all females. One of these produced the function shown in Fig. 8-23, in which a sharp maximum appears at 1500 Hz, where the curve reaches −54 db, and good sensitivity is evident over two octaves from 500 to 2000 Hz. There is a suggestion of a secondary maximum in the low frequencies.

The next figure (Fig. 8-24) shows similar results for two other animals; here the maximum point is at 1000 Hz for one animal and at 1200 Hz for the other, but the level of sensitivity is less than in the preceding figure: −32 db in one animal and −18 db in the other. Again there is the suggestion of a deflection in the low frequencies.

A fourth animal gave the results of Fig. 8-25 in which a high degree of sensitivity is shown for the upper frequency range, with a maximum point

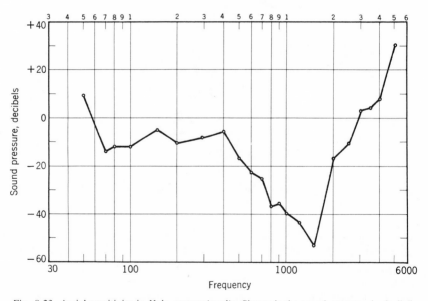

Fig. 8-23. Aerial sensitivity in *Hyla septentrionalis*. Shown is the sound pressure, in decibels relative to 1 dyne per sq cm, required for an inner ear potential of 0.1 μv.

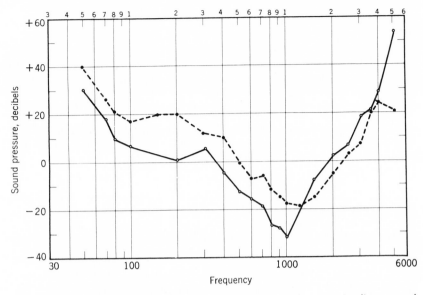

Fig. 8-24. Sensitivity functions for two additional specimens of *Hyla septentrionalis*, expressed as above.

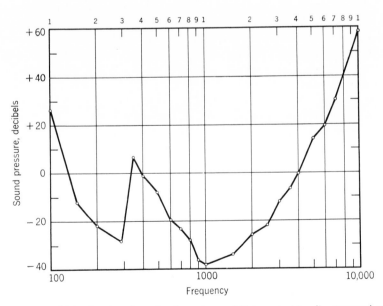

Fig. 8-25. Sensitivity functions for a fourth specimen of *Hyla septentrionalis*, expressed as in the foregoing figures.

of −38 db at 1000 Hz. Also this curve presents a striking secondary maximum in the low-frequency range, with its best point at 290 Hz.

All these results are in agreement in indicating an excellent level of sensitivity in the range about 1000–1500 Hz. The presence of a secondary maximum in the low frequencies is clear in one animal, but questionable in the others.

Hyla versicolor, the Gray Treefrog.   The common gray treefrog designated by Le Conte (1825) as *Hyla versicolor* has long been known in its wide distribution over forested areas throughout almost the whole of North America from the Plains states eastward, but only comparatively recently has it become clear that within this area there are two separate species with closely similar morphological characters. The distinction of two types was first suspected on the basis of the mating calls, which differ in pulse rate and pitch (Noble and Hassler, 1936). This subjective judgment was later borne out and made more specific by the use of sound spectrograms, which present distinctive patterns (Blair, 1962). Frogs of one type are "slow trillers," with pulse rates in their calls around 20–27 per second, and the others are "fast trillers," with rates of 47–52 per second—about twice as rapid. These calls also differ in frequency composition, with that of the "slow trillers" having a dominant frequency around 2420 Hz and that of the "fast trillers" one around 2660 Hz, a difference of more than 200 Hz (Blair, 1958).

The final proof has come from genetic studies in which artificial fertilization of eggs was carried out within and across these two calling types. The results show markedly reduced fertilization and extreme mortality of larvae when the two types are crossed; these two thus behave like distinct species.

There is a broad geographical pattern in the distribution of the two types: in the northern and eastern part of the total range there are mainly "slow trillers," for which Johnson retained the name *Hyla versicolor,* and in the southern and western areas are mainly "fast trillers," for which he used the name *Hyla chrysoscelis.* Yet within these general areas are zones in which both species occur and evidently do not hybridize to any appreciable degree. The calls themselves appear to be the critical condition that largely prevents mismatching: it was shown by Littlejohn, Fouquette, and Johnson (1960) that a female placed in the middle of a tank and exposed to sounds from two loudspeakers, that at one end of the tank playing a recording of a male call of *Hyla versicolor* and that at the other end playing a recording of *Hyla crysoscelis,* would choose the direction of her own species and swim toward it, almost without exception.

The two specimens used in the present experiments could not be positively identified, because their call notes were not observed, but tentatively are regarded as *Hyla versicolor* because they were collected in New Jersey.

The amphibian papilla in one specimen thus considered to be *Hyla versi-*

color contained 302 hair cells on the left side and 313 hair cells on the right, with tectorial connections as in Table 8-II. The total number is relatively small, and expecially reduced is the number of hair cells connected to the posterior tectorial body.

The basilar papilla in this same specimen contained 78 hair cells on the left side and 68 hair cells on the right, an ample population for this medium-sized frog.

A second specimen of this same species showed 352 hair cells in the left amphibian papilla and 355 hair cells in the right one. Again, as Table 8-II will show, the number of hair cells served by the posterior tectorial body was somewhat limited. This specimen had 74 hair cells in the basilar papilla on the left and 67 hair cells in the one on the right.

Sensitivity. — A sensitivity function for the second specimen described above is shown in Fig. 8-26. This curve indicates a poor level of sensitivity, with the best region in the medium high frequencies around 800–1500 Hz, and the maximum at 1000 Hz where +24 db is reached. This maximum is well below the dominant frequency of the mating call, determined by Blair (1958) as around 2420 Hz.

PSEUDACRIS SPECIES

The members of the genus *Pseudacris*, commonly known as chorus frogs, are noted for their clamorous heralding of the early rains of spring from their breeding sites in the vicinity of swamps and ponds, or even around temporary pools. There are 7 species, three of which were included in the study.

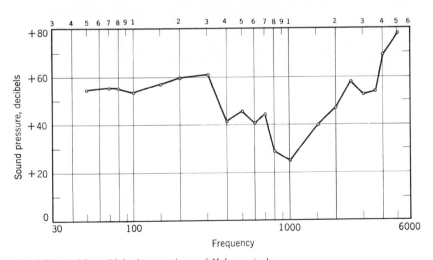

Fig. 8-26. Aerial sensitivity in a specimen of *Hyla versicolor*.

Pseudacris clarkii, the Spotted Chorus Frog. This is a small species, of about 29 mm body length in males and 31 mm in females (Duellman, 1970), found near grassland pools and in marshy areas from southern Kansas south to the Gulf of Mexico. The auditory structures have the standard form. The tympanic membrane is a broad oval, well differentiated from the surrounding skin. In a specimen with a body length of 25 mm this membrane was 1.4 mm high and 1.7 mm wide. In another specimen of 28 mm body length the tympanic membrane on one side was oval, measuring 1.5 mm by 2.0 mm, and on the other side was nearly round with a diameter of 1.8 mm.

The Amphibian Papilla. — The amphibian papilla is well developed, and was examined in detail in a specimen in which the fixation and staining were particularly favorable.

In this specimen the number of rows of hair cells increases rapidly from the anterior edge of the papilla to a maximum of 35 near the middle of the anterior region. The tectorial mass here contains thick-walled canals on the medial side and thinner ones on the lateral side. These canals extend ventrally into a region containing numerous vesicles of different sizes, mostly rather large on the lateral side and connected with a dense mass containing small vesicles on the medial side.

Then farther posteriorly, at a level a little over ⅜ of the length of the structure, an abrupt change is seen: the thin-walled canals first become elongated, extending almost all the way from the hair-cell edge to the ventral border of the tectorial body, and then this part is transformed into a thick tube with a few thin strands inside. This tube, which below and more ventrally merges into the sensing membrane, here at its dorsolateral end connects with only a few hair cells on the lateral side of the row. The remaining hair cells either have free-standing ciliary tufts or show no tufts at all. The main mass of the tectorial body does not extend this far ventrally, and only a small nodule of vesiculated tectorial material remains on the posterior end of the tube just mentioned. The nodule contains many small vesicles and a few large ones. Then a little farther ventrally this mass disappears and the tube ends in a stalk that has a small club-like expansion at its lower end. At this level there are no hair cells except the most lateral ones, which are reduced to 5 rows.

The stalk, which is the sensing membrane cut across, extends farther ventrally, expands at its foot, and attaches to the floor of the papillar recess. The dorsal part of the tectorial structure then enlarges to form a bowl-shaped body (as seen in cross section), which with the stalk has the outline of a goblet. The bowl of the goblet consists of thick-walled canals arranged in parallel and swinging medially at the bottom of the bowl; this mass is designated as the posterior tectorial body. Then the stalk portion (the sensing membrane) shifts laterally and becomes detached from the bowl portion. Thereafter the sensing membrane disappears, while a small mass of tectorial tissue representing the posterior tectorial body continues, probably going to the very end of the papilla where the wall on which it lies twists around as the amphibian recess merges with the posterior ampulla. In the region of the twist the section often includes three or four rows of hair cells. The papilla ends abruptly.

In this specimen the amphibian papilla on the left side contained 361 hair cells, and the one on the right contained 358 hair cells, with the different tectorial connections indicated in Table 8-II.

The basilar papilla on the left contained 58 hair cells and the one on the right contained 49 hair cells.

Sensitivity. — A sensitivity function for one of the specimens of *Pseudacris clarkii* is shown in Fig. 8-27. The most acute region is around 800–1000 Hz, where the curve reaches +23 db; an irregular region showing minor peaks at 2500 and 3500 appears at the high-frequency end. In general this ear is rather insensitive.

Duellman (1970) gave the fundamental frequency of the call note in this species as having a mean frequency of 78 Hz and the dominant frequency as varying between 2508 and 2652, with a mean of 2554 Hz. This dominant frequency falls in the region of the peak of sensitivity seen in the high frequencies in Fig. 8-27.

Pseudacris ornata, the Ornate Chorus Frog. This small species of chorus frog, in which the males average about 30 mm in body length and the females a little more, inhabits ponds, ditches, and wet meadows along the coastal plain from North Carolina to Mississippi and includes all but the southern tip of Florida. The specimen examined was a male 33 mm in body length.

The amphibian papilla contained 400 hair cells on the left and 394 hair cells on the right, with tectorial connections as shown in Table 8-II. The

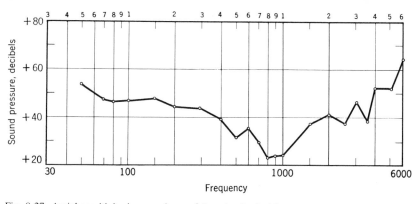

Fig. 8-27. Aerial sensitivity in a specimen of *Pseudacris clarkii*.

basilar papilla on the left side contained 38 hair cells and the one on the right contained 47 hair cells.

No satisfactory sensitivity measurements could be carried out on this specimen.

Pseudacris streckeri, Strecker's Chorus Frog. *Pseudacris streckeri* is the largest of the chorus frogs, attaining a body length of 41 mm in males and 46 mm in females (Conant, 1958). It inhabits woods, pastures, and fields in an area that includes parts of Oklahoma, Arkansas, and Texas.

The middle ear as represented in Fig. 8-28 shows a tympanic membrane that is but slightly differentiated from the surrounding skin, differing only in being thinner and largely lacking in skin glands. The notch in the columellar footplate makes an especially close fit for the anterolateral end of the operculum. Also at a level dorsal to the one shown here the middle part of the operculum is greatly thickened and has a broad depression into which the lower fork of the columella is seated. This mechanism thus contains a double notch.

The amphibian papilla is well developed, with 392 hair cells in the one on the left side, and 384 hair cells in the one on the right. Table 8-II shows details of the tectorial connections. The basilar papilla on the left contained 47 hair cells, and the one on the right contained 49 hair cells. These hair-cell populations are large for so small a frog.

Smilisca phaeota. The species *Smilisca phaeota* is a medium-sized hylid inhabiting tropical forest areas, widely distributed in lower Central America. The single specimen available was from the Canal Zone. The tympanic membrane is distinct, of oval form, with its upper edge partially covered by

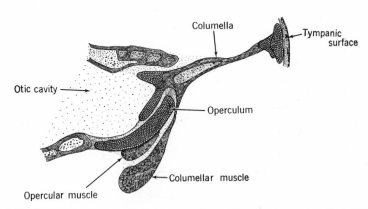

Fig. 8-28. The sound-receptive structure in a specimen of *Pseudacris streckeri*. Scale 20X.

a skin fold. In the specimen examined the tympanic membrane was 5.0 mm wide and 3.6 mm high.

In this specimen the amphibian papilla on the left side had the typical anuran structure, with a rather large population of 536 hair cells, distributed as in Table 8-II. As this table shows, 196 of these cells were connected to the anterior tectorial body, 65 were connected to a sensing membrane with its considerably expanded head portion, and 269 were served by the posterior tectorial body. There were only 6 free-standing hair cells.

The right amphibian papilla, however, was defective: the anterior tectorial body was not found in its usual location, and over the whole anterior third of the papilla the hair cells had only free-standing ciliary tufts. Then a large tectorial mass appeared and made contact with a small cluster of hair cells in 4 rows, and almost immediately sent off a sensing membrane to the floor of the papillar cavity below. This tectorial mass either was a remnant of the usual anterior tectorial body or an extraordinarily large expansion of the head of the sensing membrane. Thereafter, in the posterior region of this papilla, the usual relations were found, except that the number of hair cells connected to the posterior tectorial body was considerably smaller than on the other side.

The basilar papilla on the left side contained 88 hair cells and the one on the right contained 102 hair cells. These populations are somewhat larger than in most frogs.

Duellman characterized the call of *Smilisca phaeota* as a low growl, consisting of one or two call notes having a fundamental frequency of 110–165 Hz, with a mean at 143 Hz. He found only one harmonic to be emphasized, one with a dominant frequency that varies from 330 to 495 Hz, averaging 372 Hz. The sensitivity curve shown in Fig. 8-29 indicates a sensitive region

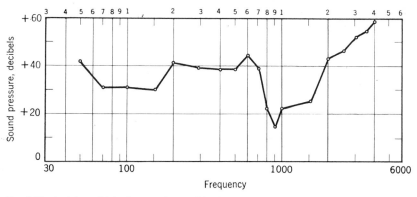

Fig. 8-29. Aerial sensitivity in a specimen of *Smilisca phaeota*.

around 800–1500 Hz, well above the range of the dominant call frequencies.

Sensitivity measurements in the single available specimen of *Smilisca phaeota* gave only the somewhat questionable results just shown. This curve indicates poor sensitivity, with a maximum region around 800–1500 that reaches +15 db at its lowest point. Duellman reported the mating call of males of this species as consisting of a low vibrant growl made up of notes of 100–130 pulses per second and a dominant frequency of 330–495 Hz with a mean of 372 Hz. The maximum region around 800–1600 Hz bears no obvious relation to the call notes in this species.

Triprion spatulatus spatulatus. This subspecies of *Triprion spatulatus* inhabits the lowland coastal areas of Sinaloa, Mexico and breeds in temporary pools after heavy rains. It is a relatively large casque-headed form: the males reach 87 mm in body length and the females 101 mm (Duellman, 1970). The two specimens studied were males of 64 and 67 mm S-V length.

The tympanic membrane is of oval form, a little wider than high. In one specimen the dimensions were 3.0×3.5 mm, in the other (the longer one) 2.7×3.1 mm. The skin of the head is co-ossified with the skull, and in placing the recording electrode it was necessary to drill away the bony surface with a dental burr until sufficient landmarks were exposed to show the location of the anterior semicircular canal. This canal then was entered in the usual way.

The amphibian papilla was examined in one of the specimens and showed some curious features. The anterior tectorial body had the usual form, except that it was much longer on the right side than on the left and contained a larger number of hair cells as Table 8-II shows. On the other hand, the sensing membrane on the left was extended and made contact with more than twice as many hair cells as the one on the right, thus evening up the relations somewhat. The posterior tectorial body also showed some peculiarities: it appeared as an expansion of the middle part of the sensing membrane rather than of its dorsal end as in other species. Then toward the posterior end of the papilla the mass of tectorial material that ordinarily rests over the hair cells at the lateral side of the papilla continued as an appendage of the sensing membrane until this membrane reached its posterior end and remained out of contact with the hair-cell layer beyond this point. This condition could be an artifact of preparation, caused by mechanical displacement or shrinkage, but careful examination of the sections did not give this impression. This tectorial mass everywhere appears to belong to the sensing membrane, and never seems to be an independent body.

Sensitivity.—Measurements on two specimens of *Triprion s. spatulatus* produced the results shown in Fig. 8-30. The sensitivity is poor in the low frequencies, grows progressively poorer as the frequency rises, and nowhere

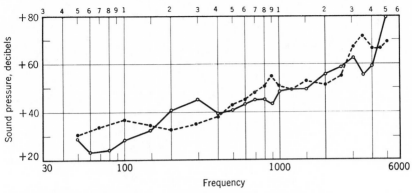

Fig. 8-30. Sensitivity functions for two specimens of *Triprion s. spatulosus*.

exhibits a prominent peak. Duellman (1970) observed that the call of this species is a single note of low pitch, but because of background conditions he was unable to obtain a measurement of the dominant frequency.

Agalychnis callidryas.   A small group of Hylidae is characterized by a vertical pupil and by peculiarities of color and the structure of the toes; these are referred to as the Phyllomedusinae, of which the species *Agalychnis callidryas* was the only available example. It is a medium-sized form inhabiting forested lowland regions scattered over an area from the southeastern end of Mexico through Central America to Panama. The four specimens studied, all males, were collected in the Canal Zone.

The tympanic membrane is distinct, though not markedly different in color from the surrounding skin. It is nearly round or slightly broader than high, and in a specimen with a body length of 49 mm it measured 2.2 mm in diameter. Another specimen of the same body size had a larger membrane, about 2.7 mm in diameter.

A study of the amphibian papilla in one of the specimens showed the standard structure, with a population of 542 hair cells in the organ on the left and 506 hair cells in the one on the right. The tectorial connections are indicated in Table 8-II and show a large proportion of the hair cells in the posterior division: about 56 percent for the left side and 52 percent for the right.

The basilar papilla contained 52 hair cells on one side and 53 on the other.

Sensitivity. — Auditory sensitivity in terms of the inner ear potentials was determined in four specimens. Results for one of these are shown in Fig. 8-31 where unusually keen sensitivity is evident: the maximum reaches −33 db for the tones 900–1000 Hz. Two other animals, as represented in Fig. 8-32, showed poorer sensitivity and much less regular functions, yet for these

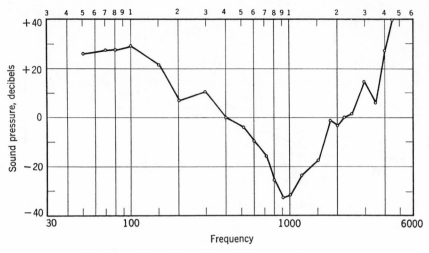

Fig. 8-31. Sensitivity in a specimen of *Agalychnis callidryas*.

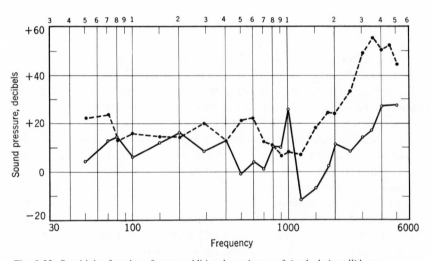

Fig. 8-32. Sensitivity functions for two additional specimens of *Agalychnis callidryas*.

the maximums are in the same general region, around 900–1500 Hz. The fourth specimen produced a maximum at 1500–2000 Hz.

These sensitivity maximums are somewhat below the dominant frequency of the mating call as reported by Duellman (1970). He found this call to have a dominant frequency in the region of 1488–2400 Hz, with a mean of 1975 Hz.

Pachymedusa dacnicolar.   This rather large hylid, the only species in the genus, inhabits arid lowland areas scattered along the entire Pacific shoreline of Mexico. A body length of 82.6 mm is reached in males and 103.6 mm in females (Duellman, 1970). These frogs are only moderately active in the dry season, and then their breeding cycle is triggered by the onset of rains, when the males begin to call from the banks of temporary pools. The call note is a single burst of short duration with a low fundamental and numerous high components, including dominant frequencies between 1120 and 2240 Hz, averaging 1727 Hz according to the observations of Duellman (1970).

The tympanic membrane is well defined, somewhat lighter in color than the surrounding skin, and nearly round, though a thick fold of skin obscures its upper edge and to some extent covers its dorsoposterior margin also. In one female of 92 mm body length this membrane was 6.0 mm in diameter, and in another of 77 mm body length the diameter measured 4.3 mm. The middle ear structure is of advanced form, with a well-developed lock mechanism as shown in Fig. 8-33. At the level shown only a few fibers of the opercular muscle are seen, as the bulk of this muscle lies more dorsally.

The amphibian papilla in this species has a large complement of hair cells, as indicated for one specimen in Table 8-II, where one ear contained 542 hair cells and the other contained 570 hair cells. The posterior tectorial body in this species arises as a gradual thickening of the dorsal end of the sensing membrane and then in the middle of the papilla becomes separated from this membrane and continues thereafter as an independent mass. The sensing membrane serves only about 14 percent of the hair cells, whereas the posterior tectorial body attaches to considerably more, to perhaps as many as

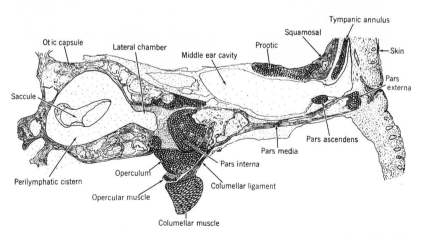

Fig. 8-33. The ear structures as seen in a transverse section of *Pachymedusa dacnicolor*. Scale 12.5X.

half of these cells. In this specimen the basilar papilla on the left side contained 53 hair cells and the one on the right contained 45 hair cells.

The results of sensitivity measurements on four specimens are shown in Figs. 8-34 to 8-36. The first of these figures presents functions for two specimens, both attaining a maximum of −33 db in the region of 1000–1500 Hz, and similar results are shown for a third animal in Fig. 8-35. A fourth specimen gave a curve of about the same form, though reaching a smaller maximum around −18 db as indicated in Fig. 8-36. The region of best sensitivity as shown agrees well with the dominant frequency of the male's call note as determined by Duellman.

CENTROLENIDAE

A small assembly of somewhat uncertain location among the advanced frogs is the Centrolenidae, a group of arboreal frogs inhabiting areas of the New World tropics from Mexico to Paraguay. They are distinguished by the presence of intercalary cartilages in two digits of the hands and feet, and the terminal phalanges of the hands have a "T" shape. This group is often placed with the hylids, but by many are given separate family status within the superfamily Bufonoidea (see Griffiths, 1963). There are only two genera, *Centrolene* and *Centrolenella*, containing about 50 species in all (Taylor, 1949,

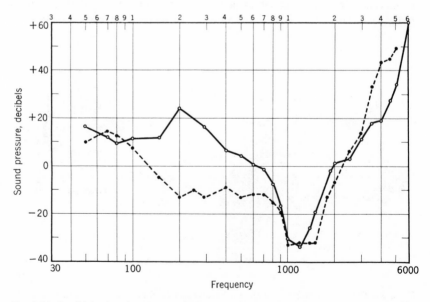

Fig. 8-34. Sensitivity functions in two specimens of *Pachymedusa dacnicolor*.

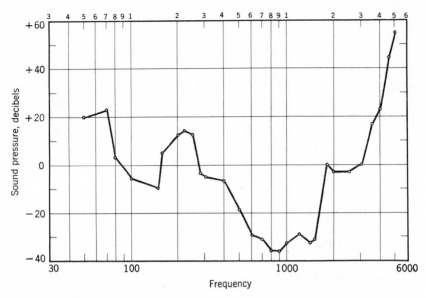

Fig. 8-35. Sensitivity for a third specimen of *Pachymedusa dacnicolor*.

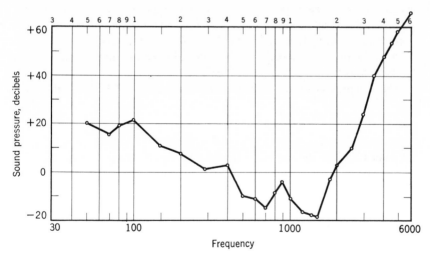

Fig. 8-36. Sensitivity for a fourth specimen of *Pachymedusa dacnicolor*.

1951, 1958; Lynch and Duellman, 1973). The two species studied were *Centrolenella euknemos* and *Centrolenella fleischmanni*.

Centrolenella euknemos. A single specimen of this species was examined in frontal sections, and the amphibian papilla on the left was found to contain 373 hair cells, distributed as shown in Table 8-III. The basilar papilla on the left had 45 hair cells and the one on the right 41 hair cells.

Only poor results were obtained from this specimen in the sensitivity tests, with the best point in the function reaching + 32 db for a tone of 1200 Hz. This animal appeared to respond adversely to the anesthetic and was in poor condition at the time of the measurements.

Centrolenella fleischmanni. In this species the tympanic membrane was not obvious on simple visual examination, but frontal sections in one of the specimens disclosed its presence as a slightly thinned area, with skin glands remaining but smaller and fewer than those found in the skin elsewhere. Beneath this area is a cartilaginous plate, the pars externa, which is especially thick posteriorly. The ascending process is reduced to a web of loose connective tissue.

The pars interna of the columella together with the operculum form a lock mechanism of a rather peculiar type, in which the operculum, which is entirely cartilaginous, is closely curled around the posterior bulbous end of the footplate. A relatively large columellar muscle sends a very slender ligament to the posterior bony process of the columella, whose anterior end is fused to the anterolateral wall of the otic capsule.

The amphibian papilla is this species has the typical form. In the speci-

Table 8-III

Hair-Cell Distribution in the Amphibian Papilla of the Centrolenidae

	To the Anterior Tectorial Body	To the Sensing Membrane	Free-standing Hair Cells	To the Posterior Tectorial Body	Total Hair Cells
	Centrolenella euknemos				
Left	163	46	8	156	373
Right	131	51	14	112	308
	Centrolenella fleischmanni				
Left	133	55	18	126	332
Right	153	42	3	121	319

men studied in detail this papilla on the left contained 332 hair cells and the one on the right 319 hair cells, distributed as in Table 8-III. In this same specimen the basilar papilla on the left side contained 35 hair cells and the one on the right 33 hair cells.

Sensitivity. — One of the specimens of *Centrolenella fleischmanni* gave the sensitivity results shown in Fig. 8-37, with a fairly good level of response for the middle tones, presenting a sharp maximum at 500 Hz where a point of − 16 db is reached.

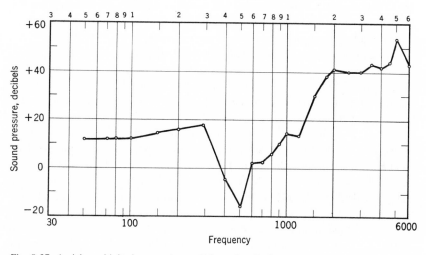

Fig. 8-37. Aerial sensitivity in a specimen of *Centrolenella fleischmanni*.

9. THE MICROHYLIDAE

The second of the three superfamilies making up the group of advanced frogs (the Neobatrachia) is the Microhyloidea, which consists of only one family, the Microhylidae. This group is considered to be closely related to the ranids and probably was derived from the same early stock.

This assembly of frogs has presented particular difficulties for the systematists. The species are highly varied in character and in earlier classifications were scattered among several families. Then Noble in 1931 proposed a new systematic scheme in which the form of the vertebral column was the principal determinant, and thus he brought together a number of species in 14 groups to form a family that he called the Brevicipitidae, characterized by the condition known as displasiocoely: the presence of two types of element in the vertebral column, including seven presacral cartilages concave anteriorly and convex posteriorly, and an eighth cartilage that is biconcave and connected posteriorly to the sacrum. Unfortunately it was soon found that this vertebral character is subject to wide variations, and Parker (1934) considered it useful but not reliably diagnostic in the determination of family status. Accordingly he found it necessary to employ a complex of characters for this purpose.

One useful identifying feature for the Microhylidae is the presence in most species of dermal folds across the palate that evidently are of assistance in grasping prey; these folds are found in no other group.

Larval development varies in three different ways within the family. Some species have free-swimming larvae that develop in pools or streams in the manner common to most frogs, others pass through the principal part of the process of metamorphosis within the egg and emerge as small but fully formed frogs. Still others reach a fairly advanced level before hatching and then continue their maturation to adult form within the nest site.

Those species with a tadpole stage present a distinctive form (Orton's Type II or Starrett's scoptanuran type). These are filter feeders; their respiration and feeding activities involve much the same mechanism. In feeding they take in water through the mouth, and the stream divides laterally to enter the two branchial chambers, passing through the filter organs in which any microorganisms are retained. Then the water flows out of each chamber through a posteromedial passage to enter an opercular tube that narrows funnel-like to a posterior spiracular outlet.

Noble found the members of this family that he called the Brevicipitidae to display such a degree of diversity as to require as many as 14 subfamilies. Parker (1934) saw difficulties in accepting several of these, considering them to be more closely related to the Ranidae, which led him to remove about half of them and to recognize only the remaining seven as properly belonging to the Microhylidae. His array of subfamilies thus included the Dyscophinae found in the Indo-Malayan region and in Madagascar, the Cophyinae of Madagascar, the Asterophryinae and Sphenophryinae of New Guinea and areas to the north, the Microhylinae of wide distribution in Ceylon, parts of India, and southern and eastern Africa, and the Melanobatrachinae in southwest India.

Parker considered the center of origin of this group of frogs as most likely the Orient and Madagascar, and Savage (1973) traced the distribution and evolution of this family from five major regions in South America, Africa, Madagascar, India, and New Guinea.

Four species of microhylids were available for study. These were *Gastrophryne carolinensis* from the southeastern United States and *Gastrophryne o. olivacea* from the Great Plains region between Nebraska and Texas, *Hypopachus c. cuneus* from south Texas and eastern Mexico, and *Kaloula pulchra* from southeast Asia. All four are closely similar in general ear structure, lacking a true tympanic membrane but with an effective substitute in the form of a cartilaginous plate (the pars externa of the columella in a flattened form) located immediately beneath the skin. These four species will now be examined.

Gastrophryne carolinensis, the Eastern Narrow-mouthed Toad. This is a small species, 22–32 mm in body length, found in a wide area of the southeastern United States; its range extends from Maryland west to the edge of Oklahoma and south to a portion of Texas. It usually lives on the borders of streams and swamps, concealed under logs and debris of various kinds, and ordinarily comes forth only at night. The pointed head, a fold of skin across the back of the head, and the absence of a tympanic membrane are distinguishing features.

The Middle Ear. — The skin continues over the side of the head, but shows the outline of the tympanic annulus beneath, from the edges of which is suspended a rounded cartilaginous plate, which is the pars externa of the columella. This plate is attached to the skin layer by connective tissue, and its inner surface is bounded by the air cavity of the middle ear. The osseous pars media of the columella attaches to the cartilaginous plate at a point somewhat posterior to its middle and then runs in a posteromedial direction through loose tissue along the lateral face of the parotic crest to the oval window where it expands as the columellar footplate. It sends an osseous shaft (the descending process) posteriorly along the lateral end of the operculum, which becomes tipped with a spike of cartilage to which the ligament of the columellar muscle is attached. Another process, which is cartilaginous though with some calcification of its lateral surface, runs through the oval window into the lateral chamber. It expands in the most lateral portion of this chamber and with the descending process forms a notch into which the upper end of the operculum enters partway. Elastic fibers connect the tip of the operculum with the lower end of the parotic crest. This structure is the usual anuran notch mechanism, though it is simpler in form than in most species: there is no suggestion of a double notch, and the lateral chamber is only partly filled.

The Inner Ear. — A study of the amphibian papilla in serial sections showed a total of 458 hair cells in the organ on the left and 442 hair cells in the one on the right. The tectorial connections for the amphibian hair cells are indicated in Table 9-I.

The basilar papilla on the left side contained 58 hair cells and the one on the right 46 hair cells.

Sensitivity. — A function for a specimen of *Gastrophryne carolinensis* is presented in Fig. 9-1. Here a fairly high degree of sensitivity is indicated in a sharp maximum at 800 Hz, where a level of −16 db is reached. After a decline there is a further maximum region at 2500 Hz, after which the response falls away at a rapid rate. A secondary maximum also appears in the low frequencies, with the best point at 290 Hz.

Results for two other specimens are presented in Fig. 9-2. One of these

Table 9-I

Hair-Cell Distribution in the Amphibian Papilla of the Microhylidae

	To the Anterior Tectorial Body	To the Sensing Membrane	Free-standing Hair Cells	To the Posterior Tectorial Body	Total Hair Cells
		Gastrophryne carolinensis			
Left	205	88	23	142	458
Right	168	90	35	149	442
		Gastrophryne olivacea			
Left	197	61	12	65	335
Right	191	77	0	72	340
		Kaloula pulchra			
Left	322	190	36	247	795
Right	363	168	48	250	829

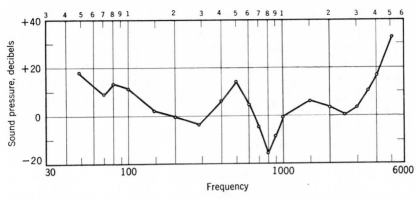

Fig. 9-1. Aerial sensitivity in a specimen of *Gastrophryne carolinensis*. Shown is the sound pressure, in decibels relative to 1 dyne per sq cm, required for an inner ear potential of 0.1 μv.

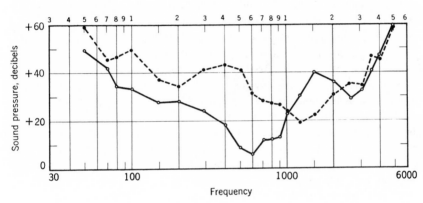

Fig. 9-2. Aerial sensitivity in two additional specimens of *Gastrophryne carolinensis*, expressed as in the preceding figure.

curves (solid line) represents a fair degree of sensitivity in the medium frequencies, with a maximum at 600 Hz where a level of +6 db is reached. This curve also shows a secondary maximum at 2500 Hz, though the level attained is very poor. The other curve in this figure represents generally poor sensitivity, with the principal maximum at 1200 Hz.

Gastrophryne o. olivacea, the Great Plains Narrow-mouthed Toad. A second species of *Gastrophryne* occurs in a more limited area west of the range of *G. carolinensis*, extending from eastern Nebraska south through Oklahoma and most of Texas to the northeastern part of Mexico, overlapping the range of the other species only to a slight extent. The two species are readily distinguished by their pigmentation patterns: the Great Plains form is largely unpigmented ventrally. In other respects—in habitat and behavior and also in auditory structure—these two species are much alike.

The Middle Ear. — The notch mechanism and the relations between the columella and its muscles are represented in somewhat schematic fashion in Fig. 9-3. At the level shown the lateral chamber does not open into the main cavity of the otic capsule, but it does so more ventrally.

A sketch of the middle ear structures based upon a dissection is presented in Fig. 9-4. Here the opercular muscle is shown running from its origin on the suprascapula to the middle of the operculum; it pulls this cartilage out of the columellar notch and permits the free transmission of vibrations inward along the columella. The columellar muscle with its long, slender tendon leading to the lower arm of the columella counteracts this operation, bringing the moving parts into the ''lock-up'' relation, and thus reduces sound transmission as discussed in more detail elsewhere.

The Inner Ear. — In a specimen of this species the amphibian papilla on the left side contained 335 hair cells, and the one on the right contained 340 hair cells. This population is noticeably smaller than that observed in *Gastrophryne carolinensis*, and as Table 9-I will show, the deficiency is mainly in the cluster of hair cells served by the posterior tectorial body.

Sensitivity. — A sensitivity function for a specimen of *Gastrophryne olivacea* is shown in Fig. 9-5. Here the best region is around 400–1000 Hz, with a maximum of −8 db at 800 Hz. No secondary maximum is evident here. Curves for two additional specimens are presented in Fig. 9-6. One of these (solid line) has a maximum at 1000 Hz and what may be regarded as a secondary maximum at 400 Hz. The other curve is similar in form, with best points at 900 and 500 Hz. However, these results are too irregular to indicate clear maximums, and the general level is very poor. Perhaps the most significant feature is the similarity to Fig. 9-5, in which the best performance of this ear is indicated for the middle frequency region around 400–1000 Hz.

Hypopachus cuneus, the Sheep Frog. This frog receives its common name from the quality of its call, which is described as a bleat like that of an abandoned lamb. It occurs in southern Texas and farther south into Mexico along the Gulf coast. It is a medium-sized frog, with males measuring 25–37 mm and females 29–41 mm in body length (Wright and Wright, 1949). It is a ground-living form, usually found well covered by litter and soil in natural hollows or in burrows made by rodents or other animals.

The single specimen obtained was a male with a body length of 32 mm. There is no eardrum; the skin is continuous over the facial area and is only slightly modified in one place by being a little thinner than elsewhere and lacking the usual skin glands.

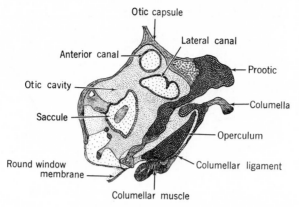

Fig. 9-3. The middle ear mechanism in *Gastrophryne o. olivacea* from a frontal section. Scale 20X.

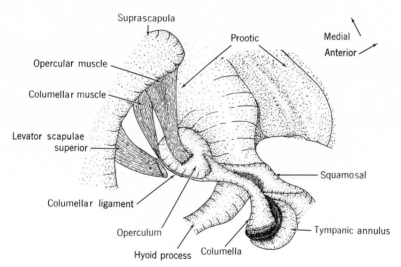

Fig. 9-4. The middle ear of *Gastrophryne o. olivacea*, drawn from a frontal section. Scale 20X.

The Middle Ear. — Although the eardrum is lacking, the columella is well developed and the pars externa consists of a flat plate of cartilage held at the edges within the tympanic annulus, as shown in Fig. 9-7. This plate in the specimen examined was a distorted oval 1.0 mm high and 1.3 mm wide, with a close contact with the skin at its midregion and a looser attachment through connective tissue toward the edges. This plate together with the skin thus forms a very small surface for the reception of aerial vibrations.

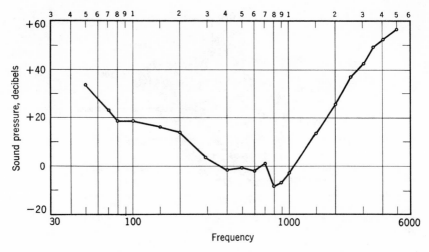

Fig. 9-5. Aerial sensitivity in a specimen of *Gastrophryne o. olivacea*. As before, the sound pressure producing an inner ear potential of 0.1 μv. is shown in decibels relative to 1 dyne per sq cm.

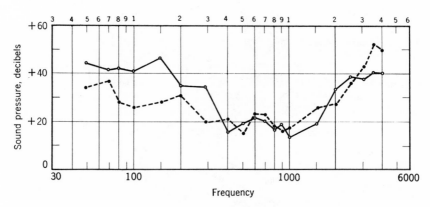

Fig. 9-6. Aerial sensitivity in two specimens of *Gastrophryne o. olivacea*, expressed as in the preceding figure.

The cartilaginous plate just described is attached to a bony pars media that runs along the outer and posterior face of the parotic crest to the columellar footplate in the oval window. This footplate has the usual complex form and makes contact with the operculum to produce a type of notch mechanism as represented in Fig. 9-8. An extended cartilaginous process runs posteriorly along the lateral surface of the operculum, and to it is attached a strand of fibrous tissue that corresponds to the usual ligament from the columellar

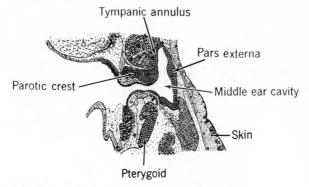

Fig. 9-7. The middle ear region in *Hypopachus cuneus*. Scale 10X.

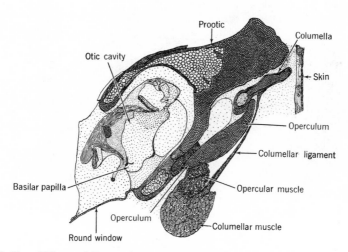

Fig. 9-8. The middle ear of *Hypopachus cuneus* in a frontal section. Scale 10X.

muscle. This strand was exceedingly thin in this specimen, and the question is raised whether it could serve in an effective manner in the control of sound transmission. This question must be held open, awaiting the examination of additional specimens to determine whether the condition here described is typical.

The Inner Ear. — The amphibian papilla as seen in frontal section is represented in Fig. 9-9. Here is shown the main body of the structure suspended from the papillar shelf, with the sensing membrane running to an attachment to the limbic pillar. This pillar contains a nerve bundle derived from the posterior portion of the eighth nerve, which also supplies the basilar papilla and the crista of the posterior semicircular canal. A window near-

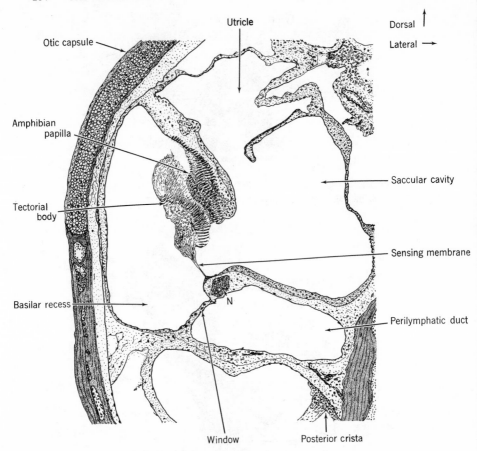

Fig. 9-9. The amphibian papilla of *Hypopachus cuneus* in a frontal section. Scale 75X. N = nerve bundle.

by leading into the perilymphatic duct provides a relief path for vibrations set up in the saccular cavity; these vibrations pass over the tectorial body and sensing membrane of the amphibian papilla and also displace the semilunar vane of the basilar papilla.

In this specimen the amphibian papilla on the left side contained 435 hair cells and the one on the right contained 472 hair cells. The distribution of these cells among the three regions of the tectorial structure could not be determined because the counting was carried out in a series of frontal sections, which do not provide this detail. The basilar papilla on the left contained 31 hair cells, and the one on the right 38 hair cells.

Sensitivity. — A response curve for this specimen is presented in Fig. 9-10. It shows a maximum region between 700 and 2000 Hz, with the best point of +33 db at 1000 Hz. The curve also shows a progressive improvement in the low frequencies, reaching +31 db at the lowest point measured at 40 Hz. This performance is generally poor, and may be explained in part by the absence of a tympanic membrane.

Kaloula pulchra. The frog *Kaloula pulchra*, one of the Microhylidae, is widely distributed in southeast Asia, including Thailand, Cambodia, Vietnam, and Malaysia on the mainland and the islands of Sumatra and Borneo.

The Middle Ear. — The 9 specimens studied varied from 56–60 mm in body length and lacked a true tympanic membrane. However, a prominent mound occurs in the ear region which has a soft, yielding area in its center. Removal of the skin in this area reveals a plate of cartilage held in the frame of a tympanic annulus as indicated in Fig. 9-11.

This plate is the pars externa of the columella and is an extension of the osseous pars media as shown; it measured 2.6 × 3.5 mm in a female specimen 64 mm in body length. Together with the skin it serves as a sound-receptive surface, as exploration with a vibrating needle readily proved.

The shaft of the columella runs to the lateral chamber of the otic capsule and forms a lock mechanism as represented in Fig. 9-12. The descending process of the footplate is unusually extended and connects through a short ligament with the columellar muscle.

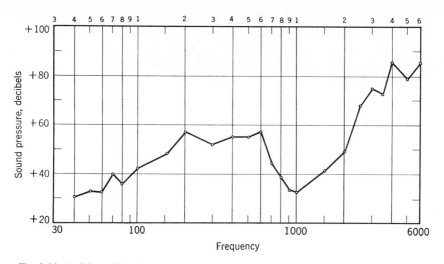

Fig. 9-10. Aerial sensitivity in a specimen of *Hypopachus cuneus*. Shown is the sound pressure in db relative to 1 dyne per sq cm giving a response of 0.1 μv.

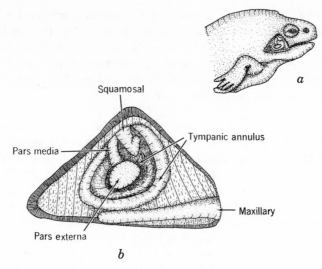

Fig. 9-11. At *a*, a dissection to reveal the ear structures beneath the skin in *Kaloula pulchra*; ¾ natural size. At *b*, the exposed ear region, enlarged 215X.

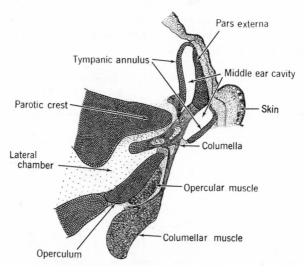

Fig. 9-12. The ear region of *Kaloula pulchra* in a frontal section.

The Inner Ear. — The amphibian papilla shows the usual form. In one specimen there were 795 hair cells in this papilla on the left and 829 hair cells in the one on the right. A little over one-fifth of these hair cells were connected to the sensing membrane, as Table 9-I shows.

The basilar papilla on the left of this same specimen contained 76 hair cells and the one on the right contained 83 hair cells.

Sensitivity. — Results on two specimens tested with aerial sounds are shown in Fig. 9-13. The function for one of these indicates a fair degree of sensitivity in the middle frequencies, with a maximum region between 600 and 1500 Hz. The function for the other specimen (broken curve) likewise shows the best response in this region, with maximum points at 700 and 900–1000 Hz, but also indicates a secondary maximum in the low frequencies, at 100–150 Hz.

Tests made on two other animals with a mechanical vibrator applied to the skin over the pars externa gave the results of Fig. 9-14. Here the curves are displaced upward somewhat relative to the aerial functions, one curve showing maximum points at 1500 and 2500 Hz, and the other showing maximums at 700 and 2000 Hz.

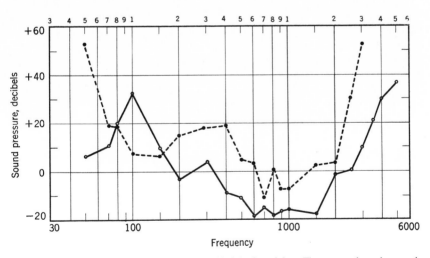

Fig. 9-13. Aerial sensitivity in two specimens of *Kaloula pulchra*. The curves show the sound pressure, in db relative to 1 dyne per sq cm, required for a response of 0.1 μv.

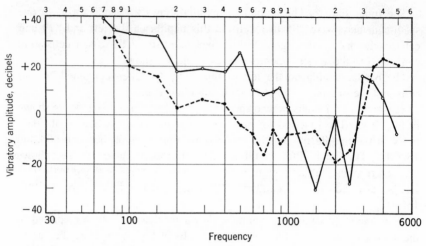

Fig. 9-14. Vibratory sensitivity in two specimens of *Kaloula pulchra*. Shown is the sound amplitude in db relative to 1 millimicron required for a response of 0.1 μv.

10. THE RANIDAE, RHACOPHORIDAE, AND HYPEROLIIDAE

THE RANIDAE

Members of the Ranidae, often referred to as the "true frogs," are worldwide in distribution. As many as ten subfamilies have been recognized, half of which occur only in Africa. Several species of the genus *Rana* were examined, and all showed an ear structure closely similar to that already described for *Rana utricularia sphenocephala* (Chapters 3 and 4). Results will be presented for additional *Rana* species that will show some of the observations made on general features and on the sensitivity to tonal stimuli, but the studies were not extensive enough to reveal species differences in full detail.

Rana adspersa. A single specimen of this large African frog was obtained, and to glean from it the maximum amount of information on the ear structures the block of tissue was split into right and left halves and these treated separately, with the right half sectioned in the frontal plane and the left half sectioned in the transverse plane. Table 10-I shows the results of hair-cell counts in these two ears, with some discrepancy between the counts made with the two planes of sectioning; those for the left ear are to be regarded as the more reliable since the transverse plane is the better for this purpose. In this specimen the free-standing hair cells could not be identified, and in the right ear the hair cells connected to the anterior tectorial body and to the sensing membrane could not be separated. There were 76 hair cells in the basilar papilla on the left and 64 hair cells in the one on the right.

A sensitivity curve for this specimen is presented in Fig. 10-1 and shows excellent acuity in the low tones, with the principal maximum at 600 Hz where a level of −59 db is reached. Above 600 Hz the decline in sensitivity is continuous and rapid, at a rate of about 40 db per octave, with the highest frequency reached at 5000 Hz.

Rana catesbeiana, the Bullfrog. This large species, known as the bullfrog because of its deep, resonant call, occurs naturally over most of the eastern and central United States, except in parts of Maine and the tip of Florida, and extends westward along a line from Nebraska through Texas and the

Table 10-I

Hair-Cell Distribution in the Amphibian Papilla of the Ranidae

	To the Anterior Tectorial Body	To the Sensing Membrane	Free-standing Hair Cells	To the Posterior Tectorial Body	Total Hair Cells
		Rana adspersa			
Left	382	123	†	229	734
Right	638ˢ	†	†	376	1014
		Rana catesbeiana			
Left	537	239	17	381	1174
Right	549	141	22	444	1156
		Rana clamitans melanota			
Left	390	110	8	367	875
*Right***	279	75	20	211	585
		Rana grylio			
Left	378	58	20	352	808
Right	324	92	17	384	817
		Rana ridibunda			
Left	231	73	15	355	674
Right	229	84	11	353	677
		Rana virgatipes			
Left	330	125	16	558	1029
Right	335	119	15	499	968

† Not determined.
ˢ Includes hair cells in sensing membrane.
** Defective ear.

edge of Mexico. It has been introduced into many other areas, where it flourishes wherever there are streams and lakes to satisfy its requirement of water in abundance.

Numerous experiments were carried out on this species, and specimens were used extensively in anatomical studies because the large size aided the dissections. Many of the results of observation on this species have already been given in Chapter 3.

The amphibian papilla contains over a thousand hair cells, with a distribution as observed in one specimen indicated in Table 10-II. In this same specimen the basilar papilla on the left side contained 99 hair cells, and the one on the right contained 95 hair cells.

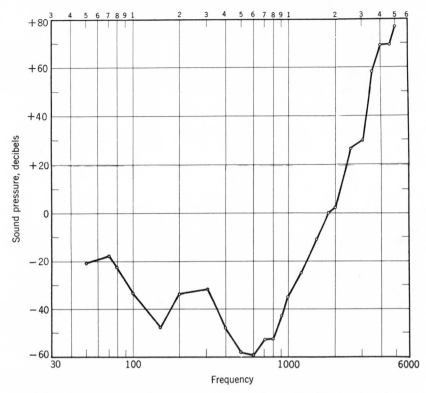

Fig. 10-1. A sensitivity curve for a specimen of *Rana adspersa*. Shown is the sound pressure, in decibels relative to 1 dyne per sq cm, required for an inner ear potential of 0.1 μv.

The sensitivity in one of the specimens is indicated in Fig. 10-2. Here are shown two regions of high acuity, one around 200–400 Hz and another around 1500–2000 Hz.

Rana clamitans, the Green Frog.   This species covers a broad area of eastern North America, occurring all along the eastern seaboard from the edge of Canada to the middle of Florida and west to a line from Minnesota to Louisiana. Conant (1958) recognized two subspecies, *Rana clamitans melanota*, north of an irregular line from North Carolina to the southern edge of Oklahoma, and *Rana c. clamitans*, south of this line, including the southern tier of states as far as eastern Texas. The specimens examined belonged to the northern group, showing much brown and black pigmentation on the dorsal surfaces. Wright and Wright (1949) indicated the maximum body lengths of males from northern areas as 95 mm and of females 100 mm, but some of our specimens exceeded these dimensions.

Table 10-II

Hair-Cell Distribution in the Amphibian Papilla of the Rhacophoridae
and Hyperoliidae

	To the Anterior Tectorial Body	To the Sensing Membrane	Free-standing Hair Cells	To the Posterior Tectorial Body	Total Hair Cells
		Rhacophorus nigropalmatus			
Left	245	154	9	284	692
Right	263	152	15	265	695
		Rhacophorus sp.			
Left	422	150	8	348	928
Right	434	153	21	400	1008
		Hyperolius viridiflavus			
Left	197	74	23	145	439
Right	206	46	15	150	417

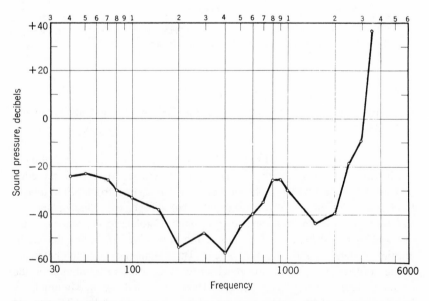

Fig. 10-2. A sensitivity curve for a specimen of *Rana catesbeiana*. Shown is the sound pressure, in decibels relative to 1 dyne per sq cm, required for an inner ear potential of 0.1 μv.

The tympanic membrane is of large size and varies from almost round to a little broader than high. It is generally larger in males than in females of comparable body length. In a male 83.0 mm in body length this membrane was 10.3 × 13.0 mm on the left side and 10.7 × 11.5 mm on the right. In a female 113 mm in body length it was 10.4 × 11.8 on the left and 10.3 × 12.2 mm on the right.

The amphibian papilla was examined in a female of 73 mm body length and on the left side was found to contain 875 hair cells, with tectorial connections as shown in Table 10-I. This papilla on the right was defective, however, with many of the hair cells shrunken and often difficult to identify. Those that could be counted are indicated in the table and total only 585. This defect cannot be accounted for as resulting from the test procedures since these were carried out on the left side; it probably represents a developmental or degenerative anomaly. The population of 875 on the left side is impressively large.

The basilar papilla in this animal appeared entirely normal on the two sides; there were 54 hair cells in the organ on the left and 47 hair cells in the one on the right.

Sensitivity. — Tests of sensitivity on a female specimen of 73 mm body length gave the results of Fig. 10-3, in which a region of very good sensi-

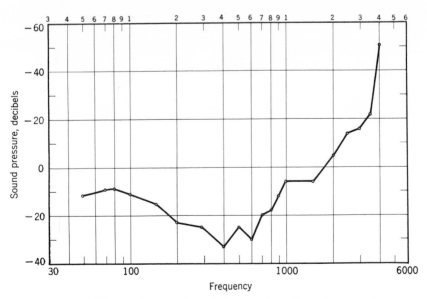

Fig. 10-3. A sensitivity curve for a specimen of *Rana clamitans*. Shown is the sound pressure, in db relative to 1 dyne per sq cm, required for an inner ear potential of 0.1 μv.

tivity appears between 200 and 700 Hz, with a peak of −34 db at 400 Hz. No secondary peak appears in this function.

A second specimen, a male of 76 mm body length, gave the results of Fig. 10-4. The solid-line curve represents the normal function, in which the primary maximum is in the region of 1000–1500 Hz where the curve reaches −39 db, and a secondary maximum appears in the lower frequencies, around 150–400 Hz. This animal was then exposed for 2 minutes to a tone of 5000 Hz at an intensity level of 1000 dynes per sq cm, after which the broken curve of this figure was obtained. There is a marked loss of response in the region of greatest sensitivity amounting to 44 db at 1500 Hz. In subsequent examination of serial sections of this specimen, no evidence was obtained of damage to the papillar structures.

Fig. 10-5 shows results of observations on two larvae of this species that were at an advanced tadpole stage with well-developed forelegs. In these tests the stimulation was with aerial sound, applied through a tube sealed over the side of the head at the level of the labyrinth. This structure was clearly evident, shining through the transparent skin covering, and the recording electrode was located on the anterior semicircular canal as usual. The sensitivity is very poor and there are large variations, so that no certain conclusions can be drawn except that a slight degree of auditory functioning is present at this stage.

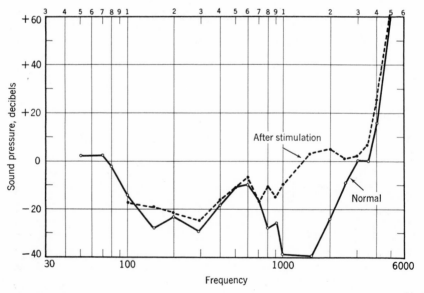

Fig. 10-4. Sensitivity in a second specimen of *Rana clamitans* under normal conditions and after overstimulation with a tone of 5000 Hz, expressed as in the previous figure.

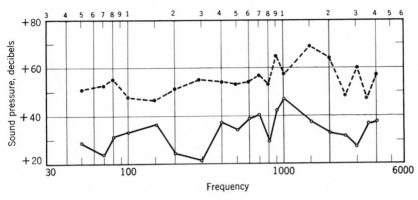

Fig. 10-5. Sensitivity functions for two specimens of *Rana clamitans* tadpoles. Shown is the sound pressure, in db relative to 1 dyne per sq cm, required for an inner ear potential of 0.1 μv.

Rana grylio, the Pig Frog. This species is known colloquially as the pig frog because its call is a deep grunt. It is highly aquatic, inhabiting lakes and marshes in the extreme southeastern part of the United States, including all of Florida and adjacent coastal regions. The distribution of hair cells in the amphibian papilla is shown for one specimen in Table 10-I. This same specimen had 83 hair cells in the left basilar papilla and 115 in the right one—a considerable number for this papilla.

The sensitivity is represented for two specimens in Fig. 10-6. There are two regions of good sensitivity, one in the low frequencies around 200 Hz and another in the high frequencies at 800 Hz in one specimen and around 1000–1500 Hz in the other.

Rana ridibunda. This is a European species of *Rana,* and the specimens examined probably belonged to the subspecies *perezi,* as their site of collection was reported as somewhere in Spain. The hair-cell distribution is shown in Table 10-I, with a total of 674 in one ear and 677 in the other. This same specimen had 72 hair cells in the left basilar papilla and 74 hair cells in the right one. A frontal section through the ear region showing the lateral chamber and portions of the lock mechanism is presented in Fig. 10-7.

Sensitivity. — The response functions shown for two specimens in Figs. 10-8 and 10-9 differ greatly in form. The first of these figures shows two regions of good sensitivity, one in the low tones around 100–400 Hz in which a maximum of −25 db appears and a second region of even greater sensitivity around 1000–1500 Hz, where the curve reaches −41 db. The second specimen, represented in Fig. 10-9, shows a similar region of low-tone sensitivity, which here reaches −51 db at 100 Hz, but the maximum in the high

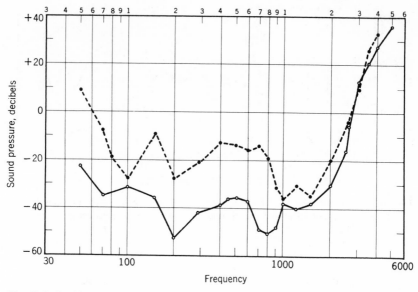

Fig. 10-6. Sensitivity curves for two specimens of *Rana grylio*. Shown is the sound pressure, in db relative to 1 dyne per sq cm, required for an inner ear potential of 0.1 μv.

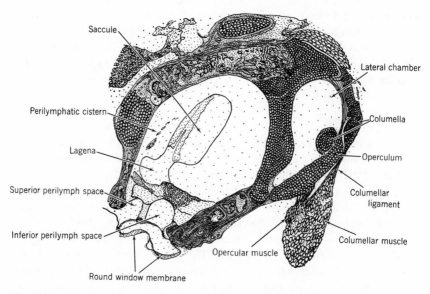

Fig. 10-7. A frontal section through the ear region of *Rana ridibunda*. Scale 20X.

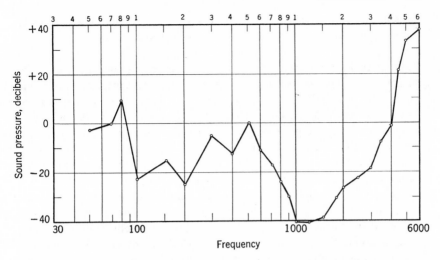

Fig. 10-8. A sensitivity function in a specimen of *Rana ridibunda*. Shown is the sound pressure, in db relative to 1 dyne per sq cm, required for an inner ear potential of 0.1 μv.

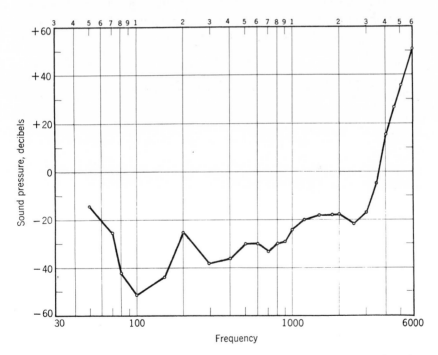

Fig. 10-9. Sensitivity in a second specimen of *Rana ridibunda*, expressed as in the foregoing figure.

tones is lacking; the sensitivity is fairly well sustained, with only a moderate decline up to 2500 Hz, after which the fall-off is very rapid.

Rana virgatipes, the Carpenter Frog.   This frog gets its common name from the character of its call, which is a clacking sound often compared to rhythmic hammering or wood-chopping. Its range is along the eastern coastal plain from New Jersey to Georgia. These frogs are often found partially submerged amidst matted grass or moss in bogs and ponds. It is a medium large frog; Conant (1958) gave the size range of males as 41–61 mm and that of females as 41–66 mm. As is usual, the tympanic membrane is somewhat larger in males than in females of a given body size.

The specimen examined was a male 62 mm in body length; the tympanic membrane was nearly round, 6.0 mm in diameter. The amphibian papilla contained 1029 hair cells on the left side and 968 hair cells on the right, with tectorial connections as shown in Table 10-I. The size of the hair-cell population in this papilla is impressive. The basilar papilla on the left side contained 64 hair cells, and the one on the right 59 hair cells.

Sensitivity. — Results of the inner ear potential tests are shown in Fig. 10-10. The sensitivity improves in an irregular fashion at the low-frequency end of the range, reaching a sharp maximum at 900 Hz, where a level of −35 db is reached. Thereafter the response falls off with great rapidity.

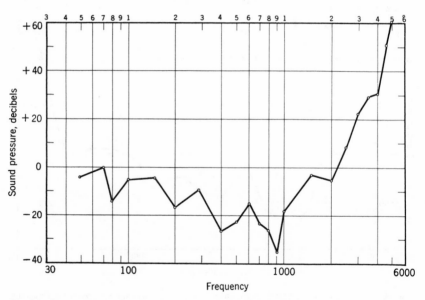

Fig. 10-10. A sensitivity function in a specimen of *Rana virgatipes.* Shown is the sound pressure, in db relative to 1 dyne per sq cm, required for an inner ear potential of 0.1 μv.

THE OLD WORLD TREE FROGS

The Ranidae are considered to have given rise to two families of tree frogs found in the Old World, the Rhacophoridae and the Hyperoliidae. It is thought that the Rhacophoridae probably developed in the Orient and then a few forms migrated to Madagascar and one genus (*Chiromantis*) to Africa. Hyperoliidae probably evolved in Africa, where all but two of the 14 genera now occur. One genus (*Heterixalus*) migrated to Madagascar and another (*Megalixalus*) to the Seychelles Islands (Liem, 1970).

The Rhacophoridae and Hyperoliidae resemble the Ranidae in many respects, notably in the presence of a firmisternal pectoral girdle: the two epicoracoid cartilages in this girdle are fused. These frogs differ from the ranids, however, in the presence of an extra joint at the ends of their fingers and toes: an intercalary cartilage occurs between the two distal phalanges. This feature is considered of great value to an arboreal frog in improving its agility in climbing through trees: it provides a more secure grasp of small twigs. This adaptation is especially useful when the tip of the digit is dilated and provided with a fleshy pad, as it is in most species, for the additional joint allows a more intimate contact between the pad and the twig being grasped. This intercalary cartilage probably was developed independently in these two derivatives of the Ranidae.

THE RHACOPHORIDAE

The family Rhacophoridae includes 10 genera, only one of which is represented in the present study. This is the genus *Rhacophorus*, of rather wide distribution in the South Seas region from India to Japan, and extending south to the Greater Sundra Islands and the Philippines. Specimens of three species were studied: *Rhacophorus maculatus*, *R. nigropalmatus*, and a third species yet unidentified.

Rhacophorus maculatus.   The middle ear structure was examined in particular detail in one of the specimens of *Rhacophorus maculatus*, and the form of its lock mechanism is indicated in Fig. 10-11. In this species the opercular and columellar muscles were easy to identify, as the opercular fibers were unusually small. Other features are of standard form. The tympanic membrane was nearly round, with a diameter of 5 mm. At the level represented in this figure the lateral chamber was completely walled off from the main cavity of the otic capsule by a thick bony septum, but ventral to this level there was a wide opening that allowed a free transmission of fluid vibrations from the footplate to the main cavity with its sense organs. The other two *Rhacophorus* species showed middle ear systems of essentially this same form.

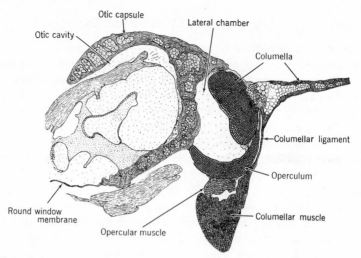

Fig. 10-11. The ear region in *Rhacophorus maculatus* in a frontal section. Scale 20X.

This animal gave the sensitivity curve of Fig. 10-12, in which two maximum regions are present, one in the low frequencies around 400–800 Hz where the sensitivity is poor, and another in the high frequencies from 1000 to 4000 Hz where the curve reaches the excellent level of −37 db for the tones 2000–2500 Hz.

Rhacophorus nigropalmatus. This species inhabits three regions of southern Asia: Siam on the mainland and the islands of Sumatra and Borneo. In the specimen examined, the tympanic membrane was well differentiated and nearly round, measuring 3.0 mm in diameter. For the tests on this animal the recording electrode was located in the saccule by an approach through the widely opened mouth. An air tube was inserted in the glottis and artificial respiration continued through the period of testing.

An examination of the amphibian papilla in this specimen showed a good-sized hair-cell population with a total of 692 hair cells on the left side and 695 on the right, distributed as shown in Table 10-II.

The basilar papilla contained 62 hair cells on the left side, and 77 hair cells on the right.

A specimen tested by means of inner ear potentials gave the curve of Fig. 10-13, in which there are two maximum regions, one around 150 Hz where the sensitivity reaches −16 db and another, more pronounced, in the region of 700–1000 Hz, where −28 db is attained.

Rhacophorus sp. Five specimens of another species of *Rhacophorus* (not further identified) were studied; these varied in body size from 60 to 66 mm.

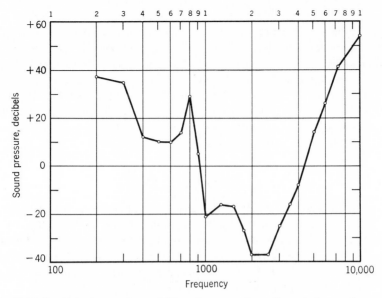

Fig. 10-12. Sensitivity in a specimen of *Rhacophorus maculatus*. Shown is the sound pressure, in db relative to 1 dyne per sq cm, required for an inner ear potential of 0.1 μv.

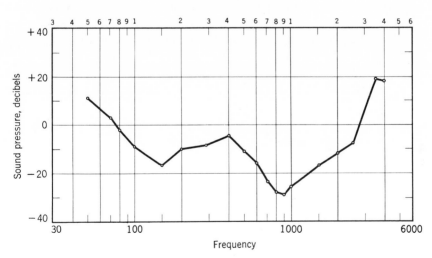

Fig. 10-13. A sensitivity function for a specimen of *Rhacophorus nigropalmatus*, expressed as the sound pressure, in db relative to 1 dyne per sq cm, required for an inner ear potential of 0.1 μv.

The tympanic membrane, oval in form, measured 4.5 mm wide by 5.0 mm high. In the amphibian papilla of one of these specimens there were 928 hair cells on the left side and 1008 on the right, with tectorial connections as shown in Table 10-II. The basilar papilla in this specimen contained 96 hair cells on the left and 90 hair cells on the right.

Sensitivity in this species is represented for two animals in Fig. 10-14, in which the curves agree in showing best sensitivity for the middle tones between 400 and 2000 Hz, with a maximum indicated for one of the curves at 800 Hz, where the level reached was −36 db.

THE HYPEROLIIDAE

As mentioned above, the Hyperoliidae include 14 genera of which 12 occur in Africa. Specimens were available representing two of these, *Hyperolius* and *Kassina*, both of which are found south of the Sahara Desert. Laboratory bred specimens of *Hyperolius viridiflavus* were obtained from the Amphibian Facility of the University of Michigan, and wild specimens of *Kassina maculata* came from a commercial source.

Hyperolius viridiflavus.   Six specimens were examined, all males, with body lengths of 24–27 mm.

The Middle Ear. — On superficial examination a tympanic membrane appears to be absent in this species, for the skin is continuous over the side of the head. Careful scrutiny, however, shows a shallow depression in the re-

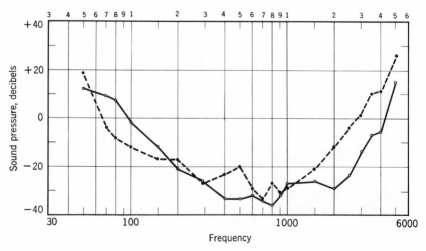

Fig. 10-14. Sensitivity curves for two specimens of *Rhacophorus* sp. expressed as in the foregoing figures.

gion just behind the eye at the edges of which the outline of the tympanic annulus shines through; then on removal of the skin, which comes away easily, a thin, almost transparent membrane may be seen stretched over the annular opening. This membrane corresponds to the fibrous middle layer of the typical tympanic membrane, as readily shown in serial sections through this region. Just medial to this fibrous layer is a cartilaginous plate that represents the outer surface of the pars externa of the columella. Finally, covering this plate medially, and extending over the more peripheral portions of the fibrous layer, is a sheet of mucous membrane that is common to the whole middle ear cavity. These relations are shown in Fig. 10–15. The diameter of this tympanic membrane was measured as 0.9 mm.

A lock mechanism is present as in the middle ears of other frogs, though there are a number of peculiar features. Thus the operculum is particularly thin and is attached to the otic capsule at its posteromedial end by a definite joint. The opercular muscle is relatively large and runs well forward to its attachment to the operculum on its midlateral surface. The columellar footplate sends off a cartilaginous process posteriorly that connects with a ligament from the columellar muscle running alongside the operculum but essentially independent of it. An opercular muscle, recognizable by its relatively fine fibers, attaches to the operculum along the middle of its lateral surface.

This structure, partly reconstructed by combining features of three adjacent sections, is presented in Fig. 10-16. At the level shown here the lateral chamber is separated from the main portion of the perilymphatic cistern by a wall of cartilage, but dorsally of this level the posterior portion of this wall opens up to provide a free path between the columella and this cistern, and thus to the amphibian and basilar papillae contained therein.

The sensitivity of the ear is represented for two specimens in Fig. 10-17,

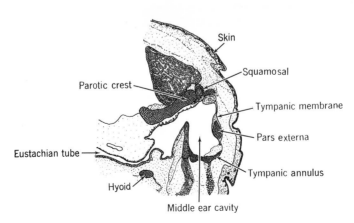

Fig. 10-15. The ear region in a specimen of *Hyperolius viridiflavus*. Scale 15X.

where a sharp maximum appears in the middle of the frequency range, reaching −20 db at 1000 Hz, and there is the suggestion of a secondary region of sensitivity in the higher tones around 2000–4000 Hz. Two additional specimens gave the results of Fig. 10-18, in which one shows a somewhat similar function, with the principal maximum again in the mid-frequency region, here peaking at 900 Hz, and a definite secondary maximum around 4000 Hz. The other animal shown here appears to lack the primary maximum in the middle of the range, though there is a suggestion of a maximum in the mid-frequencies and a definite one in the region of 3000–5000 Hz.

The Inner Ear. — An examination of the amphibian papilla in one of the specimens showed a total of 439 hair cells on the left side and 417 hair cells on the right, with tectorial connections as indicated in Table 10-II. The basilar papilla in this same specimen had 71 hair cells on the left and 78 hair cells on the right. These are good-sized populations, but are significantly smaller than those observed in rhacophorids.

THE SOUND STIMULATION PROCESS
IN THE ANURAN EAR

At this point, now that the detailed anatomy of the anuran ear has been surveyed, it is well to review the particular processes by which sounds are effective in stimulating the action of this type of ear. Most generally among the species of this order a tympanic membrane is present as a thin disk of modified skin stretched over a round or oval opening in a cartilaginous tympanic annulus, a disk distinguished from the surrounding skin by being relatively thin and lacking the subcutaneous layers and marked at the edges by a deep furrow. Sound waves striking this surface (either aerial or aquatic waves) exert an acoustic pressure and set the membrane into a vibratory motion that is transmitted through the columella to the footplate in the oval window, where alternating pressures are exerted on the fluid of the lateral chamber (as represented in Fig. 3-17). As already noted, this internal pressure has two principal routes of outflow to the round window and through this window to the middle ear and mouth cavities and thus to the outside, one route passing through the amphibian papilla and the other passing through the basilar papilla.

In both these papillae the alternating pressures produce movements of masses of tectorial tissue in which the ciliary tips of the auditory hair cells are embedded, and thereby stimulate these cells. The speed of propagation of sound in the ear tissues is of the order of 1500 meters per second and thus is so rapid relative to the distances involved within the ear that the actions can be regarded as essentially in phase: the entire mass of tectorial tissue with its attached hair tufts vibrates as a whole, and the responses of these cells reproduce closely the acoustic variations of the stimulus.

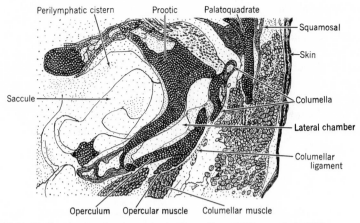

Fig. 10-16. Further details of ear structure in *Hyperolius viridiflavus*. Scale 30X.

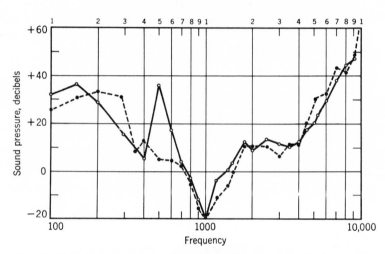

Fig. 10-17. Sensitivity in two specimens of *Hyperolius viridiflavus*, expressed as the sound pressure in db relative to 1 dyne per sq cm required for an inner ear potential of 0.1 μv.

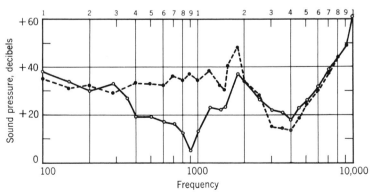

Fig. 10-18. Sensitivity curves for two additional specimens of *Hyperolius viridiflavus*, expressed as in the foregoing figure.

Further consideration of the anuran middle ear shows that it serves as a mechanical transformer in matching the impedance of the inner ear to that of the aerial medium. The tympanic membrane receives the sound pressures over its surface and transmits them in summated form through the ossicular chain to the inner ear fluids and through these to the auditory papillae. This transformer action affords a considerable gain in sensitivity, depending upon the dimensional characteristics, which vary with species but can be estimated as of the order of 20–30 db.

When a frog is in the water, it often floats with the head only partially submerged, so that the tympanic membranes are directly accessible to aerial sounds, and then the effectiveness of the system appears to be maximal. When the animal dives and the tympanic membranes are water-covered, the reception of aerial sounds will be greatly reduced. Then the hearing will be limited to the fraction of acoustic energy entering the water itself and now brought to bear on the submerged tympanic membrane. Sound produced in the water (as by another frog in the same pool) will then be readily received through a direct conduction through the tympanic surface and the columella.

PART III. THE URODELES

11. THE SALAMANDER EAR

As was done for the anurans, a consideration of the anatomy of the ear in the urodeles will begin with the skull structures for purposes of general orientation and then will include the labyrinth as a whole. Finally, with these topographical features in mind, a series of figures will be presented to show the forms of the auditory receptor organs and their relations to other structures and especially the pathways by which these organs may be stimulated by sounds.

GENERAL STRUCTURE

First to be examined in this orienting survey are two species of *Ambystoma*, which are among the more advanced of the salamanders.

Figure 11-1 presents a dorsal view of the skull of *Ambystoma gracilis*, drawn from a dissected specimen. The auditory end-organs are contained in the prootic, with which the exoccipital is fused, and the squamosal and quadrate are closely adjacent. Further details are shown in Figs. 11-2 and

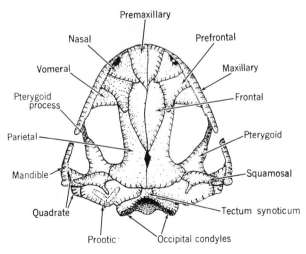

Fig. 11-1. The skull of *Ambystoma gracilis*, from a dissected specimen. Dorsal view, Scale 3.5X.

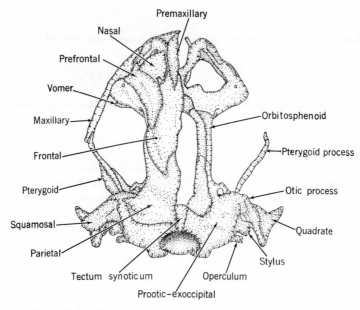

Fig. 11-2. The skull of *Ambystoma maculatum*. Dorsal view, after Theron, 1952. Scale 4.2X.

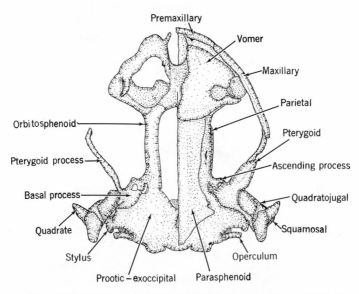

Fig. 11-3. The skull of *Ambystoma maculatum*. Ventral view, after Theron, 1952. Scale 4.2X.

11-3 representing the species *Ambystoma maculatum*, redrawn from Theron (1952). These figures were produced by a graphic reconstruction method, and are unusual for their detail and precision. In both of these some of the more superficial elements have been omitted on one side to reveal the deeper structures.

The stylus and operculum are indicated, extending laterally out of an opening (the oval window) in the prootic-exoccipital complex. These two elements have their basal portions held in the window by soft tissue (ligament and cartilage), so that they are able to move over short distances in response to sound vibrations. At the larval stage the outer end of the stylus is attached to the squamosal, but in the process of transformation to the adult this connection shifts largely to the quadrate as shown here.

THE SALAMANDER LABYRINTH

Lateral and medial views are given of the labyrinth of the alpine newt, *Triturus alpestris* (Figs. 11-4 and 11-5). This salamander is widely distributed in Europe, not only in the high mountains as its name implies, but also in hilly and even lowland areas. These two figures, taken from Birkmann (1940), show the typical elongated form of this structure, with general locations of the usual endorgans. As Fig. 11-5 shows, the amphibian papilla has an inferior and central position and the lagenar papilla, though not visible here, is close by. These and other structural features of the inner ear will now be examined.

THE SALAMANDER INNER EAR

The inner ear of the salamander presents many features in common with that of frogs. There is an amphibian papilla and in a number of species a second inner ear receptor also, which unfortunately has heretofore been called a basilar papilla on the supposition that it corresponds to the one given that name in the anurans. These organs are formed of hair cells and supporting cells of the same kinds as in anurans (and other vertebrates) and contain tectorial tissues whose general character and function in the stimulation process are closely parallel. There are differences, however, and the problem of their relationships will be given attention later on.

DETAILED ANATOMY OF THE SALAMANDER EAR REGION

THE AMPHIBIAN PAPILLA

As in the anurans, the primary auditory receptor is one to which DeBurlet gave the appropriate name of amphibian papilla; indeed it is the only one present in a number of salamander species. This organ has a location in the head that corresponds to that of the anurans, suspended below a limbic shelf arising low on the cartilaginous or bony septum between cranial and otic

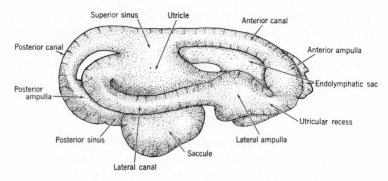

Fig. 11-4. The labyrinth of *Triturus alpestris*. Lateral view, after Birkmann, 1940. Scale 40X.

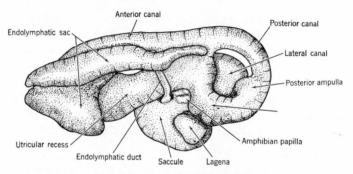

Fig. 11-5. The labyrinth of *Triturus alpestris*. Medial view, after Birkmann, 1940. Scale 40X.

cavities, and it takes a form that can be regarded as a simplification of the anuran type. This structure will be followed in detail in a series of transverse sections running from anterior to posterior ends as was done for the anurans.

The following series is taken from a specimen of *Ambystoma maculatum* in which the papillar structure may be regarded as typical of the more advanced urodeles.

The amphibian papilla is first seen anteriorly as a little notch at the dorsolateral corner of a pillar of limbic tissue that arises alongside the lower portion of the otocranial septum, and to which the saccule is closely attached, as shown in Fig. 11-6 and more clearly in Fig. 11-7. The first of these figures is a transverse section showing almost the entire head, and as usual the skin and muscle tissues had been removed over the skull roof on one side for the placement of a recording electrode. The section includes the cranium above with the mouth cavity below and also a portion of the lower jaw region. Of interest for general orientation and for further reference are the frontoparietal and parasphenoid bones that enclose the cranium above and

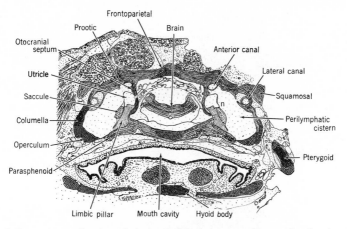

Fig. 11-6. Transverse section through the head of *Ambystoma maculatum*, showing the anterior ear region. *n* represents a notch where the amphibian papilla begins. Scale 20X.

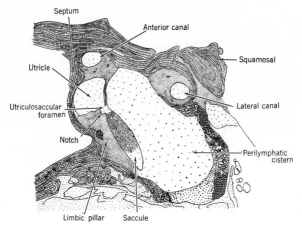

Fig. 11-7. Transverse section through the right ear region of *Ambystoma maculatum*, at a level posterior to the preceding. Scale 20X.

below and the squamosals and prootics covering the sides; gaps are left in the bony enclosure at the outer ventral ends that are filled by the columella and operculum. Further details concerning these elements in the oval window will be noted later.

The notch that forms the most anterior indication of the amphibian papilla lies between the upper end of the saccule and the lower part of the utricle and in Fig. 11-8 extends more deeply into the limbic pillar where it is cov-

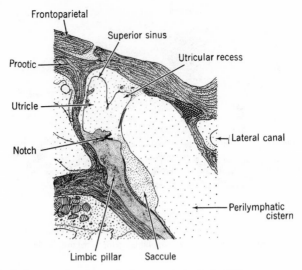

Fig. 11-8. As in the preceding figure, farther posteriorly. Scale 20X.

ered by epithelial lining cells. A little farther posteriorly will appear the first of a series of hair cells. The notch opens in the limbic bed and is divided by a thin membrane as seen in Fig. 11-9; this membrane is the perilymphatic window and the cavity below is the perilymphatic duct that leads eventually into the cranial cavity. The limbic shelf overhanging this area has a typical form and bears the papillar organ on its undersurface.

The amphibian papilla may be examined further in the enlarged view of Fig. 11-10. The shelf consists of a core of limbic tissue enclosed above by large epithelial cells, and within the core are scattered cells and a good many nerve fibers that supply the sensory layer below. This sensory layer consists of closely packed supporting cells, seen in cross section as a continuous row. The nucleated ends of these cells are attached to the limbic floor of the shelf and send fine pillar processes downward. These processes expand to clasp the lower ends of the hair cells, whose cilary tufts extend downward through openings between the processes. The ciliary tufts connect with a tectorial body as shown.

The tectorial body is a structure that in cross section resembles a lattice but perhaps is better described as a series of closely aligned canals leading into a mass of irregularly vesiculated material below. The vesicles are mostly globular, sometimes ovoid, of various sizes, and the mass evidently has the function of taking up vibrations from the surrounding fluid, transmitting them to the canaliculate structure above, and thereby producing a bending of the ciliary tufts of the hair cells through which these cells are stimulated.

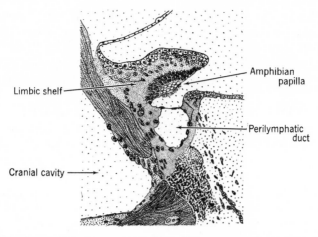

Fig. 11-9. As above, still farther posteriorly. Scale 20X.

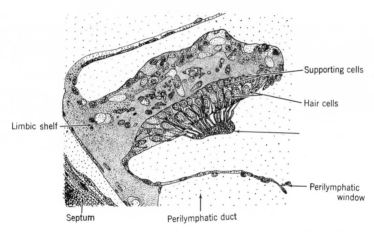

Fig. 11-10. Detailed view of the amphibian papilla of *Ambystoma maculatum*, in transverse section. Scale 200X.

Let us examine this papilla further. Its form is roughly that of a hemisphere, or in some species is more flattened so as to resemble about a third of a sphere, and consists of three nesting portions: a layer of supporting cells at the periphery forming a containing cup or outer shell, a layer of hair cells forming a second shell within this one, and the tectorial body making up a middle core that faces outward into the papillar recess. The organ is contained in a mass of limbic tissue and forms the inner wall of a small cavity (the papillar cavity) that is separated by an extremely thin membrane (the perilymphatic window) from the large space of the perilymphatic duct.

This papilla has just been shown in frontal sections in Figs. 11-9 and 11-10, but the form as described is more evident, and is presented in further detail, in sketches based on sagittal sections of a specimen of *Ambystoma texanum*. Figure 11-11 shows the general form of the structure in this lateral view, with further detail given for a portion in the enlarged picture of Fig. 11-12. The lower part of this figure shows a portion of the tectorial body with the channels into which the ciliary tufts of the hair cells protrude. Drawings of an individual supporting cell and a hair cell are presented in Fig. 13 and Fig. 11-14 shows how the ends of the hair cells fit into openings in the perforated membrane separating the hair-cell layer from the tectorial body.

As Figs. 11-11 and 11-12 show, the supporting cells have their cell bodies at the periphery of the cup and send radially inward their long, slender processes that expand at the ends to form wide feet that are interlocked so as to produce round openings between them. In these openings are secured the outer ends of the hair cells whose main bodies seem loosely held between the middle portions of the extended processes. The ciliary tufts of the hair cells extend outward toward the center of the cup; their ends enter the radiating canals and attach to their walls. These canals end in a mass of tectorial material that consists of numerous vesicles of varying sizes. It is this vesiculated tissue (the tectorial body) that extends out into the papillar recess and is set in motion by the mass movement of the inner ear fluids produced by an acoustic stimulus.

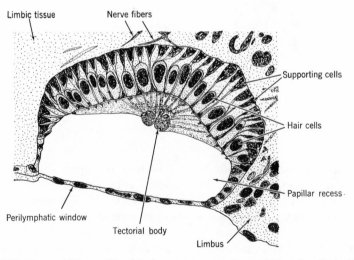

Fig. 11-11. The amphibian papilla of *Ambystoma texanum*, in a sagittal section. Scale 400X.

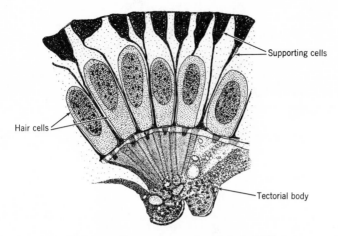

Fig. 11-12. A portion of the preceding specimen, enlarged. Scale 750X.

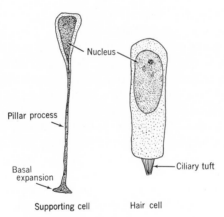

Fig. 11-13. Isolated views of a supporting cell and a hair cell, from the amphibian papilla of *Ambystoma texanum*. Scale 1500X.

There seem to be two forms of ciliary connection. The usual one, seen for all the middle hair cells, is an attachment of the end of the ciliary tuft to the wall of a canal close to its inner end. The second form, shown by the more peripheral hair cells, consists of a tectorial strand that runs out directly from the central mass to attach to one or two ciliary tufts. These peripheral hair cells are not enclosed in well-defined spaces or tubes as the more central hair cells are, but are located singly or in groups of two or three in areas that are somewhat vaguely outlined by webs of tectorial tissue. The hair cells of the

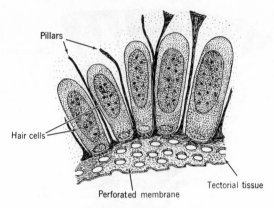

Fig. 11-14. The perforated membrane of a specimen of *Ambystoma texanum*, with a few hair cells and pillar processes in place. Scale 2000X.

central group, making up perhaps half of the whole, are enclosed by tubes with noticeably thicker walls than the others.

The Acoustic Pathways.—The path of sound transmission through the amphibian papilla is now to be followed. Vibrations communicated to the water in which the animal is immersed (or perhaps coming through the air if the animal is on land) will exert alternating pressures on the oval window elements and on the inner ear fluids beyond, tending to set this entire structure into vibratory motion. However, because fluids and fluid-filled tissues are for all practical purposes incompressible, there can be no movement unless somewhere beyond these is a relief route. As experiments later to be described have proved, this route in the salamander is distinctive: it is a passage to the opposite side of the head. It runs first through a foramen in the otocranial septum, then through the cranial cavity to the opposite ear, and on through the same structures in reverse sequence to the other oval window.

The particular route of the fluid movements will be traced in a number of sectional views. Figure 11-9 has shown the presence of a thin membrane (the perilymphatic window) leading into the perilymphatic duct, but at the level of that section the duct is closed below; a foramen is beginning to open in the bony septum but is still obstructed by limbic and cellular tissues. Then in the next and several successive sections as represented in Fig. 11-15 this foramen is widely open. Vibratory currents pass through the perilymphatic duct and enter the cranial cavity, following the pathway indicated by the large arrows in Fig. 11-16. These currents pass over the tectorial mass protruding into the channel below the amphibian papilla and thus set in motion the embedded ends of the ciliary tufts of the hair cells.

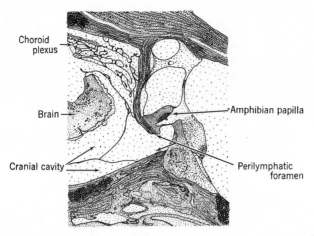

Fig. 11-15. The amphibian papilla in *Ambystoma maculatum* in relation to the perilymphatic foramen and brain cavity. Scale 20X.

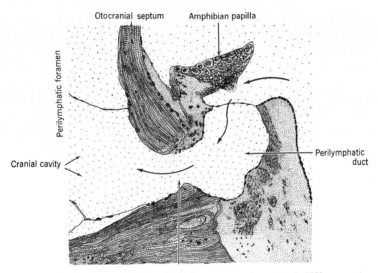

Fig. 11-16. Paths of vibratory fluid flow in *Ambystoma maculatum.* Scale 80X.

In the section shown in Fig. 11-17 the amphibian papilla is coming to an end; only four rows of hair cells are still to be seen, and in a short distance farther posteriorly this papilla terminates and the foramen that here is already occluded by soft tissue becomes completely closed by bone (Fig. 11-18).

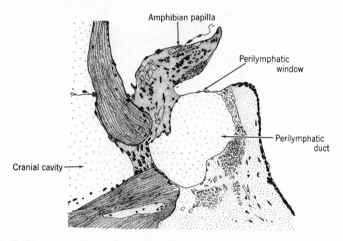

Fig. 11-17. The pathway from papillar region to cranial cavity in *Ambystoma maculatum*, occluded by soft tissue. Scale 80X.

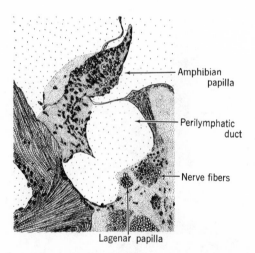

Fig. 11-18. A level posterior to the above; the path to the cranial cavity is closed by bone. Scale 80X.

In this species as in most salamanders a second auditory endorgan is present; the one commonly called the "basilar" papilla soon follows in this posterior progression through the head. The name "basilar papilla" is quite inappropriate for this endorgan; because it is clearly a derivative of the lagena, it will be referred to as the "lagenar papilla."

THE SECOND AUDITORY PAPILLA: THE LAGENAR PAPILLA

The most anterior trace of this second auditory papilla is seen in Fig. 11-18 as a small patch of epithelial cells located on a ridge that partially divides the perilymphatic cavity into two parts. A mass of nerve fibers will also be found laterally to this patch of cells; these elements innervate the second papilla. At a short distance farther posteriorly the picture represented in Fig. 11-19 is obtained, in which the main perilymphatic duct remains above and a lower portion of this perilymph space is displaced downward. Intruding between and separating these two spaces at this level is a third cavity that has pushed out from the roof of the lagena; this cavity contains the second auditory papilla, bathed in endolymph. Farther posteriorly the lower perilymph space comes to an end, and the condition is as represented in Fig. 11-20.

The lagena as seen here is contained in a relatively large cavity, well enclosed in the limbic tissue of the region, with its macular organ running along the medial wall. The tectorial material overlying the hair cells of the lagena is heavily loaded with statolithic crystals. The lagenar papilla is not seen in this section because it lies on a more anterior plane; its location is thus above the septum that separates the upper perilymph space from the lagenar cavity below, at a position indicated in this figure by a small arrow. This papilla is shown in Fig. 11-21 in a frontal section in which the acoustic pathway to this second papilla is readily followed: sound enters through the oval window, passes across the perilymphatic cistern to the small lagenar recess at the end of the saccule, and then exits through a thin membrane into the brain cavity. The lagenar papilla hangs from the roof of this recess and is shown in further detail in Fig. 11-22 in a transverse section and again in Fig. 11-23 in a frontal section in which its appearance is much the same. There are about 10 rows of hair cells adjacent to which is a little hillock of tectorial material into which the ciliary tufts of these cells protrude. The nerve fibers supplying these hair cells are shown in Fig. 11-22, running through the limbic tissue above the hair-cell layer. A further view representing this structure in a frontal section is given in Fig. 11-24.

The pathway as shown in Figs. 11-16 and 11-21 contains no serious obstacles; there are only three thin membranes between the oval window region and the brain cavity: two membranes bounding the papillar recess and one between the perilymphatic duct and the brain cavity proper. Such membranes, having a density equal to that of the surrounding fluid and practically lacking in tensile strength, are almost perfectly transparent to fluid vibrations; they merely serve to separate the different fluids, and acoustically can be neglected. The only significant factors in this sound path are the mass of the fluid that must be moved and the frictional resistances along the way.

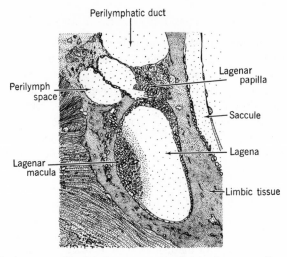

Fig. 11-19. The lagenar papilla of *Ambystoma maculatum* lying between perilymphatic duct and the lagenar cavity, seen in transverse section. Scale 80X.

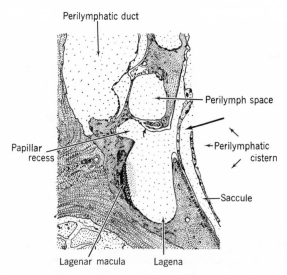

Fig. 11-20. Sound vibrations (heavy arrow) reach the papillar recess through the lagena. Scale 50X.

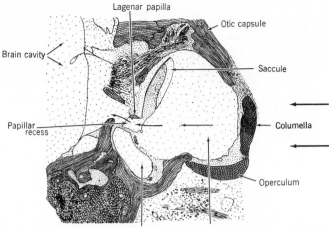

Fig. 11-21. The sound pathway from the columella to the brain cavity, in a frontal section of *Ambystoma maculatum*. The heavy arrows on the right represent the incident sound stimulus, and the smaller arrows the course through the ear tissues. Scale 20X.

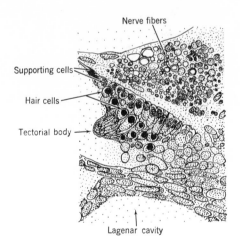

Fig. 11-22. The lagenar papilla of *Ambystoma maculatum* in a transverse section. Scale 350X.

COMPARISONS WITH ANURAN RECEPTOR ORGANS

The Amphibian Papilla. — This papilla in the urodeles is relatively simple. As followed from anterior to posterior ends it arises from a shelf on the octocranial septum within the saccular cavity, increases rapidly in size until it contains a fairly large number of rows of hair cells (around 30 in one spec-

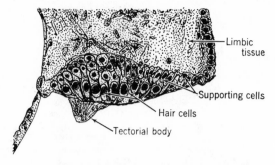

Fig. 11-23. The lagenar papilla of *Ambystoma maculatum* in a frontal section. Scale 250X.

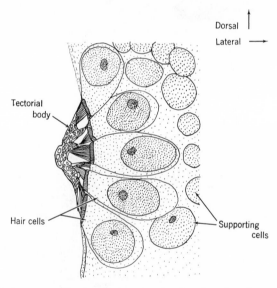

Fig. 11-24. The lagenar papilla of *Ambystoma texanum* in a frontal section.

imen examined), and then gradually declines. In the specimen referred to there was a total of 202 hair cells in this papilla.

The general form of this sensory structure corresponds closely to the anterior portion of the amphibian papilla in anurans: the limbic shelf is similar in design, the hair cells are suspended in the same manner, and the tectorial body has a corresponding form. However, there is no sensing membrane attaching to certain of the hair cells in this sensory body as is found in anurans, and the long posterior division present in the amphibian organ of the advanced frogs is lacking also.

The number of hair cells in this papilla falls far below that occurring in the amphibian papilla of most frogs, which may amount to a few hundred and in some species exceeds a thousand. The number in salamanders is more comparable with that in the anterior division of this papilla in some of the smaller frog species, such as certain of the hylids.

In other respects, however, the resemblances between these papillae in salamanders and frogs are such as to give strong assurance that these are corresponding organs and are properly designated by the same name.

The Lagenar Papilla — The resemblance just found between the amphibian papillae of anurans and urodeles fails to hold for the second type of auditory organ. Although this organ in the salamander has without much questioning been referred to by previous investigators as a basilar papilla, corresponding to the second papilla in anurans, this extension of the name cannot be justified. Indeed, the differences between these organs in anurans and urodeles could hardly be more profound: these two have different locations in the head, different relations to other labyrinthine structures, and they appear to operate by different physical principles.

This second papilla in salamanders is not a derivative of the inferior saccule (as is the second papilla of anurans) but forms an extension at the dorsal end of the lagena. Instead of appearing in a deep inferior position it lies close beside the amphibian papilla and makes use of the same route as that structure for the exit of the sound waves.

This papilla employs a protruding tectorial mass for the sensing of fluid vibrations just as do the caudate amphibian papilla and the main part of the amphibian papilla of anurans. The tectorial body protrudes into the vibratory path and is set in motion along with the inner ear fluids through the action of sound waves. There is no sensing membrane to assist in vibratory reception as in the basilar papilla of anurans. Thus it is clearly inappropriate to employ the same name for this organ as given to it in anurans: there is indeed no correspondence of any kind except that in both groups it constitutes a second auditory receptor in addition to the amphibian papilla. As mentioned above, this organ will be designated as the ''lagenar papilla'' because of its location and apparent derivation. (Still another type of papilla will be encountered when we come to the caecilians.)

It would be difficult to conceive that this organ could have had the same origin as the basilar papilla of anurans. These two are clearly separate developments, probably arising quite independently out of primitive equilibratory organs.

THE SALAMANDER EAR AS AN AQUATIC RECEPTOR

In anurans, as we have seen, except for a few species that evidently have suffered degenerative changes, the ear is well adapted to the reception of aerial sounds: there is a tympanic membrane stretched over an air chamber (the middle ear cavity) at the side of the head, and sound pressures acting on this membrane produce a vibratory motion that passes through an ossicular chain to the inner ear fluids and thence to the auditory papillae whose hair cells produce the final sensory response.

In salamanders, however, this peripheral receptor mechanism is absent; the skin, with heavy muscle layers beneath, extends without interruption over the sides of the head. There are inner ear structures below the surface, but clearly these are not well adapted to the direct reception of aerial sounds.

Kingsbury and Reed considered this problem near the beginning of the present century, and after a careful inquiry into the nature of the sound-reception mechanisms of urodeles, in which they covered 23 species, including representatives of all eight of the then recognized families, they set forth a series of hypotheses in an effort to account for the operation of these mechanisms in the different types. This effort was followed a little later by the observations of Reed (1915, 1920) on additional species that further supported the position reached in the first reports. It is a tribute to the quality and completeness of these studies that no comparable treatment of this subject has appeared since that time; the observations, along with the hypotheses advanced in explanation of them, have been widely accepted with relatively little question or criticism. The most extensive treatment of this area in later years was that of Monath in 1965, in which the problem was carefully examined and the position of Kingsbury and Reed was generally upheld.

Kingsbury and Reed described in great detail the structures of the salamander ear that have to do with the reception of sounds, with consideration of the embryonic development of these structures and the changes that occur in the course of this development. They gave particular attention to the alterations that appear at metamorphosis when the larval stage transforms into the adult condition.

Through an insight that was well ahead of its time, these authors considered that because of the limitations of middle ear structure the urodeles are insensitive to aerial sounds and responsive only to sounds in the water or to the vibrations of solid objects, such as the ground or the floor of their pool, with which their bodies may come in contact. They then developed three hypotheses concerning the mode of transmission of sound to these ears, with the choice for a given animal depending on the particular ear structure present at its current stage of development.

1. *The Mouth-floor Hypothesis, for Larval Types.* The first hypothesis applies to aquatic forms, including larvae, that remain in the water until they transform, and the adults of certain species like *Necturus* and *Cryptobranchus* that continue to be aquatic in habit throughout their lives and in which the larval form of the auditory structures persists. Kingsbury and Reed con-

sidered that in these aquatic forms the sounds are transmitted to the ear through the floor of the mouth. The complete chain of sound conduction includes the firm structures of the mouth floor, mandible, palatoquadrate, squamosal, stylus, and columella, which are pictured in Fig. 11-25, and in addition there are inner ear fluids forming the final segment of the pathway: the perilymph that occupies a major part of the otic capsule and the endolymph within one or two vesicles on the wall of this capsule containing the auditory papillae with their hair cells; see Figs. 11-16 and 11-21. There are usually two of these vesicles containing endorgans that will be referred to as the amphibian and lagenar papillae; in a number of species the amphibian papilla alone is present.

2. *The Foreleg and Muscle Hypothesis, for Terrestrial Animals.* A second hypothesis dealt with sound reception in terrestrial forms. When the animal reaches the adult stage and leaves the water for a largely terrestrial existence, it commonly stands with the head raised, out of direct contact with the substratum. Kingsbury and Reed pointed out that sound transmission from the substratum to the jaw is no longer effective when the animal maintains such a posture and a different route must be employed. In their theory the path of sound entry is then through the forelegs. The complete chain of transmission, pictured in Fig. 11-26, involves forelegs and shoulder girdle, and then the opercular muscle and operculum, so that finally it is the vibrations of the opercular cartilage occupying the oval window at the posterior end of the otic capsule that are transmitted to the inner ear fluids and reach the auditory papillae.

3. *Siren Type.* A third hypothesis was found necessary for the Sirenidae, which lack the connection between mandible and columella that appears in

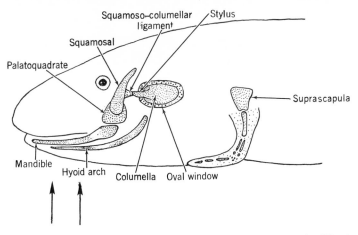

Fig. 11-25. Sound reception in salamanders: the mouth-floor hypothesis. After Kingsbury and Reed, 1909 (their Fig. 21a).

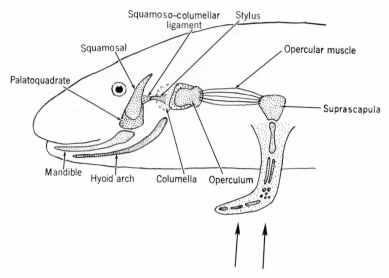

Fig. 11-26. Sound reception in terrestrial salamanders, as represented by Kingsbury and Reed, 1909 (their Fig. 21b).

the other species and also lack the operculum. In members of this family the chain of connections for sound transmission, as formulated by Kingsbury and Reed, resembles that for larvae, except that the columella is reached through a ligament from the hyoid arch rather than through the jaw suspension. Thus the chain of transmission includes the floor of the mouth, hyoid arch, and columella, so that finally the auditory papilla is stimulated as in the others by way of the inner ear fluids.

Evaluation of the Kingsbury-Reed Hypotheses

Critically considered, these hypotheses presented by Kingsbury and Reed have a number of drawbacks. The mouth-floor hypothesis appears to be overly specific: a more general route is available. Sounds transmitted through the water and reaching the body of a salamander will pass through the skin and muscle layers at the side of the head with little reflection, because these tissues have an acoustic impedance closely similar to that of the aqueous medium, and thus all these structures will vibrate with the same amplitude. The bone structures beneath, specifically the squamosal and palatoquadrate, also will be involved in the acoustic vibration because these have densitites and elasticites of the same general order of magnitude as the superficial tissues and will be sufficiently well matched to them to take up a considerable fraction of the vibratory motion. These bones cover a broad area at the side of the head, with the squamosal extending well forward and the palatoquadrate covering middle and posterior regions, so that the acoustic forces impressed upon the side of the head are well integrated as they are conveyed inward.

These forces are transmitted through a contact between the columella and either the palatoquadrate or squamosal, or both, usually provided by an extension of the columella known as the stylus or else by a simple fusion between one of these superficial bones and the fenestral plate. The floor of the mouth, as considered by Kingsbury and Reed, may also convey the aquatic vibrations, but not with the effectiveness of the lateral route just described. The special route suggested by them for *Siren* also seems secondary in importance, as the lateral route is available for this group also.

The special hypothesis proposed for terrestrial animals, in which sounds are considered to pass through the ground to the forelegs, and then by an elaborate route to the ear, can hardly be defended. In the first place there is the improbability that any useful information exists in the substratum and is to be derived from it. Predators, the creatures that the salamander must detect and escape from, like such animals as snakes, turtles, and water birds, are either naturally cautious and slow-moving, or have learned to be so in their search for prey, and hardly set the ground into vibration in their movements. Even the larger animals, such as otters and skunks, that feed upon amphibians on occasion are wont to move cautiously, and the sounds that they produce in making their way through thickets and foliage along the watercourses are likely to be aerial sounds, and these do not enter the ground to any appreciable extent. Of course large, heavy-bodied mammals such as the proverbial herd of elephants might readily be detected through ground vibrations, but such animals are not common in the areas inhabited by salamanders. In any event, if the salamander is in a depression of some kind and in close contact with the ground, it would probably be safer to remain there rather than to move and signal its presence. Thus it is difficult to discover any value to the salamander in a perception of sounds through the substratum.

Further, the idea that sounds can be transmitted from the substratum along the legs to scapula and suprascapula and then through the opercular muscle to the ear is hardly credible. Experimental tests have shown that a muscle under tension will indeed transmit sound vibrations, but it fails to do so when relaxed. The opercular muscle therefore would have to be maintained in almost continual contraction to be of value in providing emergency signals. Moreover it would have to be kept tensed while the other muscles close around it were relaxed, or else it would lose a large part of its vibratory energy to these others, as well as to the still firmer skeletal tissues in the vicinity. In short, the conception of a muscle as a suitable path for the transmission of sound is far from attractive, and it is remarkable that this idea has been taken seriously, without any positive evidence to support it, for the better part of a century. Most likely the salamander when on land has no efficient way of receiving sounds from either the air or the substratum and is effectively deaf; its survival under these circumstances must depend on its natural wariness, its visual capabilities, concealment in dense vegetation, and the practice in most species of venturing forth only under cover of darkness. Of course these animals may be able to perceive sounds of extraordinary intensity, of the order of thunder claps, by the aerial route, but this reception can be only of marginal value to survival.

If we agree with Kingsbury and Reed that the salamander ear is poorly equipped for the reception of aerial sounds, as the absence of a tympanic membrane clearly testifies, and we reject their mouth-floor and substratum hypotheses for the reasons given, there remains only one practical manner of sound reception, through the water. This conception is reasonable from a practical standpoint since these animals are aquatic in habit at important stages in their life cycle. For the great majority of species the eggs are laid in water and hatch out there, and the larvae maintain an aquatic existence until they reach maturity and metamorphose to the land stage. Then later in the season when these animals become sexually active many of them reenter the water for breeding purposes. Among certain species a male selects a particular site in a pool or stream and seeks to entice females to this place; the female deposits her eggs in a special packet that is attached to some fixed object underwater, and the male then approaches this packet and fertilizes the eggs within.

It is commonly supposed that the male attracts females to his chosen site in a pond or stream by the production of a hedonic secretion; but there is the further possibility that he uses a vocal call as a general summons. Some salamanders are known to produce sounds, though in air these are usually described as faint and unimpressive, and ordinarily no important function has been ascribed to them. Some investigators, however, have suggested that these vocalizations may have a significant function in the mating process, and here is a possibility requiring further investigation (see Maslin, 1950; Blair, 1963).

If indeed underwater acoustic signals are produced by the male, these would have an extensive range throughout the pool as compared with a chemical substance and would better promote the success of the breeding process.

As is well known, sound waves are readily transmitted from one substance to another when the acoustic qualities of density and elasticity of the two substances are similar; and on the contrary the waves are reflected from the interface between the two materials when these qualities are markedly different. Thus aerial sounds on striking ordinary body tissues are largely reflected, and the ears of terrestrial animals like ourselves are serviceable only because they are provided with a mechanical transformer, the middle ear apparatus, which consists of a thin, light membrane, the tympanic membrane, that moves easily under the air pressures that strike its surface and concentrates these pressures on a much smaller area (the stapedial footplate) and with a lever reduction provided by the ossicular chain that increases the force; by this mechanical device the inner ear fluids can be set in motion so as to involve the auditory hair cells within.

The salamander lacks such a transformer mechanism, and the transmission loss from the air to its sensory cells must be very great, estimated as of the order of 30 to 40 db.

When in the water, however, there is no such loss. The salamander's body tissues, the skin and muscle layers at the surface, and also the ear structures beneath them, have acoustic properties closely similar to those of water; and though there will be some loss because the match is not perfect, a sound in the water will transmit a considerable part of its energy to the head and its contents. The effect of this transmission then will depend upon a number of conditions.

If the sound frequency is low, so that the wavelength of the aquatic vibrations is of the order of magnitude of the animal's head, or larger, then the action of the sound will be such as to vibrate the head as a whole, much as would be the case if the sound source were a vibrating bar applied to the head's surface.

The effects of such a stimulus then will depend upon the internal constitution of the driven mass: upon the axis of the imposed motion relative to the disposition of the internal structures, upon the inertial properties of these structures, and upon their degree of coupling with one another. Let us consider the effects of aquatic vibrations on a specimen of *Hynobius nebulosus*, as represented in Fig. 11-27.

The columella lies deep below the surface at the side of the head and constitutes a large bony and cartilaginous plate closely covered by muscle tissues and the outside layer of skin. These soft tissues are well matched to the acoustic impedance of water, and when the head is immersed will readily receive and transmit inward any sound vibrations passing through the water medium. The large mass of the depressor mandibulae muscle will vibrate along with the skin layer and will convey the movements to the bone and cartilage layer immediately below formed by the squamosal and palatoquadrate, which in the picture presented here extend the receptive layer forward and in general cover a large area along the side of the head.

As seen in sections this configuration of ossicular structures differs according to the level at which the cut is made, as these structures vary in their particular shapes and interrelations. A different picture is presented in Fig. 11–28 for another specimen of this same species. Here a series of arrows indicates the principal pathway from the fenestral plate through the perilymphatic cistern to the amphibian papilla and across the perilymphatic window into the brain cavity.

Sounds transmitted through the water convey vibratory energy to all these

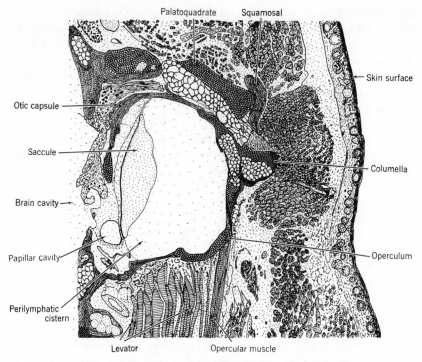

Palatoquadrate   Squamosal

Skin surface

Otic capsule

Saccule

Columella

Brain cavity

Papillar cavity

Operculum

Perilymphatic cistern

Levator   Opercular muscle

Fig. 11-27. The ear region of *Hynobius nebulosus*, in a frontal section. Scale 125X.

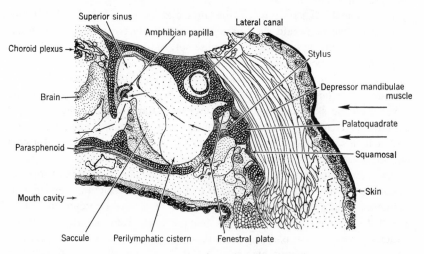

Superior sinus   Lateral canal

Amphibian papilla

Choroid plexus

Stylus

Depressor mandibulae muscle

Brain

Parasphenoid

Palatoquadrate

Squamosal

Mouth cavity

Skin

Saccule   Perilymphatic cistern   Fenestral plate

Fig. 11-28. The ear region of *Hynobius nebulosus*, in a transverse section. Scale 25X.

structures, but differentially, depending on their properties of inertia, stiffness, and mutual attachments. The more superficial tissues, the skin and muscle layers, have densities so much like the water medium as to be essentially transparent to the water waves. The bony and cartilaginous structures beneath these layers, because of their similar construction and because they are closely joined by connective tissues, probably respond as a unit. Also the tissue mass next in order, consisting of the otic capsule and its contents, which is somewhat flexibly attached to the cranial framework above and below provided by parietal and basisphenoid bones, can be expected itself to act as a unit. The movements of these two masses, the superficial acoustic elements and the otic capsule, in response to alternating sound pressures can be expected to be somewhat different in both amplitude and phase because of variations in inertia and in the security of their attachments to adjoining structures. The difference in motion between these two structures becomes significant at their place of contact at the stylus, shown in Fig. 11–28 as a link between fenestral plate and palatoquadrate. The differential action is the equivalent of a vibratory stimulus applied to the fenestral plate itself; this resultant stimulus gives rise to fluid displacements within the perilymphatic cistern, and these displacements when transmitted through the passages leading to the auditory papillae bring about an excitation of the auditory hair cells. The final pathway through the amphibian papilla is indicated by the line of arrows in Fig. 11-28.

The picture just presented holds for the action on the salamander ear of low-frequency vibrations in the aquatic medium, for which the movements of the fluid particles, and their forces acting on the ear structures, are essentially in phase. When the frequency rises, and the wavelengths become small relative to the structures involved, this pattern becomes increasingly complex, as the several parts of the mechanism no longer vibrate in phase. The resultant forces acting on the fenestral plate or columella are then reduced and finally become altogether ineffective. Observations on the sensitivity of the salamander ear fully bear out this expectation: this sensitivity is well maintained along the low-frequency range and then falls progressively as the frequency is raised.

This ear is well suited to the reception of aquatic sounds, but can be expected to be of little value for aerial reception. The mismatch in impedance between the aerial medium and the tissues of the head will produce almost total reflection of the sound energy; only extremely loud sounds can produce an appreciable effect. Thus it seems likely that the salamander ear is essentially an aquatic receptor.

A limiting condition to this aquatic reception of sound is the small size of the head (specifically the distance between the two ears and hence the difference in sound paths from the source to the two sides of the head) in relation to the wavelength of the sounds in water. The smaller this difference the greater is the similarity of the sound waves reaching a given receptor organ from right and left sides so as to produce a mutual cancellation of response. The orientation of the head will have a significant effect on this reception; an animal directly facing a source of sound will be in an unfavorable position to perceive it, and a more advantagous position will be with a body position at right angles to the direction of the sound.

The reception of sounds thus will be affected by the animal's own behavior with respect to the sound stream, and also with respect to any obstacles in the medium that might obstruct and reflect the sound waves. The animal might assume certain positions with respect to the sound source and objects in the area that would greatly affect sound reception. A favorable position would be with one side of the head toward the sound source and the other side close to a solid object that would serve as a barrier to the sound waves. To what extent salamanders assume definite listening attitudes remains to be discovered.

Further consideration of these conditions of underwater hearing, and especially the long wavelengths of the sounds to which the salamander ear is mainly sensitive, leads to the suggestion that for continuous pure tones what the salamander hears is essentially the onset of the train of waves: this wave front reaches one ear ahead of the other, and then a little later both ears are about equally stimulated. The priority of action in the nearer ear makes this stimulus distinctive and also gives an indication of the direction of the source. Then as the sound is sustained both ears are stimulated about equally and the priority cue is no longer available. It follows that the salamander is able to perceive the onset of each of a series of sound bursts, or abrupt changes in a continuous sound, but not the sustained quality of a train of waves.

THE ROUND WINDOW PROBLEM

Before the action of sounds on the salamander ear is considered further, it is necessary to examine what may be called the round window problem. All ears employing fluid movements to convey sound vibrations from the outside to the deep-lying sensory cells require a relief route of some kind for these fluid movements to occur. Fluids for all practical purposes are incompressible, so that a certain inward displacement of the fluid of the ear is possible only if somewhere there is a corresponding outward displacement to make room for it. Most vertebrate ears—those of mammals, birds, and many reptiles—employ a round window for this purpose: in an inner wall of the cochlea is an opening covered by a thin, flexible membrane beyond which lies an air space (the middle ear cavity), and as air is readily compressible a certain displacement of the columellar footplate at the entrance to the cochlea can be matched by a displacement of equal volume produced by a bulging of the round window membrane.

A number of species among the reptiles, such as the snakes and turtles, lack a middle ear air cavity, and in these a different method, the reentrant fluid circuit, is employed. In these ears a fluid pathway leads from an area in the cochlear wall (an area corresponding to the usual location of a round window) and follows a circuitous path to the outer surface of the columellar footplate; thus when under the action of a sound this footplate moves inward, displacing a certain quantity of cochlear fluid ahead of it, the same quantity of fluid flows around the reentrant circuit to the front of this footplate. Accordingly, as the footplate vibrates in response to sound, the fluid in the reentrant pathway surges back and forth. This mode of mobilization of the cochlear fluid serves well enough at low frequencies, but becomes less effective as the frequency rises because of the considerable mass of fluid that must be moved at high rates, and the friction thereby generated.

Surprisingly, an examination of the three existing orders of amphibians reveals three different solutions to this problem of cochlear fluid mobilization: the two just mentioned that occur in other vertebrates and a third that is unique.

The Anura (frogs and toads) possess a round window and employ the first, most common method of fluid mobilization. The Gymnophiona (caecilians) use a reentrant fluid circuit like many of the reptiles. The Caudata (salamanders), however, have resorted to another and quite remarkable expedient: the sounds literally go in one ear and out the other. This means further that hearing in these animals is invariably binaural: every sound passes through both ears.

The first evidence that this condition exists was simply anatomical. As shown in Fig. 11-29 for the newt *Taricha granulosa*, a firm structure of bone and cartilage, the otocranial capsule, encloses the two ears and the brain between, with only two openings, the oval windows, one on each side, each

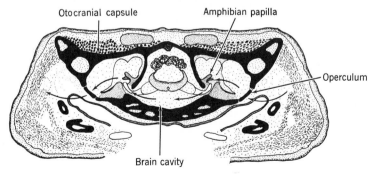

Fig. 11-29. A cross section through the ear region in a newt, *Taricha granulosa*. Scale 5X. From Wever, 1978a.

loosely covered by a cartilaginous plate, the operculum. To be sure, there are a number of small perforations in the walls of the otic capsule that transmit nerves and blood vessels, but they are well occluded by these tissues and can be neglected as acoustic leakage paths. Thus, in the many species represented by this figure, sound pressures exerted at one operculum can produce fluid displacements only by traversing the path indicated by the arrows, leading to the operculum on the other side. As shown, this path goes through three different fluids separated by thin membranes; from the right operculum the course is first through the perilymph of the perilymphatic cistern here, then through a membrane into the saccule, which contains endolymph and also bears within it, attached to the lateral wall of the otocranial septum, the amphibian papilla that is the principal auditory receptor. Contained within a recess in the lagena, which can be regarded as a diverticulum of the saccule, and not shown here because it lies at a more ventral level, is a second auditory organ, the lagenar papilla, that is also traversed by a branch of the fluid pathway much as the amphibian papilla is. From the saccule the passage continues through a thin membrane, the perilymphatic window, and becomes the perilymphatic duct that enters the brain cavity; this duct ends at another thin membrane, the arachnoid, separating the perilymph of the duct from the cerebrospinal fluid that fills the brain cavity. From here on the course continues to the opposite side, with the same fluids, membranes, and sensory structures encountered in reverse order, until the opposite operculum is reached.

Most salamanders in the adult stage present the general pattern here displayed: an oval window on either side contains a single fenestral element; one of these windows receives the initial sound and is able to set a long, complex fluid pathway in vibratory motion because of the presence of the corresponding window and responsive element on the other side. In many species, including all the Salamandridae, to which the present example belongs, and also the Ambystomatidae and Pleurodelidae, the element in the oval window appears to be an operculum. In several other species, however, such as the Proteidae (*Necturus* and *Proteus*), Cryptobranchidae, Amphiumidae, and Sirenidae, this element is considered to be a columella. In still others, including the large group of Plethodontidae, the fenestral element most likely represents a fusion of columella and operculum. Only the Hynobiidae and the larvae of Ambystomatidae, and perhaps a few others, have clearly separate columellas and operculums in this window.

The identification of the fenestral element depends almost altogether on developmental studies like those of Kingsbury and Reed in which the structures are carefully followed in their successive stages. The form of the adult structure alone gives no reliable indication of this identity. Fortunately this identification does not seem to have any significant functional implications: the fenestral element, regardless of its derivation, serves its purpose in responding to the sound pressures imposed from the outside and setting up a vibratory pathway through the two ears involving their auditory papillae.

The presence of a separate columella and operculum in *Hynobius* and many larvae might give concern if it were considered that these two elements could act independently so that one would move inward under the influence of a sound while the other immediately adjacent would serve as the yielding site, moving outward. This action would produce a local eddy and would greatly reduce the propagation inward of the sound pulses to the auditory papillae. That this does not happen is evidenced by the high level of sensitivity of *Hynobius* species; these compare favorably with other forms in which a single element occupies the oval window. Thus it appears that the sound acts similarly on these two fenestral elements: both columella and operculum respond together and combine their effects in the transmission of vibrations inward to the sensory endings. This compatibility of function probably has brought about the widespread fusion of these two elements or portions of them.

Further evidence that the course of sound transmission in the salamander ear is as just described, and that the stimulation is invariably bilateral, was obtained in the course of a series of three experiments, as follows:

1. *The bilaterality of stimulation.* Responses to sounds were obtained by inserting fine needle electrodes on both sides through holes drilled in the otic capsules so as to enter the dorsal part of the perilymphatic cistern. Stimulation was produced with a vibrating needle applied to the right operculum, and potentials were obtained from both ears. In general, the responses were larger for a given stimulus when recorded from the side on which the vibrations were applied. The sensitivity curves, which show the sound pressure required for a constant response of 0.1 μv, reflect this difference and are presented in Fig. 11-30. The ipsilateral responses exhibit a significantly greater sensitivity over the entire range, with the difference in this instance averaging about 19 db. Evidently there are leakages of acoustic energy along the fluid pathway: not all the vibratory motion that passes over the right auditory papilla is able to reach the left one.

2. *Loading and immobilizing one operculum has a bilateral effect.* It was observed that if the vibrating needle were applied to the operculum on the right side, and the recording electrode was on the right, a blocking of the left operculum by pressing over it a mass of bone wax or modeling clay had a profound effect, greatly reducing the responses to some tones and often increasing them for others. In general, as Fig. 11-31 shows, the reduction was greatest for those tones to which the ear was normally the most sensitive, and smaller changes, which were sometimes increases, occurred for tones outside this sensitive range. The loading and blocking of the operculum as described evidently adds both mass and stiffness to the vibrating fluid column and alters the resonances of the system in complicated ways.

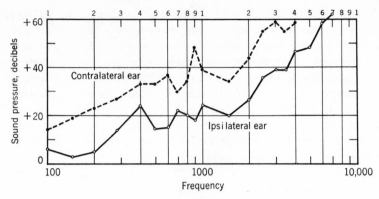

Fig. 11-30. Sensitivity functions for the two ears of *Ambystoma tigrinum* for vibratory stimulation applied to the operculum on the right side. From Wever, 1978a. Shown is the vibratory amplitude, in db relative to 1 millimicron, required for a response of 0.1 μv.

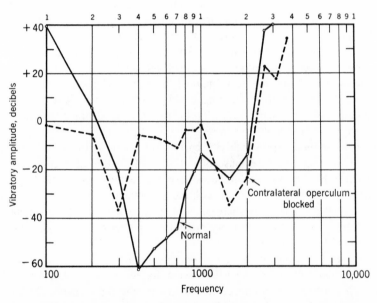

Fig. 11-31. The effects as seen in a specimen of *Ambystoma maculatum* of blocking the left operculum while applying a vibratory stimulus to the right operculum. The recording was from an electrode in the otic capsule on the right. From Wever, 1978a.

3. *Binaural stimulation produces additive effects that depend on phase relations.* A vibratory stimulus was applied to one operculum and the intensity adjusted to produce some arbitrary level of inner ear potentials as recorded from an electrode on the right side, and then the same procedure was carried out when the stimulus was applied to the other operculum, with the electrode position unchanged. An application of both stimuli simultaneously then produced an increased response, one greater by $\sqrt{2} = 1.414$ than either stimulus by itself, at a certain phase relation between the two stimuli that had to be discovered by trial. Then when these two stimuli were delivered at a phase relation just 180° from the former one, the result was an absence of response; the two stimuli exactly cancelled one another. Intermediate values of phase between the two used in the above tests gave intermediate amounts of response. This is precisely what is to be expected in the relations of sinusoidal responses coming from two (or more) sources: they combine like vector quantities.

The evidence that the two ears in salamanders are never independent, but always are interactive, will be found of great significance in our consideration of the general problem of sound stimulation.

PROTECTIVE DEVICES IN THE SALAMANDER EAR

I. The Function of the Opercular Muscle

The widely accepted view of the function of the opercular muscle in the salamander ear, as already mentioned, is that propounded by Kingsbury and Reed in which it was supposed that this muscle serves as a path of sound transmission to the inner ear. For the reasons already given this muscle seems quite unsuitable for this purpose, and a radically different conception of its function is now offered: this muscle is one of two that have a protective action and serve to reduce the transmission of sounds inward to the auditory papillae and thus to guard against excessive stimulation that might damage these delicate receptive structures.

Contraction of the opercular muscle pulls the opercular disk outward in its window and probably has two mechanical effects: to jam the disk in this window so as to impair its mobility and at the same time, by an outward displacement of all fluids and membranes in the sound path from columella on, to increase the stiffness of this entire conductive mechanism.

That a reduction of sound transmission to the auditory papillae is the result of applying tension to the operculum has been proved experimentally. In one series of tests a fine thread was tied around the end of the opercular muscle near its attachment to the suprascapula and run over a light, smooth-acting pulley to a scale pan on which weights could be placed. Inner ear potentials were recorded in the usual way, with an active electrode in the otic capsule, introduced through a fine hole drilled into the lateral semicircular canal, with an indifferent electrode located elsewhere. Sounds were presented and adjusted in intensity for some convenient level of response well within the normal range of this ear, first in the absence of tension on the opercular muscle and then when such tension was applied. The scale pan was made as light as practicable (0.7 grams) and its weight was always included in the totals; but a zero tension was easily obtained by lifting the pan. The experiment was carried out on specimens of *Ambystoma maculatum* and *Ambystoma tigrinum*. The tones used were usually 400 and 1000 Hz; no special study was made of frequency relations.

The observations showed great lability of the responses to sounds in the presence of any kind of manipulation of the opercular mechanism. Evidently this system is a delicately balanced one and is under some degree of neural or dynamic control even in these anesthesized animals.

The application of small tensions, such as the weight of the scale pan alone, or total weights of the order of 1 or 2 grams, usually produced slight increases in response initially, and then in a few seconds the response returned to its original level or showed a moderate decrease. Then if the weight was removed, the response at once increased, usually to a higher level than its initial value, and thereafter slowly declined to this initial level.

The application of larger tensions, up to 10 grams or so, produced immediate declines in response, often of the order of 30 to 40 percent. Still stronger tensions were applied only occasionally, toward the end of an experiment, because of the danger of damaging the muscle; these tensions reduced the response sometimes to as little as one-third (− 10 db) or one-fourth (− 12 db) of the original level.

The degree of reduction, up to 12 db, as demonstrated in these tests may seem very limited in extent in view of the wide range in the sound levels that an animal might encounter, but it must be borne in mind that in this experimental situation, with the animal under deep anesthesia, the muscle is without its normal tonus and is greatly weakened in consequence. Thus extreme caution was necessary to avoid its injury. A number of these muscles were torn in the preliminary trials before the safe limits of the manipulations were learned. There is no doubt that in the active, waking animal the muscle is capable of a much greater range of sound control than could be demonstrated here. Still, even a fourfold (12 db) reduction as observed in these tests would serve in significant degree in the protection of this ear against acoustic damage. This damage probably consists of a tearing away of the connections between the ciliary tufts of the hair cells and their cell bodies or a separation of these tufts from the tectorial tissues that produce their deflections under the influence of sounds, and this sort of displacement presents an injury threshold: a level beyond which recovery does not occur. Even a moderate degree of protection therefore can have a critical significance.

It must be borne in mind also that the opercular muscle is not alone in its role of ear protection: as will now be indicated, the quadrate muscle also appears to have a similar and in most species an even more significant role in this regard.

It is surprising that in the salamander ear a contraction of the opercular muscle produces a reduction in sound transmission, whereas in anurans it has a contrary effect, that of freeing the lock mechanism of the middle ear and thus of improving transmission. It is evident that in these two amphibian orders the opercular mechanism has taken two widely divergent evolutionary courses.

II. THE FUNCTION OF THE QUADRATE MUSCLE

In 1904 Drüner described a muscle in the salamander to which he gave the name "inter ossa quadrata" and mentioned Ruge (1897) as having seen it previously. Drüner indicated the origin of this muscle as on the ventral end of the palatoquadrate, with its insertion on fascia close beneath the skin in the throat region. He found this muscle to be innervated by the jugular branch of the facial nerve and conceived its function to be that of a general constrictor for the hyoid area. No more specific function has been assigned to this muscle, but it is now suggested that it is one of two protective muscles for the ear mechanism of salamanders.

This new conception requires a reversal of Drüner's designations of origin and insertion for the muscle, with the attachment on the fascia along the midline of the throat now to be considered as the origin, from which the two fan-like bodies of fibers rapidly converge to their dorsolateral insertions, often through a terminal ligament, on the palatoquadrates on either side. The location of this pair of muscles in a specimen of *Ambystoma tigrinum* is indicated in Fig. 11-32, in which they were exposed by first removing the skin

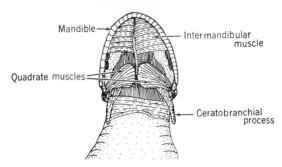

Fig. 11-32. The quadrate muscles in a specimen of *Ambystoma tigrinum* in a ventral view, exposed by removing the skin and portions of the intermandibular muscle. About natural size.

of the throat and then detaching some of the more posterior fibers of the intermandibular muscle, which cover a portion of the quadrate muscle in this region. The quadrate muscle in this species is spatulate in form at its origin and bends around dorsoposteriorly along the first ceratobranchial cartilage on its way to the palatoquadrate.

Because of its oblique course this muscle is not easily made out in sections, and it has to be followed closely along the series. Also, because it was a regular practice in our histological preparation of specimens to remove much of the skin of the head in order to facilitate the infiltration of fixatives and other fluids, the portions of these muscles close to the skin along the midline were often damaged or torn away. Indeed, it was not anticipated that structures so remote from the ear as the throat region would have significance for this organ's functioning. Fortunately in most of the specimens enough of these muscles remained to show their general course and connections, and in a few the whole extent could be determined. Figure 11-33 shows one of the more complete preparations in a specimen of *Hynobius nebulosus* where the quadrate muscle on the right side runs from its midline position in fascia beneath the skin to its dorsolateral attachment to the palatoquadrate. Also shown here is the connection of the palatoquadrate to the columella, made mainly by a bony stylus that extends laterally from the main plate portion of the columella.

Özeti and Wake (1969) pictured this and other throat muscles in a number of salamander species and showed that in most of them the quadrate muscles fan out more broadly than the ones represented here for *Ambystoma tigrinum;* in many of these the fibers from right and left sides were found to meet in the midline more closely and over a greater extent than seen here. In the species *Hypsetriton wolterstorffi* they found these muscles meeting closely along the midline and together occupying more than half the throat area.

These muscles are present in all salamander species examined and in most of them appear to be well suited for the role of protecting the ear against overstimulation by sounds. In many species, as shown for *Hynobius nebulosus* in Fig. 11-33, a contraction of the muscle will exert tension on the palatoquadrate in a ventral and somewhat medial direction, and through the connection between the dorsal end of the palatoquadrate and the outer face of the columellar footplate, partly through a direct fusion and partly through an interconnection made by the stylus, this footplate is pulled in a largely ventral direction. Such a displacement, besides adding tension to the whole vibratory system, will bring the footplate into contact with the ventral edges of the oval window and thereby damp its movements. This mechanism appears to be an even more effective one for the protection of the ear against damage by excessive sounds than that provided by the opercular muscle system.

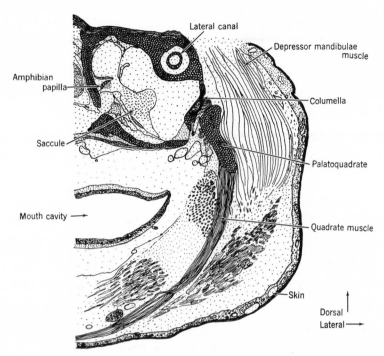

Fig. 11-33. A transverse section through the right side of the head of a specimen of *Hynobius nebulosus*. Scale 25x.

12. THE HYNOBIIDAE AND
CRYPTOBRANCHIDAE

INTRODUCTION

Now to be considered in respect to the structure of the salamander ear and its performance in the reception of sounds are the various species available for study, disposed in a series that follows the Dowling and Duellman arrangement of this group in four suborders according to their conception of a progressively increasing level of specialization.

This series begins with the Hynobiidae in the suborder Cryptobranchoidea, which are regarded as the most primitive of salamanders and probably the best living representatives of the basal stock from which all others were derived; the series then continues with the Cryptobranchidae in this same suborder that are closely similar to this basic group.

Next come the Sirenoidea containing the single family Sirenidae, a group of eel-like forms that lack the pelvic girdle and hind limbs. These species undergo only an incomplete metamorphosis, which somewhat obscures their relationships to the other salamanders.

The third suborder is the Salamandroidea, a highly varied assemblage including the Salamandridae, the newts and salamanders of wide distribution in both hemispheres, the Proteidae, consisting of aquatic forms in the two genera *Necturus* in eastern North America and *Proteus* in Italy and Yugoslavia, and the Amphiumidae, often called Congo eels, found in the southeastern portion of North America.

The fourth suborder is the Ambystomatoidea, including the two families Ambystomatidae and Plethodontidae, considered to have arisen from the hynobiids but now exhibiting a number of advanced characters; they are widely distributed in North America.

THE HYNOBIIDAE

The family Hynobiidae includes the Asiatic land salamanders of which there are 5 genera and about 20 species, most of these species in the genus *Hynobius*. These salamanders are mostly found on the islands of Japan, with a few on the Kurile and Sakhaline islands to the north, and others are scattered along the mainland in Asiatic Russia, Manchuria, Taiwan, and South China.

This family has been placed in the suborder Cryptobranchoidea, in company with the family Cryptobranchidae, as the most primitive group of salamanders, distinguished primarily by their practice of external fertilization; in all others the fertilization is internal, involving the transfer from male to female of a sperm packet. In this more primitive group the transfer of sperm is indirect, not requiring contact between the sexes. The particular procedure varies with species and is only of incidental interest here, but will be described briefly as observed in *Hynobius nebulosus* by Rehberg (Thorn, 1968, pp. 46–48).

Females are enticed into a particular area of a stream or pool largely (it has been supposed) by the male's introduction into the water there of hedonic secretions; the female responds to these substances and produces clusters of eggs contained in a pair of gelatinous bags that she attaches to some fixed object, such as a twig or water plant, located a little below the water surface. The male, which has remained in the area that he has adopted as his own, repeatedly visits these bags and manipulates them, fertilizing the eggs in so doing.

The type species for the genus *Hynobius* is *nebulosus*, and 6 specimens of these were available for study, along with a single specimen of *H. okiensis*. These are salamanders of typical body form and moderate size, averaging about 15 cm in length. *Hynobius nebulosus* inhabits areas adjacent to ponds and pools of stagnant water on the islands of Kyushu and Shikoku in Japan; *H. okiensis* is from the island of Oki.

Hynobius. A frontal section showing the ear region in a specimen of *Hynobius nebulosus* is presented in Fig. 12-1. Here the columella and operculum are clearly indicated as separate structures in the lateral wall of the otic capsule, well below the skin and muscle layers. Shown also are related structures, including the squamosal bone and palatoquadrate cartilage in a location anterolateral to the otic capsule. The amphibian papilla lies deep in the medial portion of this capsule adjacent to the brain.

Further details of otic structure appear in Fig. 12-2, representing a section from the same specimen as the foregoing but a little farther ventrally. The columella is cut through the middle showing the prominent headpiece extending into the large muscle mass of the depressor mandibulae. An enlarged view of the columella as seen here is presented in Fig. 12-3. The footplate is a thin bony shell containing a large marrow cavity and surrounded by a ring of cartilage, and the headpiece is similarly constructed, with marrow within and a cap of cartilage extending laterally.

The Quadrate Muscle. — A muscle to be called the quadrate (Drüner in 1902 named it the "inter ossa quadrata" muscle), which at the level shown runs largely longitudinally, extends anteriorly to the region of the columella, but does not make a firm, direct connection with it. The ligamentary end of

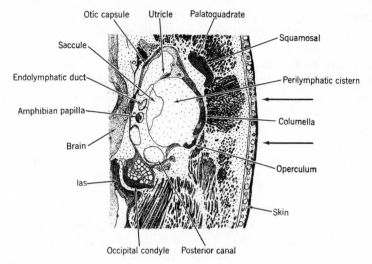

Fig. 12-1. The ear region of *Hynobius nebulosus* in a frontal section. Scale 12.5X.

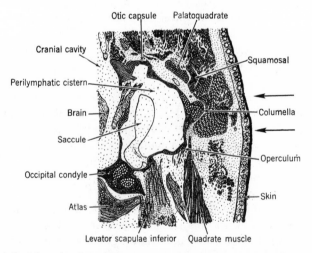

Fig. 12-2. A frontal section through the same specimen as in the preceding, at a more ventral level. Scale 12.5X.

this muscle passes through the fascia surrounding the columellar headpiece and continues below the level shown here to connect with the palatoquadrate beyond at a position anterolateral to the otic capsule. This insertion is represented in Fig. 12-4, which shows a transverse section from another specimen of this same species. At the level indicated the palatoquadrate is fused

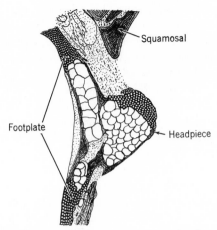

Fig. 12-3. Detailed view of the columella as seen in the preceding figure. Scale 41.5X.

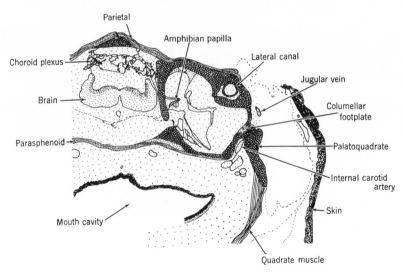

Fig. 12-4. The ear region of *Hynobius nebulosus* in a transverse section. Scale 20X. The lateral muscle masses are indicated only in outline.

with the columellar headpiece and thus firmly attached to the footplate. The connection of quadrate muscle to columella is largely indirect, but nevertheless firm.

Incidentally to be noted is the location of the columella with respect to the jugular vein (lateral head vein) and internal carotid artery: the columella is above the artery and below the vein; this relation is typical of salamanders and is often a useful guide in the identification of these parts.

A further representation of the quadrate muscle, cut rather obliquely, appears in Fig. 12-5 and shows the relation to the suprascapula. The posterior portion of the muscle runs close along the lateral surface of this cartilage, between it and the skin, but no connections to the suprascapula have been seen. Only the posterior end of the opercular muscle is shown here; the main part of this muscle runs forward at a more dorsal level to its attachment to the operculum. The opercular and quadrate muscles thus run closely alongside in a portion of their courses, but are quite independent.

Figure 12-6 shows the further course of the quadrate muscle in a transverse section through almost the whole head in the species *Hynobius okiensis*. This muscle extends ventrally and then medially below the mouth cavity, and finally the right and left portions come together and meet immediately beneath the skin in the throat region (not shown here). Most of the fibers in these two broad, fan-like bundles terminate on skin and fascia along their courses, so that the size of the bundles as seen in sections rapidly dwindles

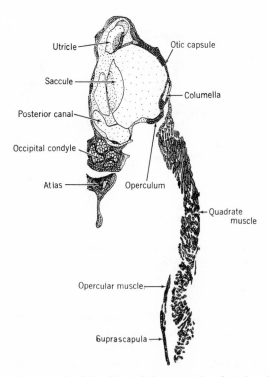

Fig. 12-5. The quadrate muscle of *Hynobius nebulosus*, seen in a frontal section. Scale 15X.

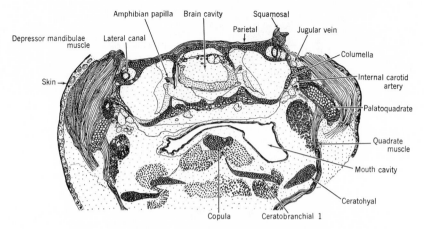

Fig. 12-6. A transverse section through the head of *Hynobius okiensis*. Scale 10X.

as we pass ventrally, but a few fibers appear to be continuous through the throat region.

It will be apparent from the figure that a contraction of these muscles exerts tension on both columellas, acting through the palatoquadrates. As mentioned earlier, this is a means of reducing the mobility of the columellas and thus of protecting the inner ear against overstimulation by sounds.

The Opercular Muscle. — As has long been known, the operculum in the salamander ear is provided with a muscle, commonly called the opercular muscle as a general term, though its derivation and form vary considerably in the different species. The anterior end of this muscle in *Hynobius nebulosus* is represented in Fig. 12-7; it consists of a bundle of relatively small fibers that runs posteriorly to an attachment on the lateral face of the suprascapula. Other muscle fibers of larger cross section just medial to these constitute the main part of the levator scapulae and have their attachments to the exoccipital bone.

The access of sound vibrations to the amphibian papilla in *Hynobius nebulosus* is evident in Fig. 12–4, as this papilla lies in the pathway from the columellar footplate through the perilymphatic cistern into the brain cavity. This pathway is clearly indicated by the series of arrows in Fig. 12-8. The vibratory flow through the lagenar papilla takes a closely similar course as indicated in Fig. 12-9, in which the complete circuit from right to left sides is portrayed. Details of the lagenar papilla in this species are represented in Fig. 12-10, and it is evident that the ciliary tufts of the hair cells in this papilla, embedded in a tectorial body, will be washed over and thus stimulated by the acoustic waves.

The anatomical relations for *Hynobius okiensis* are closely similar. Sounds

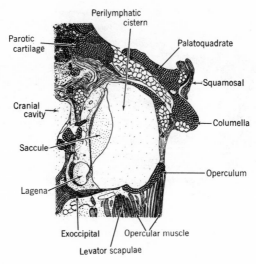

Fig. 12-7. A frontal section showing the right ear region and the anterior end of the opercular muscle in *Hynobius nebulosus*. Scale 20X.

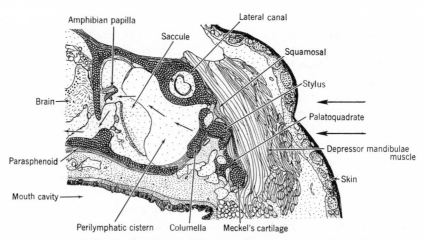

Fig. 12-8. The sound pathways through the amphibian papilla in *Hynobius nebulosus*. Scale 25X.

acting on the side of the head involve the skin and muscle mass, with the palatoquadrate beneath as shown in Figs. 12-11 and 12-12, and these acoustic effects thereby reach the columellar plate. From this plate the vibrations are conducted across the perilymphatic cistern to the amphibian papilla as may be seen in Fig. 12-13. They then continue through the perilymphatic window into the perilymphatic duct, finally reaching the brain cavity.

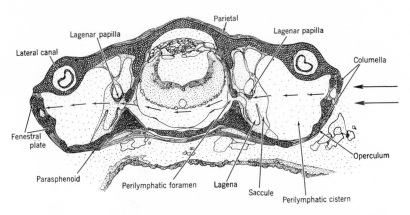

Fig. 12-9. The complete acoustic path through the head in *Hynobius nebulosus*. Scale 25X.

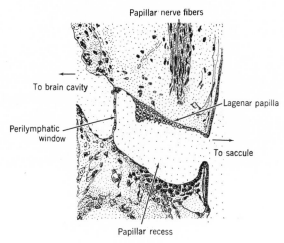

Fig. 12-10. The lagenar papilla in *Hynobius nebulosus* in a frontal section. Scale 125X.

Numbers of Hair Cells. — In one specimen of *Hynobius nebulosus* the number of hair cells in the amphibian papilla on one side was determined as 155 and on the other side as 170, and in the lagenar papilla of this specimen the numbers were 46 and 40. In another specimen of this species there were 100 hair cells in the amphibian papilla on the right and 90 hair cells in the one on the left, and in the lagenar papilla the corresponding numbers were 21 and 19. In the specimen of *Hynobius okiensis* there were 77 hair cells in the amphibian papilla on the left side and 70 hair cells in the one on the right, and in the lagenar papilla there were 51 and 52 hair cells on the two sides.

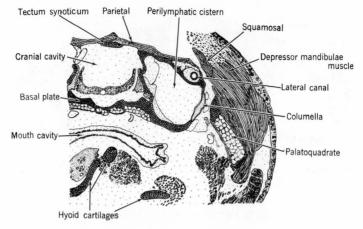

Fig. 12-11. The ear region in *Hynobius okiensis* in a transverse section.

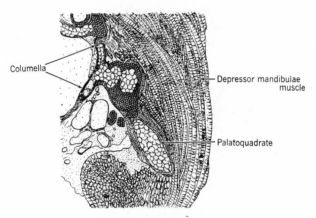

Fig. 12-12. A transverse section through the columellar region in *Hynobius okiensis*.

Sensitivity Functions. — The sensitivity of these ears was tested in the usual way, with an active electrode inserted through a hole drilled into the perilymph space of the lateral semicircular canal, with results shown in Figs. 12-14 and 12-15 for two specimens of *Hynobius nebulosus* and in Fig. 12-16 for the specimen of *Hynobius okiensis*. These functions for the *nebulosus* species are similar in form, with a broad region of moderate sensitivity in the low frequencies around 100 to 500 Hz, reaching a level of +2 and +4 db for one of these but only around +15 for the other. The specimen of *H. okiensis* shows an irregular curve at a much poorer level of performance, only reaching +34 db at the best point at 700 Hz.

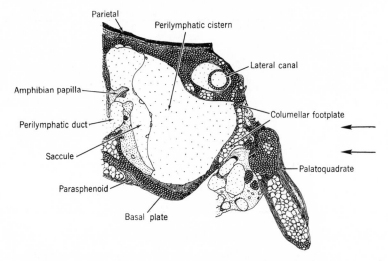

Fig. 12-13. The peripheral receptive structures in *Hynobius okiensis* in a transverse section. Scale 20X.

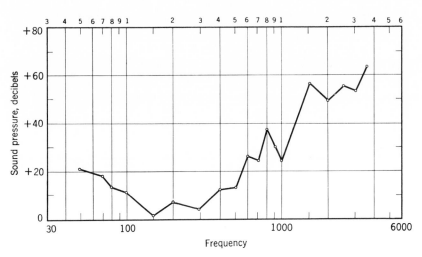

Fig. 12-14. Aerial sensitivity in a specimen of *Hynobius nebulosus*. Shown is the sound pressure, in decibels relative to 1 dyne per sq cm, required for a potential of 0.1 μv.

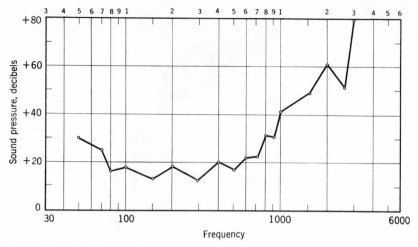

Fig. 12-15. Aerial sensitivity in a second specimen of *Hynobius nebulosus*, represented as in the foregoing figure.

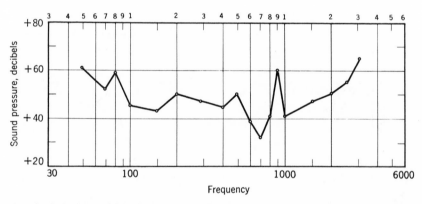

Fig. 12-16. Aerial sensitivity in a specimen of *Hynobius okiensis*, represented as in the foregoing figures.

These ears clearly are poorly adapted to respond to aerial sounds, but can be expected to respond reasonably well to aquatic stimuli that most likely are involved in the mating process.

CRYPTOBRANCHIDAE, THE GIANT SALAMANDERS

Two species of the genus *Cryptobranchus* are recognized, *C. alleganiensis*, which is by far the widest in distribution, and *C. bishopi*, occupying a small area in the Ozark region of Missouri.

Cryptobranchus alleganiensis. This is a large, full-bodied salamander with a greatly flattened head, known generally as the hellbender (see Fig. 12-17). It is about 50 cm long on the average, but females are often larger, up to 70 cm.

The head is broad, about 20 cm wide in a large specimen. The legs are short and stumpy, with 4 toes on the fore feet and 5 toes on the hind ones. The tail is compressed, with a distinct keel.

This species occurs in the eastern, central, and southern states, in tributaries of the Susquehanna and Allegheny river systems in the East, and in numerous tributaries of the Mississippi system through Missouri, Arkansas, and Georgia. *Cryptobranchus* is a further species, along with *Necturus, Amphiuma*, and larval forms in general, in which the columella has its footplate in the oval window and sends forth a headpiece or stylus that makes contact with the squamosal, or with this element in combination with the palatoquadrate. This structure is shown first in a lateral view (Fig. 12-18) based upon a dissection and then in a transverse section (Fig. 12-19) that indicates in detail the connection between the head of the columella and the inner end

Fig. 12-17. A specimen of *Cryptobranchus alleganiensis*. Drawing by Anne Cox.

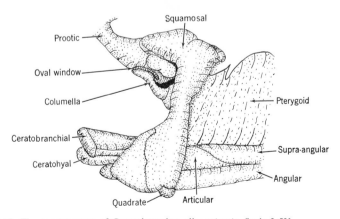

Fig. 12-18. The ear structures of *Cryptobranchus alleganiensis*. Scale 2.5X.

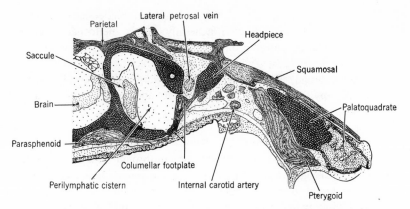

Fig. 12-19. A transverse section through the ear region of *Cryptobranchus alleganiensis*. Scale 5X.

of the squamosal. The palatoquadrate, and also the pterygoid, form a coherent mass of firm material alongside the squamosal, all of which is subject to vibratory pressures from the aquatic medium outside, and which transmits, mainly through the squamosal connection but in some degree by way of the adjacent soft tissues, the imposed vibrations from the outside.

The hair-cell populations in amphibian and lagenar papillae were determined in two specimens of *Cryptobranchus*, as Table 12–I shows. This ear is well provided with hair cells.

A further figure shows these ear structures in a frontal section (Fig. 12-20). The fenestral plate formed by the base of the columella fills the oval window and bears a bony extension, the stylus, which at a different level (already indicated in the foregoing figure) makes a firm contact with the squamosal.

Sensitivity to aerial sounds is shown for three specimens of *Cryptobran-*

Table 12-I

Hair-Cell Populations in the Cryptobranchidae

In the Amphibian Papilla		In the Lagenar Papilla	
Left	*Right*	*Left*	*Right*
#1 *Cryptobranchus alleganiensis*			
491	437	112	120
#2 *Cryptobranchus alleganiensus*			
361	332	82	84

chus alleganiensis in Fig. 12-21. The best of these three curves shows good sensitivity to sounds in the low-frequency range, up to about 250 Hz, and a progressive decrease in sensitivity for the high tones. This ear can be expected to operate as a very effective receiver of water vibrations at the low end of the sound spectrum.

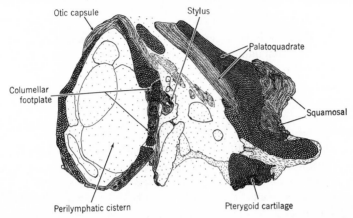

Fig. 12-20. A frontal section through the ear region of *Cryptobranchus alleganiensis*. Scale 8X.

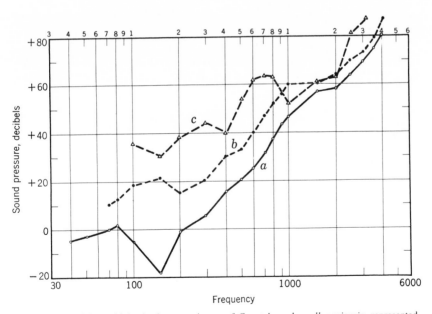

Fig. 12-21. Aerial sensitivity in three specimens of *Cryptobranchus alleganiensis*, represented as the sound pressure, in decibels relative to 1 dyne per sq cm, required for an inner ear response of 0.1 μv.

13. THE SIRENIDAE

The family Sirenidae includes two genera, *Siren* and *Pseudobranchus*, with the first of these containing two species, one of them with two subspecies; specimens were available of the greater siren, *Siren lacertina*, which occurs along the Atlantic coastline from Virginia through all of Florida, and of the lesser siren, *Siren intermedia intermedia*, which is found in the coastal half of South Carolina and southward through southern Georgia and Florida. Specimens were also obtained of *Pseudobranchus striatus spheniscus*, a subspecies that occurs in north central Florida and southwest Georgia.

These salamanders are fully aquatic, eel-like in body form, and live in a variety of water habitats from ditches to streams and lakes, usually in areas choked with water plants and plant debris, where the animals find food such as worms, crustacea, and larvae of various kinds, and (perhaps incidentally) engulf large amounts of vegetable material.

Siren lacertina.   This species has an average length of about 60 cm; a sketch of the head portion of a smaller specimen, 25 cm long, is shown at *a* of Fig. 13-1. There are three pairs of external gills, and forelegs only are present. The sketch at *b* shows the ear region after removal of the superficial tissues, and this view is presented in detail in Fig. 13-2. To the right in Fig. 13-1

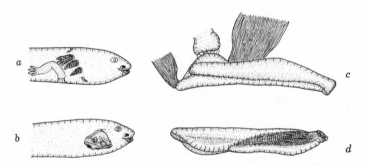

Fig. 13-1. The head region in a specimen of *Siren lacertina*. *a*, a view of the right side showing mouth, gills, and a foreleg; *b*, an anterior area dissected to expose the ear structures, shown in detail in the following figure; *c*, the mandible in a lateral view, together with its articulation and muscles; and *d*, this element in a medial view. Scales: *a*, *b*, about natural size; *c*, *d*, 5X.

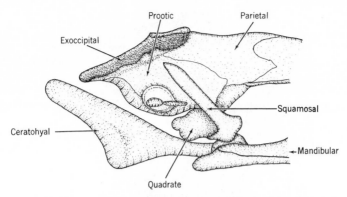

Fig. 13-2. The ear structures of *Siren lacertina* represented in detail in a lateral view. Scale 6X.

are lateral and medial views of the mandible, with the principal mandibular muscles and the articulatory element shown in part *c*.

A frontal section through the ear region is shown in Fig. 13-3. The otic capsule is cartilaginous, and the rather large oval window is filled by a columellar plate with a headpiece that is extended anterolaterally as a stylus to which a hyocolumellar ligament is attached. There is an operculum with an opercular muscle that mainly arises from one of the vertebral aponeuroses, but has a tenous attachment also to the suprascapula.

At another level, as shown in Fig. 13-4, the fibers of the hyocolumellar ligament appear to be continuous all around the outer border of the cerato-hyal and connect with a quadrate muscle coming from skin and fascia in the neck region. The ceratohyal evidently serves as a sort of pulley to change the direction of pull from posterior to lateral; this no doubt makes the action of the quadrate muscle more effective in reducing the mobility of the colu-mella.

The protection of this ear against overstimulation by sounds is thus provided by two muscles: by the quadrate muscle as just indicated and by the opercular muscle that in its contraction pulls the operculum posteriorly and by its attachment to the columella tends to lock this element in place.

Only the amphibian papilla is present in sirens, and hair-cell counts made in three specimens of *Siren lacertina* are given in Table 13-I.

Sensitivity to aerial sounds was determined in two specimens as shown in Fig. 13-5. In the region from 100 to 1000 Hz there is a good level of response, indicating that this ear should serve well as a receiver of auditory signals in the water.

Siren intermedia. The Eastern lesser siren, *Siren intermedia*, presents an ear structure closely similar to that of the preceding species; a transverse section

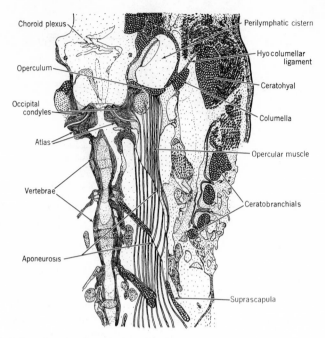

Fig. 13-3. A frontal section through the ear region of *Siren lacertina*. Scale 10X.

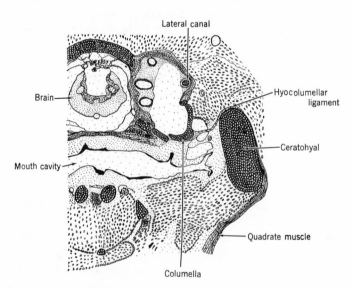

Fig. 13-4. A transverse section in a specimen of *Siren lacertina*. Scale 10X.

Table 13-I
Hair-Cell Populations in *Siren lacertina*

In the Amphibian Papilla	
Left	*Right*
#1 *Siren lacertina*	
349	350
#2 *Siren lacertina*	
384	238
#3 *Siren lacertina*	
375	375

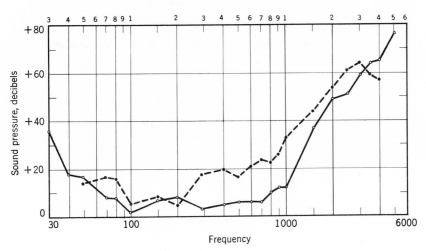

Fig. 13-5. Sensitivity to aerial sounds in two specimens of *Siren lacertina*. Shown is the sound pressure, in decibels relative to 1 dyne per sq cm, required for an inner ear potential of 0.1 μv.

through the greater part of the head is shown in Fig. 13-6. The quadrate muscle originates in the neck region, attaches firmly to the lateral surface of the ceratohyal, and its force is continued to the hyocolumellar ligament that is joined to the columella. (A few ligamentary fibers may may run continuously around the lateral surface of the ceratohyal, but the muscle fibers do not do so.) Details of ear structure, and especially the relations of columella,

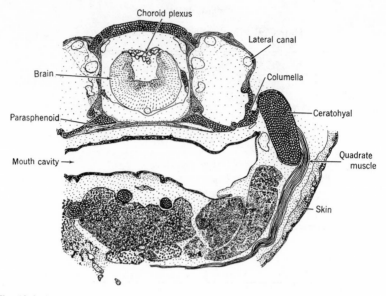

Fig. 13-6. A transverse section through the ear region of *Siren intermedia*. Scale 15x.

hyocolumellar ligament, and ceratohyal, are shown in Fig. 13-7. A section taken from another specimen shows a particularly broad hyocolumellar ligament (Fig. 13-8).

Another specimen, shown in Fig. 13-9, in which the skin had been preserved over the ventral portion of the head was used to show the continuity of the right and left quadrate muscles. These muscles thin out in the middle of the neck region, but many of the fibers continue to the opposite side, as shown by a close tracing under the microscope.

Sensitivity to aerial sounds is indicated for three specimens of *Siren intermedia* in Fig. 13-10. This sensitivity is poor, with the maximum in the region of 290 to 600 Hz, and there is a rather rapid decline for lower and higher tones. In one specimen the number of hair cells in the amphibian papilla (the only one present in this species) was 350 on the left side and 332 on the right. In a second specimen these numbers were 236 and 239, and in a third specimen 323 and 325.

Pseudobranchus striatus spheniscus. In this subspecies of the dwarf siren the columella, as Fig. 13-11 shows, is a simple plate filling the ventrolateral opening in the capsular wall (the oval window) and lacks a definite stylus. A hyocolumellar ligament connects with both the columella and otocranial

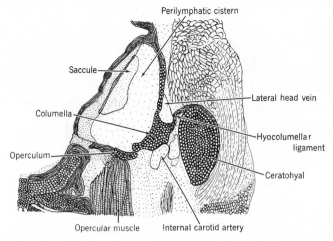

Fig. 13-7. Details of the columellar region in *Siren intermedia*. Scale 20X.

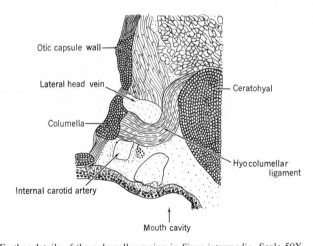

Fig. 13-8. Further details of the columellar region in *Siren intermedia*. Scale 50X.

floor and runs to a firm attachment on the ceratohyal. The cross section in Fig. 13-12 shows this attachment and also a separate attachment of the quadrate muscle to the ceratohyal. It is clear that contractions of the quadrate muscle, acting through the ceratohyal, will exert tension on the columella and tend to fix it in its window. Thereby the ear is protected against excessive sounds.

Sensitivity functions in response to aerial sounds are presented for two

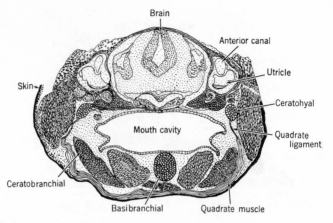

Fig. 13-9. A transverse section through the head in *Siren intermedia*. Scale 12.5X.

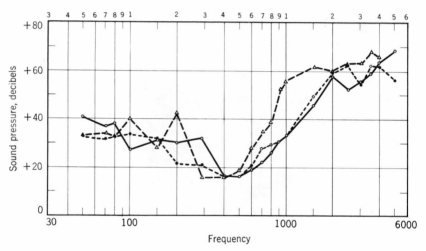

Fig. 13-10. Aerial sensitivity in three specimens of *Siren intermedia*. Shown is the sound pressure, in decibels relative to 1 dyne per sq cm, required for an inner ear potential of 0.1 μv.

specimens in Fig. 13-13. This sensitivity is poor, with the tones below 1000 Hz somewhat more effective than the others.

EAR PROTECTION IN THE SIRENIDAE

The columella in *Siren lacertina* as seen in Fig. 13-4 is a plate of cartilage that remains attached to the otic capsule wall along its anterior margin, but elsewhere is free, connected only by a ring of ligamentary tissue. This plate

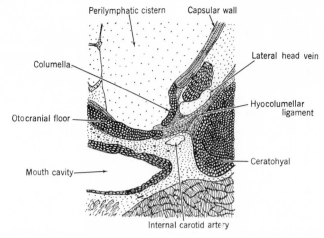

Fig. 13-11. The middle ear structures in *Pseudobranchus striatus spheniscus*. Scale 50X.

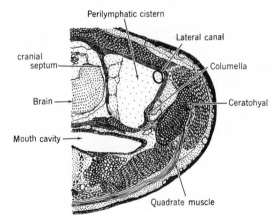

Fig. 13-12. A transverse section through the right half of the head in *Pseudobranchus striatus spheniscus*. Scale 20X.

therefore can respond as a vibrating flap under the influence of sound pressures transmitted inward through the skin and muscle layers at the side of the head. This transmission encounters the large ceratohyal cartilage, which is embedded in the muscle mass in this region and which probably moves bodily with the sound waves, in some degree unifying their force.

The columella does not bear a stylus in any true sense, though it has a thickened region over which lies a large ligamentary mass extending lat-

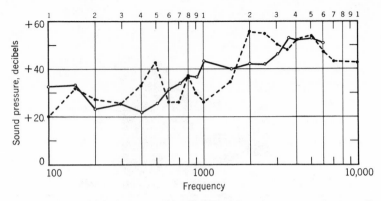

Fig. 13-13. Aerial sensitivity in two specimens of *Pseudobranchus striatus spheniscus*. Shown is the sound pressure, in decibels relative to 1 dyne per sq cm, required for an inner ear potential of 0.1 μv.

erally into the area covered by the ceratohyal and connecting to this cartilage by a heavy ligament.

As usual, two muscles serve to protect this ear against overstimulation by sounds:

1. The quadrate muscle, with its origin in the throat region, fans out and sends forth a terminal band-like ligament that reaches the columella over a considerable area. Along a large part of this course it passes around the ceratohyal, which appears to serve as a sort of pulley to change the direction of pull from posterior to lateral; this no doubt makes the action more effective. Perhaps this tension also serves to stiffen the ceratohyal and make it more of an obstacle to the inward transmission of sound.

2. An operculum is present (see Fig. 13-3) as an independent nodule of cartilage at the posterior end of the otic capsule, rather loosely connected by fascia to the medial wall formed by the exoccipito-prootic bone, and attached by ligament laterally to the lower edge of the columella.

The opercular muscle consists of a limited group of fibers attached to the operculum, along with fibers on either side that go to fixed structures—the exoccipito-prootic and a nearby part of the otic capsule wall. Contraction of this muscle draws the operculum posteriorly, stiffening the contents of the otic capsule and reducing the motion in response to sounds. The attachment of the more peripheral fibers of the opercular bundle to adjacent immobile tissues serves to fix the operculum further.

The small number of fibers involved in this opercular action suggests that this is a minor mode of ear protection and that this function is mainly served by the quadrate muscle, acting through the ceratohyal as shown in Fig. 13-12.

14. THE SALAMANDRIDAE:

THE NEWTS

The family Salamandridae is the most widespread of all the urodele groups; it includes the dominant species of Europe, Asia, and North Africa and is well represented in North America also. An important genus in this family, *Salamandra*, will first be examined and then the discussion will turn to a number of other genera for which specimens were available, including *Triturus* (also known as *Triton*), *Taricha*, *Notophthalmus* (also called *Diemictylus*), *Salamandrina*, *Pleurodeles*, *Cynops*, and *Euproctus*. In these last-named genera the columella has been reduced still further than in the *Salamandra* species, and the structure that remains, as seen in the adults, has been incorporated into the fenestral border to such an extent that it is hardly discernible. Kingsbury and Reed (1909) by following in *Triturus* the course of development through the embryonic stages to the adult condition were convinced that vestiges at least of the columellar tissue persist.

Salamandra salamandra. The species *Salamandra salamandra*, commonly known as the European salamander, is a heavy-bodied form occurring in eight subspecies distributed over Europe and extending into west Asia and north-west Africa. The subspecies examined was *S. s. salamandra*, the spotted salamander, with a wide range in Europe, including the Alpine and Carpathian regions and extending west to Rumania, Bulgaria, and the edge of Turkey. An example is shown in Fig. 14-1.

In this subspecies mating occurs over a long period, most actively in spring but extending into autumn, and involves an act of amplexus in which a spermatophore is taken up by the female; then after a gestation period of several months she bears living young that though small have well-developed legs and a long tail; these larvae are deposited in water and are active swimmers, fully able to make their own way. After two or three months of growth, or sometimes if they were born late in the summer during the following spring, these larvae change to the adult form, leave the water, and thereafter remain terrestrial (except for the females who at the end of the breeding season enter the water temporarily to give birth).

The ear structures are shown in Fig. 14-2 in a lateral view drawn from a dissected specimen. Here the stylus of the columella is seen, just posterior

Fig. 14-1. A specimen of *Salamandra salamandra*. Drawing by Anne Cox.

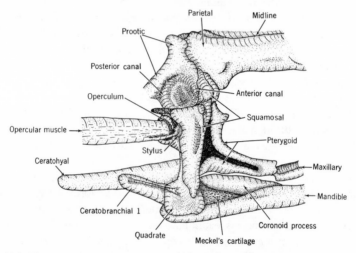

Fig. 14-2. The ear region in *Salamandra salamandra* as seen in a dissected specimen. Scale 5X.

to the squamosal and partly covered by it, and immediately adjacent is the operculum, to which a heavy opercular muscle is attached. Above in this figure is the prootic, with posterior and anterior semicircular canals well defined by bony ridges, and the otic capsule deep within the region enclosed by these canals.

A transverse section through the region of the amphibian papilla is presented in Fig. 14-3. Here the columella occupies the oval window and is fused with the ventrolateral portion of the otic capsule. At this level the perilymphatic duct is seen leading from the perilymphatic window just below

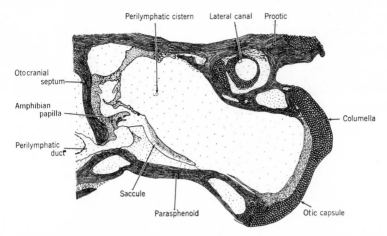

Fig. 14-3. A transverse section through the right side of the head of *Salamandra salamandra* at the level of the amphibian papilla. Scale 25X.

the amphibian papilla and passing through the otocranial foramen into the brain cavity.

A transverse section at a point well posterior to that in which the amphibian papilla can be seen is presented in Fig. 14-4. Here the operculum is present, cut across at a level at which it more than half fills the oval window, and is separated from the columella by connective tissue. The lagena is shown, opening medially into a papillar recess, with the lagenar papilla dependent from the dorsal wall. An enlarged view of this lagenar area is given in Fig. 14-5, and further details of the lagenar papilla itself appear in Fig. 14-6.

Auditory sensitivity in this species was measured with aerial sounds applied through a tube over the skin surface in the facial area. Results for two animals are given in Fig. 14-7. In one of these specimens (broken curve) the maximum sensitivity was at 300 Hz, declining for lower and higher tones, especially rapidly above 1500 Hz. For the other animal (solid curve) the best point was at 100 Hz, with a very rapid decline for lower and higher tones. Results for a third specimen are presented in Fig. 14-8; this curve represents the mean of three successive series of measurements. Here the maximum is at 400 Hz, with a moderate decline for lower tones, and a rather rapid one for higher tones, with some irregularities in the upper frequencies.

Triturus SPECIES

The *Triturus* group contains the dominant salamander species of Europe, covering an extensive range not only in the mountains but also in hills and lowland areas. These salamanders are always found near water sources, but in habit are mainly terrestrial. They are active at night and spend the day-

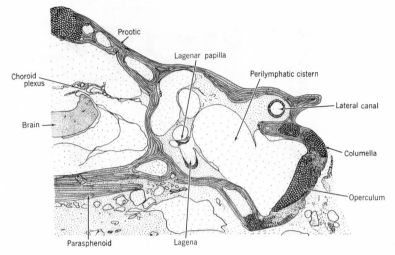

Fig. 14-4. A transverse section at a level posterior to the preceding, passing through the lagena and lagenar papilla. Scale 20X.

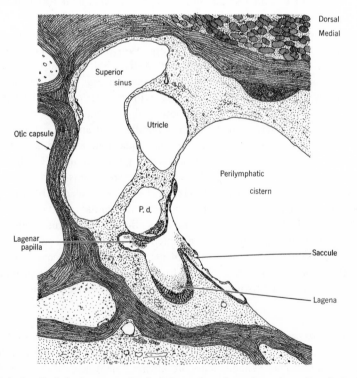

Fig. 14-5. Detail of the lagenar region in *Salamandra salamandra*. pd = perilymphatic duct. Scale 50X.

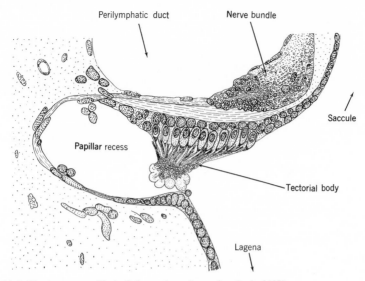

Fig. 14-6. The lagenar papilla in *Salamandra salamandra*. Scale 250X.

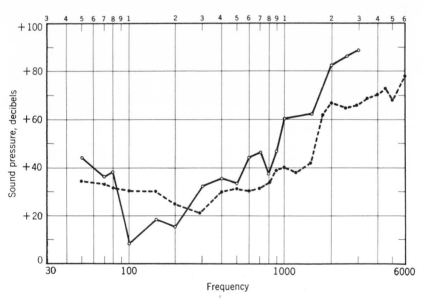

Fig. 14-7. Sensitivity to aerial sounds in two specimens of *Salamandra salamandra*. Shown is the sound pressure, in db relative to 1 dyne per sq cm, required for an inner ear potential of 0.1 μv.

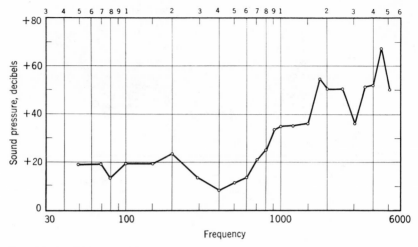

Fig. 14-8. Sensitivity to aerial sounds in a third specimen of *Salamandra salamandra*, represented as in the foregoing figure.

time hours hidden away under protective litter such as leaves, moss, and twigs. Their lung capacity is limited, and when in the water they periodically come to the surface to take a gulp of air. The eggs are laid singly in the water, attached to some submerged object. The female gives careful attention to the handling of each one; often it is stuck to a leaf and the edges of the leaf bent around to form a little pocket.

There are at least five subspecies, of which two were studied.

Triturus cristatus cristatus. *Triturus c. cristatus* is the principal subspecies widely distributed in the northern regions of Europe from Britian and France to Russia as far as the Ural mountains. A transverse section through the ear region is shown in Fig. 14-9; here is seen the large operculum, loosely connected at its anterolateral end to a small stylus, which is all that is recognizable of the columella. These structures are presented in an enlarged drawing in Fig. 14-10 which shows the connection between operculum and stylus as formed by connective tissue. No specific function for the stylus is evident here; and it appears that sound from the outside reaches the fluid of the perilymphatic cavity simply by bulk conduction by the peripheral skin and underlying muscle and other tissues.

The numbers of hair cells in amphibian and lagenar papillae were determined for two specimens of *Triturus c. cristatus* and are shown in Table 14-I.

Sensitivity in terms of the inner ear potentials was measured in several specimens, with results shown by the curves of Figs. 14-11, 14-12, and

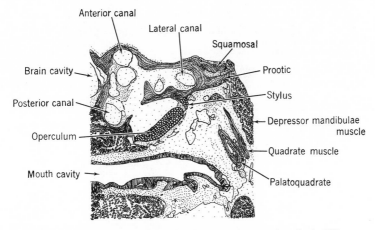

Fig. 14-9. The ear region of *Triturus c. cristatus*, in transverse section. Scale 15X.

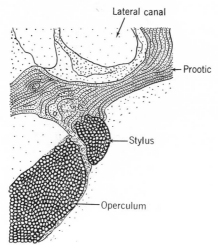

Fig. 14-10. The ossicular area of *Triturus c. cristatus* showing stylus and operculum. Scale 60X.

14-13. In general these functions show best sensitivity in the middle frequencies from 150–800 Hz and rapidly reduced sensitivity for the higher tones. The two specimens represented in Fig. 14-13 are exceptional in that the sensitivity extends well into the low frequencies.

Triturus cristatus carnifex.   The alpine crested newt inhabits mountainous areas of northern Italy, Austria, and northern Yugoslavia. A figure showing

Table 14-I

Hair-Cell Populations in the Salamandridae

In the Amphibian Papilla		In the Lagenar Papilla	
Left	Right	Left	Right
	#1 *Triturus c. cristatus*		
177	146	15	14
	#2 *Triturus c. cristatus*		
174	158	10	26
	#1 *Triturus c. carnifex*		
145	135	24	19
	#2 *Triturus c. carnifex*		
248	256	15	21
	#3 *Triturus c. carnifex*		
193	169	14	16
	#1 *Taricha rivularis*		
166	160	34	34
	#2 *Taricha rivularis*		
166	190	56	44
	#3 *Taricha rivularis*		
199	201	40	49
	#1 *Taricha granulosa*		
177	149	21	38
	#2 *Taricha granulosa*		
190	163	33	20
#1 *Notophthalmus viridescens (mature adult)*			
123	120	*	
#2 *Notophthalmus viridescens* (mature adult)			
129	142	*	
#3 *Notophthalmus viridescens* (mature adult)			
102	88	*	
#4 *Notophthalmus viridescens* (mature adult)			
187	159	*	
#5 *Notophthalmus viridescens* (red eft)			
98	112	*	
#6 *Notophthalmus viridescens* (red eft)			
†	109	*	
#7 *Notophthalmus viridescens* (red eft)			
119	115	*	

Table 14-I

Hair-Cell Populations in the Salamandridae

In the Amphibian Papilla		In the Lagenar Papilla	
Left	Right	Left	Right
#1 *Pleurodeles waltl*			
113	132	63	77
#2 *Pleurodeles waltl*			
172	161	45	57
#1 *Euproctus asper*			
121	112	23	25
#2 *Euproctus asper*			
114	100	32	31

*Lagenar papilla lacking in species.
†Not determined.

the ear structures, based on a dissection, is presented in Fig. 14-14. Neither the columella nor its stylus is evident here, but a study of serial sections through this region disclosed an irregular body of cartilage in the lateral wall of the otic capsule lying between the border of the operculum and the rather thick wall forming the anterolateral enclosures of the otic capsule; this is without question a remnant of the columella.

An operculum is present, not visible in Fig. 14-14 but shown in Fig. 14-15 in a view from a more posterior position. The opercular muscle is divided into superior and inferior portions, as indicated.

Hair-cell populations for both amphibian and lagenar papillae were observed for three specimens and are shown in Table 14-I. As will be seen, the amphibian papilla is well developed, whereas the lagenar papilla contains only a few hair cells.

Triturus marmoratus. The marbled newt, *Triturus marmoratus*, inhabits a wide area including all of Spain and Portugal and the southwestern half of France. Within its area it avoids high elevations and prefers the valleys and lower altitudes generally.

A cross section through the ear region, given in Fig. 14-16, shows a well-defined columella, connected loosely by ligamentary fibers to a large operculum posteriorly and fused anteriorly to the bony wall of the otic capsule. Immediately anterior to the columella is the lateral head vein, and posterior

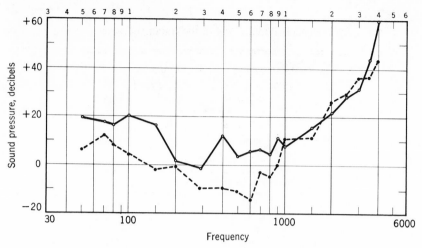

Fig. 14-11. Sensitivity curves for two specimens of *Triturus c. cristatus*, for aerial sounds. Shown is the sound pressure, in db relative to 1 dyne per sq cm, required for an inner ear potential of 0.1 μv.

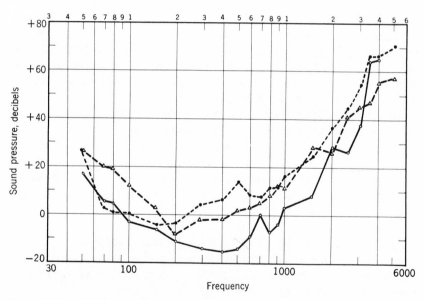

Fig. 14-12. Sensitivity curves for three additional specimens of *Triturus c. cristatus*, for aerial sounds, represented as in the foregoing figure.

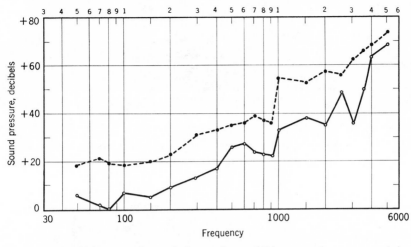

Fig. 14-13. Sensitivity curves for two further specimens of *Triturus c. cristatus*, represented as in the foregoing figures.

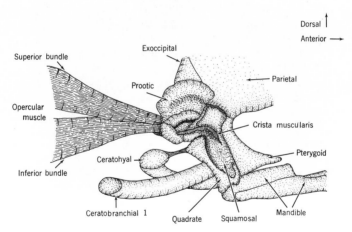

Fig. 14-14. The ear region in *Triturus cristatus carnifex* in a lateral view. Scale 7.5X.

to it is the internal carotid artery—landmarks that assist in identifying this element. Lateral to the oval window area is a large mass of tissue made up of skin at the outer surface, next a relatively thin muscle layer, and finally a somewhat loose mass of fascia. The ceratohyal cartilage lies in this fascial layer, at a level slightly posterior to the otic capsule. Extending laterally into the muscle layer are the palatoquadrate and the closely attached squamosal bone. Two heads of the opercular muscle were present, the one shown in

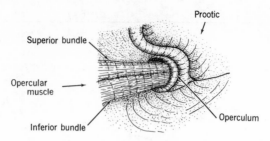

Fig. 14-15. A posterolateral view of the opercular region in *Triturus cristatus carnifex*. Scale 10X.

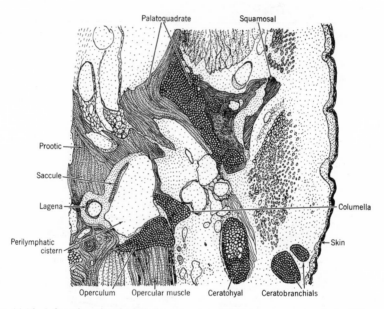

Fig. 14-16. A frontal section through the ear region of *Triturus marmoratus*. Scale 25X.

the figure and another medial to this one that followed a similar course to the suprascapula in a more ventral plane.

Taricha torosa, the California Newt. The genus *Taricha* contains three species, all occurring along the Pacific coast of North America from Alaska to California. The species *Taricha torosa* is restricted to California and is found in two principal areas, one, the subspecies known as the Coast Range newt, occupying a strip along the coastline except its extreme northern and southern ends, and another, the Sierra newt, that lives well inland in the Sierra mountains.

The ear structures of *Taricha torosa* are represented in Fig. 14-17. The oval window is filled by the operculum, from which runs a well-developed opercular muscle to its origin on the suprascapula. A columella is not obvious and is considered to be incorporated in the otic capsule adjacent to the oval window. A somewhat enlarged view of some of these structures is presented in Fig. 14-18.

The lagenar papilla is shown in a transverse section in Fig. 14-19, dependent from the roof of a small recess extending medially from the lagenar cavity. Immediately above this papilla, in limbic tissue lateral to the perilymphatic duct, is a small bundle of nerve fibers supplying this papilla.

In one specimen of this species the amphibian papilla contained 234 hair cells on each side, whereas the lagenar papilla on the left contained 13 hair cells and the one on the right 26 hair cells. In another specimen the amphibian papilla showed 126 hair cells on the left and 151 on the right, whereas the lagenar papilla had only 13 hair cells on the left side and 10 cells on the right. Still another specimen had 155 hair cells in each amphibian papilla and 11 hair cells in each lagenar papilla. The lagenar papilla appears to be poorly developed in this species.

An aerial sensitivity curve for one specimen of *Taricha torosa* is shown in Fig. 14-20. This curve attains a rounded maximum in the region of 290–400 Hz and shows a rather regular decline in sensitivity for lower and higher tones. Another specimen, represented in Fig. 14-21, showed somewhat poorer

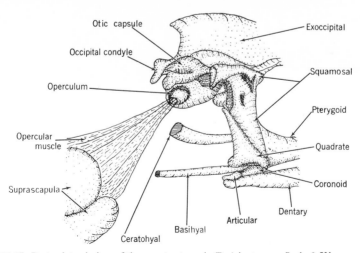

Fig. 14-17. Posterolateral view of the ear structures in *Taricha torosa*. Scale 6.5X.

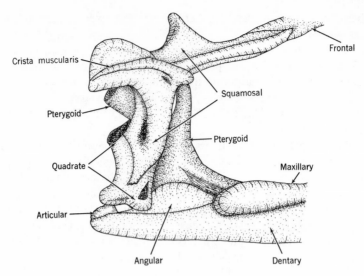

Fig. 14-18. An enlarged view of the ear structures in *Taricha torosa*. Scale 9X.

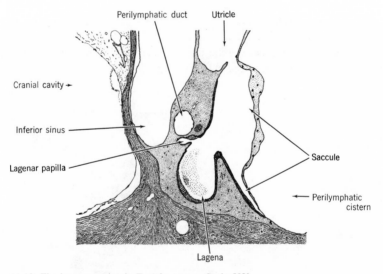

Fig. 14-19. The lagenar region in *Taricha torosa*. Scale 50X.

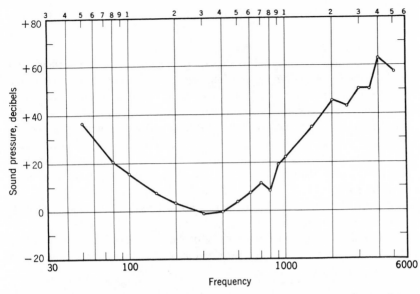

Fig. 14-20. Aerial sensitivity in a specimen of *Taricha torosa*, represented as the sound pressure, in db relative to 1 dyne per sq cm, required for an inner ear potential of 0.1 μv.

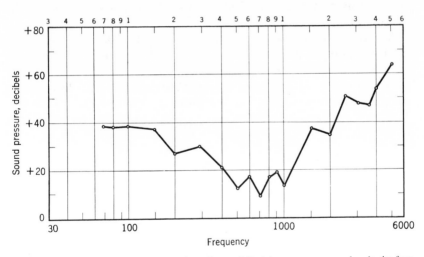

Fig. 14-21. Aerial sensitivity in a second specimen of *Taricha torosa* expressed as in the foregoing.

sensitivity, with considerable variations with frequency in the best region from 500–1000 Hz.

Taricha rivularis. A second species of *Taricha* occurs in the northern part of California, inhabiting streams and rivers in woodland areas near the coastline. Shown in Fig. 14-22 is a side view of the head of *Taricha rivularis* with a dissection to reveal the ear structures. The operculum is shown, with its muscle removed. A somewhat enlarged representation of the ear structures as seen in a dorosolateral view is presented in Fig. 14-23. Only the extreme anterior end of the opercular muscle is indicated. The hair-cell populations were determined for both papillae in three specimens, as indicated in Table 14-I. Both auditory papillae in this species are well developed, with the lagenar organ containing about one-fourth as many hair cells as the amphibian papilla.

An aerial sensitivity function for one of the specimens is presented in Fig. 14-24. This ear shows a somewhat irregular maximum region around 150 to 600 Hz, with peaks at 200 and 400 Hz.

Taricha granulosa. This third species of Pacific newt occurs along the coastline from British Columbia to the northern part of California, inhabiting a variety of streams, pools, and ditches in the aquatic phase and in wooded areas usually concealed under forest debris or even in grasslands in the terrestrial phase.

A transverse section through the amphibian papilla is shown in Fig. 14-25, and the sizes of the hair-cell populations for both papillae in two specimens are indicated in Table 14-I.

The sensitivity appears to be somewhat poorer than in the other *Taricha* species studied; curves for three specimens are given in Figs. 14-26 and 14-27. For the best of these three ears the sensitivity curve reaches its most

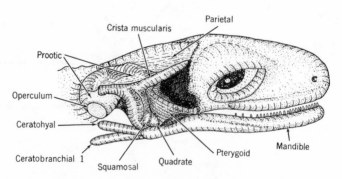

Fig. 14-22. A lateral view of the head of *Taricha rivularis*. Scale 5X.

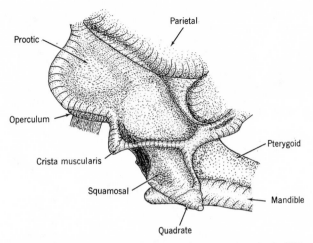

Fig. 14-23. A dorsolateral view of the ear region in *Taricha rivularis*. Scale 10X.

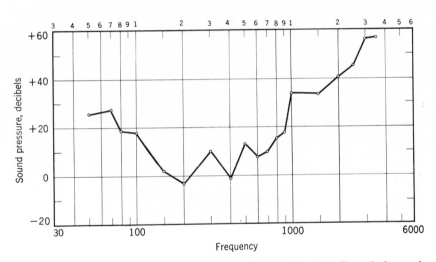

Fig. 14-24. Sensitivity to aerial sounds in a specimen of *Taricha rivularis*. Shown is the sound pressure, in db relative to 1 dyne per sq cm, required for an inner ear potential of 0.1 μv.

effective level in the region of 150 to 300 Hz, but plainly these ears do not perform well in response to aerial sounds.

Notophthalmus viridescens. The family Salamandridae contains two genera in North America, *Notophthalmus*, the Eastern newts, and *Taricha*, the

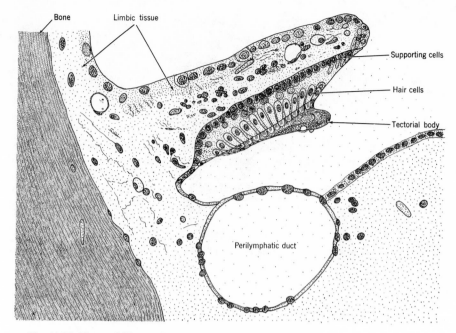

Fig. 14-25. The amphibian papilla in *Taricha granulosa*. Scale 250X.

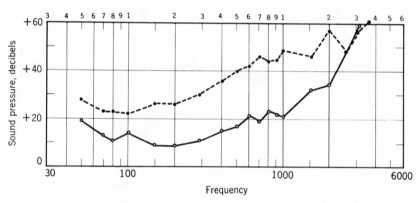

Fig. 14-26. Aerial sensitivity in two specimens of *Taricha granulosa*. Shown is the sound pressure, in db relative to 1 dyne per sq cm, required for an inner ear potential of 0.1 μv.

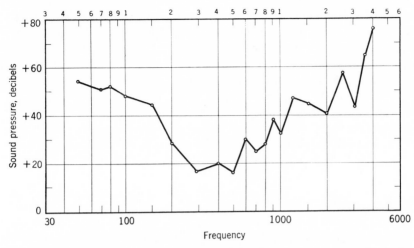

Fig. 14-27. Aerial sensitivity in a third specimen of *Taricha granulosa*, represented as in the preceding figure.

Western newts. *Notophthalmus* was formerly known as *Diemictylus*, and some have used the now inappropriate names of *Triturus* or *Triton*.

The Eastern newts include three species: *viridescens*, *perstriatus*, and *meridionalis*, of which only the first was available for study, and indeed only one among the four subspecies that are included in that group; this subspecies was *Notophthalmus v. viridescens*.

Notophthalmus v. viridescens. This subspecies, known as the common spotted newt, occupies a wide area of the northeastern United States, reaching northward into Ontario and Quebec and including the Great Lakes area, then extending south all along the eastern coastline halfway across Georgia and Alabama and as far west as Indiana and most of Kentucky and Tennessee.

These salamanders as adults inhabit pools, swamps, and the quiet bayous of streams, usually near wooded areas. The larvae are initially aquatic but then pass into a terrestrial stage, when they come out on land and live in woodlands; these are known as "red efts" and differ strikingly from the aquatic forms. The usually brown or green color changes to a bright orange-red, and these animals continue a terrestrial life for periods varying from one to three years, after which they reach the breeding stage. These breeding individuals gather in the spring on the banks of ponds and streams, and, after mating, the females deposit their eggs singly on plants below the water surface. The eggs hatch out in a month or so, and another aquatic stage begins. In some localities the red eft form is omitted, and the mature larvae transform directly into breeding adults, without going through a terrestrial stage.

The ear region is represented in a frontal section in Fig. 14-28. The section shows a relatively thin bony wall covering most of the lateral face of the otic capsule, giving way posteriorly to a small cartilage that evidently represents a remnant of the columella. This cartilage connects by ligament to a rather large operculum, to which are connected two heads of the opercular muscle.

The hair-cell population for several specimens of *Notophthalmus viridescens* is represented in Table 14-I. Note that this salamander lacks a lagenar papilla.

Sensitivity functions for two adult specimens of this species are presented in Fig. 14-29. For both animals the best sensitivity is in the region of 600–800 Hz and falls away for both lower and higher tones. These two specimens were mature adults; a sensitivity curve for a red eft is shown in Fig. 14-30. This curve shows a broad form, with the best region around 290-500 Hz, and the sensitivity declines progressively for lower and higher tones.

Salamandrina terdigitata. This small salamander occurs only in Italy, in the Apennine mountains, most abundantly in the region of Genoa, and is usually concealed in leafy litter in wooded areas adjacent to small streams. It moves about mainly by night in search of prey and thus is seldom seen. It measures about 7.5 cm in body length.

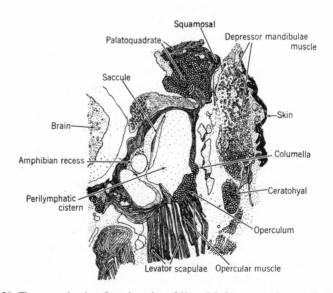

Fig. 14-28. The ear region in a frontal section of *Notophthalmus v. viridescens*. Scale 20X.

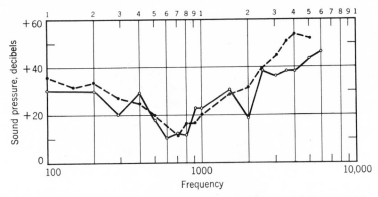

Fig. 14-29. Sensitivity to aerial sounds in two adult specimens of *Notophthalmus v. viridescens*. Shown is the sound pressure, in db relative to 1 dyne per sq cm, required for an inner ear potential of 0.1 μv.

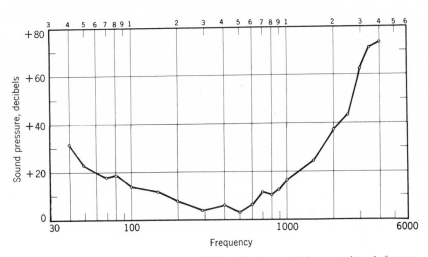

Fig. 14-30. Aerial sensitivity in a specimen of *Notophthalmus v. viridescens* at the red eft stage. Represented as in the foregoing figure.

A frontal section through the ear region is shown in Fig. 14-31. The lateral wall of the otic capsule is completely osseous; anteriorly it is fused with the palatoquadrate and squamosal. There is a large operculum, of cartilage as usual, with a well-developed opercular muscle. This muscle appears to have only a single head.

In one specimen the amphibian papilla on one side contained 187 hair cells, and the one on the other side 185 hair cells. A lagenar papilla was present

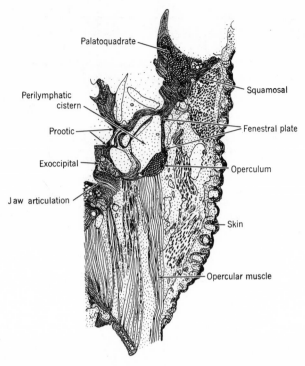

Fig. 14-31. A frontal section representing the ear region in *Salamandrina terdigitata*. Scale 20X.

containing only 19 hair cells on the left side and 20 hair cells on the right.

The sensitivity of this ear is poor as indicated for two specimens in Fig. 14-32. One of these curves indicates a rather uniform range from 290 to 2000 Hz and falls progressively for higher and lower tones. The other curve follows the same course in the low frequencies, but shows progressively less sensitivity as the frequency rises.

Pleurodeles waltl. This large salamander inhabits the southern and western areas of Spain and Portugal and also parts of Morocco. It is aquatic in habit and rarely emerges onto the land, preferring ponds, lakes, and swamps with abundant vegetation. It is the largest of the European tailed amphibians, averaging about 20 cm in length and often reaching twice that.

A cross section through the ear region is shown in Fig. 14-33. There is a distinct columella, though it is small compared with the operculum, that fills a wide segment of the otic capsule on its posterolateral side. There is a strong opercular muscle.

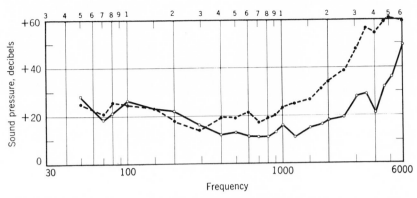

Fig. 14-32. Aerial sensitivity in two specimens of *Salamandrina terdigitata*. Shown is the sound intensity in db relative to 1 dyne per sq cm required for an inner ear potential of 0.1 µv.

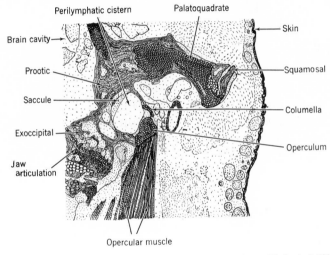

Fig. 14-33. A frontal section through the ear region in *Pleurodeles waltl*. Scale 7.5X.

The number of hair cells in the two auditory papillae was determined in two specimens, as shown in Table 14-I. As will be noted, the hair-cell population of the lagenar papilla is comparatively large, on the average approaching two-fifths that of the amphibian papilla.

A sensitivity curve for one specimen is shown in Fig. 14-34. The best region is in the extreme low tones; from 100 to 400 Hz upward the curve rises steeply. This is an excellent level of low-tone sensitivity for a salamander ear.

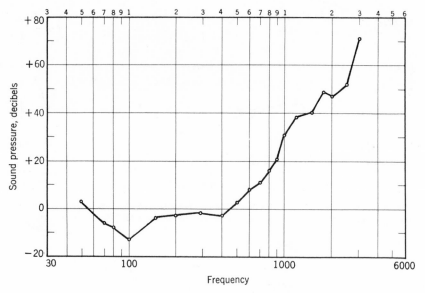

Fig. 14-34. Aerial sensitivity in a specimen of *Pleurodeles waltl*, indicated as the sound pressure, in db relative to 1 dyne per sq cm, required for an inner ear potential of 0.1 μv.

Cynops pyrrhogaster. The fire-bellied newt of Japan is a rather large type and occurs in five or more "races," which differ both in color markings and in body form. The body length is around 11–14 cm, sometimes greater in females.

The ear structures are shown in frontal section in Fig. 14-35. Here the otic capsule is mainly osseous and incorporates the cartilaginous columella on the lateral side. Continuous with the columella is the operculum, to which is attached an opercular muscle of good size. The palatoquadrate is very large and connects through the squamosal to the great mass of the depressor mandibulae muscle lying beneath the skin surface.

The number of papillar hair cells was determined in five ears; on the average there were 137.6 hair cells in the amphibian papilla and only 12.2 hair cells in the lagenar papilla.

Sensitivity curves for two specimens of *Cynops pyrrhogaster* are shown in Fig. 14-36. These indicate a fair level of sensitivity in the low tones up to 400 Hz, and progressively poorer response for higher tones.

Euproctus asper. This salamander occurs in the Spanish Pyrenees and extends also into the southwest border of France. It is confined to swift mountain streams and torrents; the lungs are reduced, and cutaneous respiration is

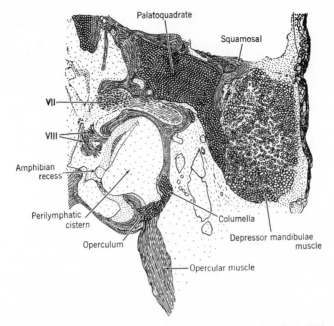

Fig. 14-35. The ear region of *Cynops pyrrhogaster*, in a frontal section. Scale 20X.

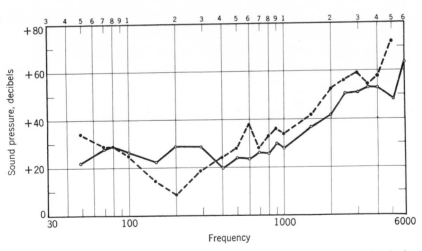

Fig. 14-36. Aerial sensitivity in two specimens of *Cynops pyrrhogaster*, represented as in the foregoing figures.

largely relied upon, which is fully adequate in rapidly moving and well-aerated waters.

This species shows two other adaptations to its torrential habitat; it creeps along the bottom rather than swimming freely in its search for prey and has developed a definite act of copulation in which the sperm packet is transferred directly to the female. The female then attaches her eggs to the underside of stones located in the swift current.

After the eggs hatch in a month or so, the larvae remain in the water for 8–9 months, then transform to the adult stage and leave the water; they finally return to the water three or four years later when they have reached sexual maturity. The general anatomy, including the structure of the ear, is closely similar to that in *Triturus* species. A sketch of middle and inner ear structures, based upon a dissection, is shown in Fig. 14-37. A lagenar papilla is present, with a fair level of development as indicated by the size of the hair-cell population, as Table 14-I shows.

Sensitivity to aerial sounds, as indicated for two specimens in Fig. 14-38, is definitely poor, at least at the adult stage in which these measurements were taken. Of course this ear would be expected to be much more serviceable in the water, yet when we consider the background of a mountain torrent, it seems doubtful that this ear provides any useful information to its possessor, at least at the adult stage.

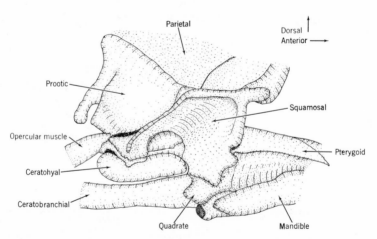

Fig. 14-37. A lateral view of the ear structure in *Euproctus asper*. Scale 10X.

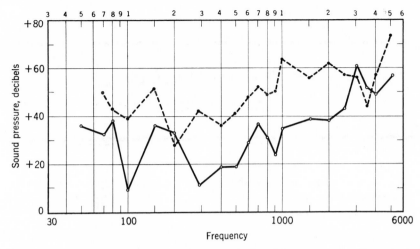

Fig. 14-38. Sensitivity to aerial sounds in two specimens of *Euproctus asper*. Shown is the sound pressure, in db relative to 1 dyne per sq cm, required for an inner ear potential of 0.1 μv.

15. THE PROTEIDAE AND

AMPHIUMIDAE

Of somewhat uncertain status, but generally considered to be related to the Salamandridae of the foregoing chapter, are the families Proteidae and Amphiumidae, now to be described.

FAMILY PROTEIDAE

The genus *Necturus*, whose members are known colloquially as mudpuppies or sometimes as water dogs, includes a group of American salamanders that are thoroughly aquatic in habit and are permanent larvae; the three pairs of external gills persist through life, though these animals also possess lungs. This genus and the European genus *Proteus*, a subterranean lake-dweller that it closely resembles, constitute the small family of Proteidae.

Several species of *Necturus* are widely distributed over the eastern half of North America, inhabiting all kinds of permanent waters from streams and rivers to ponds and lakes, and even muddy, weed-infested sloughs. This salamander is relatively large, of eel-like form, with four toes on both fore and hind legs. The most widespread species, specimens of which were used in this study, is *Necturus m. maculosus* and inhabits chiefly the Mississippi and Ohio river systems and their tributaries, including waterways that have been artificially linked to the original systems by canals.

Necturus m. maculosus.　　The species *Necturus maculosus* is the largest in the genus, with adult lengths averaging about 29 cm, though sizes up to 43 cm have been recorded. A drawing of an adult specimen is shown in Fig. 15-1.

A lateral view of the head is given in *a* of Fig. 15-2 and shows the prominent external gills. In *b* of this figure a dissection has exposed the ear region beneath the skin and muscle layers, the details of which are shown in part *c*.

The columella has a footplate of oval form filling an opening in the prootic (the oval window), and extending anterolaterally is a conical process, the stylus, which is connected by a ligament (the squamosocolumellar ligament) to a short projection of the squamosal.

Fig. 15-1. A specimen of *Necturus m. maculosus*. Drawing by Anne Cox.

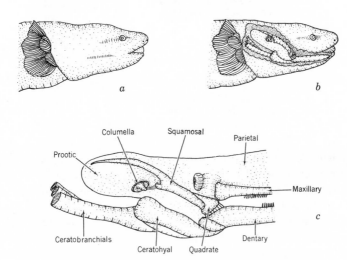

Fig. 15-2. The head of *Necturus m. maculosus* with a dissection to show the ear structures. Scales: *a*, *b*, 0.5X; *c*, 1.5X.

A transverse section through the ear region is shown in Fig. 15-3. Seen here is the complex structure of the columella consisting of a bony framework that includes pockets of cartilage and marrow spaces; this structure is attached dorsally to the prootic by ligamentary tissue and similarly attached medially to the cartilaginous floor of the perilymphatic cistern. Lateral to the columella is a dense ligamentary mass that merges with another enclosing the ceratohyal cartilage, and still farther laterally is the great bulk of the depressor mandibulae muscle and finally the skin layer.

Sound passing through the water medium will reach the columella through the mass of peripheral tissues, with the squamosal serving in a measure to summate and integrate the wave motion.

The size of the hair-cell population in the amphibian papilla was determined in four specimens, with the results shown in Table 15-I. These num-

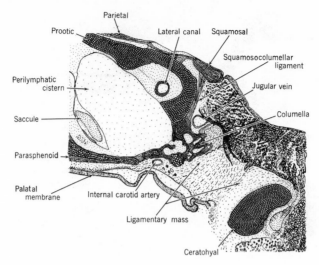

Fig. 15-3. The ear region of *Necturus m. maculosus*. Scale 10X.

bers of hair cells compare favorably with those found in the amphibian papilla of other salamander species. There is no lagenar papilla.

Sensitivity curves in response to aerial sounds are shown for two specimens in Fig. 15-4. This ear responds poorly to such sounds, though one specimen showed a fair degree of sensitivity in the region of 400 Hz. This ear is probably well adapted for the reception of aquatic sounds, especially those of low frequencies.

FAMILY AMPHIUMIDAE

There are two species of *Amphiuma*, one with two subspecies; these are *Amphiuma means means*, *A. means tridactylum*, and *A. pholeter*; available for study were the two subspecies of *A. means*. These animals are commonly known as Congo eels.

Amphiuma means means. These salamanders are almost completely aquatic, though they are known to come out on land at times during rainy periods. They have long sturdy bodies, often attaining a length of 85 cm and a diameter of around 2 cm, and have both forelegs and hindlegs so small as to appear useless; a distinctive feature of this subspecies is the presence of only two toes on each of the four feet.

These animals live in a variety of water places from ditches to ponds and streams, and usually have a sort of nest or haven that they construct of plant debris, and in which they are usually concealed.

Table 15-I

Hair-Cell Populations in the Proteidae and Amphiumidae

In the Amphibian Papilla		In the Lagenar Papilla	
Left	Right	Left	Right
	#1 *Necturus maculosus*		
403	427	*	
	#2 *Necturus maculosus*		
280	248	*	
	#3 *Necturus maculosus*		
382	368	*	
	#4 *Necturus maculosus*		
366	359	*	
	#1 *Amphiuma m. means*		
248	237	60	58
	#2 *Amphiuma m. means*		
255	231	74	75
	#3 *Amphiuma m. means*		
262	267	65	64
	#1 *Amphiuma m. tridactylum*		
305	285	104	101
	#2 *Amphiuma m. tridactylum*		
294	320	118	105

*Lagenar papilla lacking in species.

This species occurs in a wide arc along the southeast coast of the United States from the edge of Virginia to the eastern half of Louisiana, including all of Florida and extending north to a portion of Alabama.

The skull structure of this subspecies was studied early by Wiedersheim (1877), and Fig. 15-5 follows a revision of his drawing made by Baker (1945). The strong squamosal is seen entering into the jaw articulation by way of the quadrate, and extending posteroventrally from beneath this bone is the columella, whose footplate is hidden in the wide oval window of the otic capsule. Figure 15-6 shows these cochlear features in detail in a transverse section, which also indicates the large perilymphatic space through which vibrations of the columellar footplate can make their way inward to the amphibian papilla. The exit passage, shown in Fig. 15-7, is through a foramen

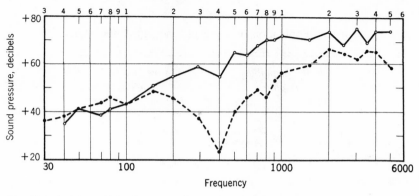

Fig. 15-4. Aerial sensitivity in two specimens of *Necturus m. maculosus*, represented as the sound pressure, in decibels relative to 1 dyne per sq cm, required for an inner ear potential of 0.1 μv.

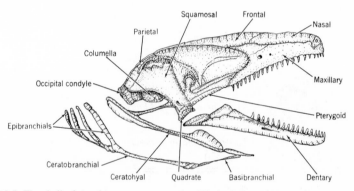

Fig. 15-5. The skull of *Amphiuma m. means*, after Wiedersheim, 1877.

in the otocranial septum ventral to the brain and across to the other side as usual in salamanders.

As shown here, a lagenar papilla is present in this salamander genus and is located along the path of fluid vibrations from perilymphatic cistern to brain cavity.

Observations on three specimens of *Amphiuma m. means* showed the size of the hair-cell populations in both papillae as represented in Table 15-I. The average number of hair cells in the amphibian papilla for these six ears is 250 and in the lagenar papilla is 66; these numbers indicate a good level of development for both papillae.

The sensitivity to aerial sounds is represented for two specimens in Fig. 15-8. The better of these two animals responds fairly well in the low fre-

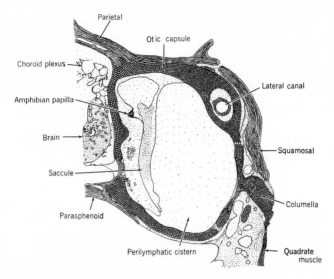

Fig. 15-6. The ear region of *Amphiuma m. means*. Scale 20X.

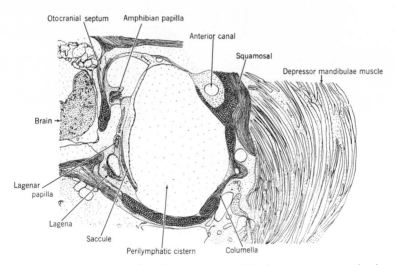

Fig. 15-7. A transverse section through the ear region of *Amphiuma m. means* at a level more posterior than in the preceding figure. Scale 20X.

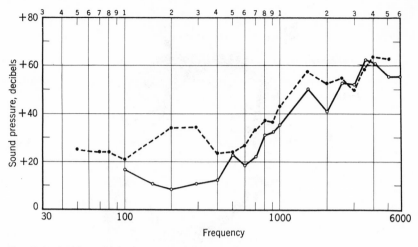

Fig. 15-8. Aerial sensitivity in two specimens of *Amphiuma m. means.* Shown is the sound pressure, in decibels relative to 1 dyne per sq cm, required for an inner ear potential of 0.1 μv.

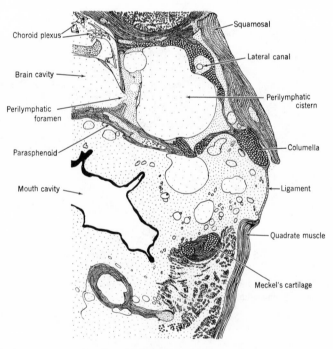

Fig. 15-9. A transverse section through the ear region in *Amphiuma m. tridactylum.* Scale 8X.

quencies from 100-400 Hz and shows progressively poorer responses for higher tones.

Amphiuma means tridactylum.   This second subspecies of *Amphiuma means* overlaps the area of the first in Alabama and extends westward to the edge of Texas and northward through Mississippi and Arkansas almost to the southern tip of Illinois. This form runs somewhat larger than the other subspecies, with differences also in coloration and the presence of three toes on each foot. The general structure in the ear region is closely similar, as may be seen in Fig. 15-9. Both amphibian and lagenar papillae are present, and these appear to be well developed, as Table 15-I indicates. The numbers of hair cells present in these two specimens exceed those observed in the first subspecies by amounts that seem significant.

Sensitivity to aerial sounds as indicated for two specimens in Fig. 15-10 is generally comparable to that obtained for the other subspecies: best sensitivity is shown for the lower frequencies (though with considerable variability), and there is a rapid, progressive decline in sensitivity for the higher tones. Though poor in response to aerial sounds, these ears can be expected to be fully serviceable for the reception of low-frequency vibrations in the water.

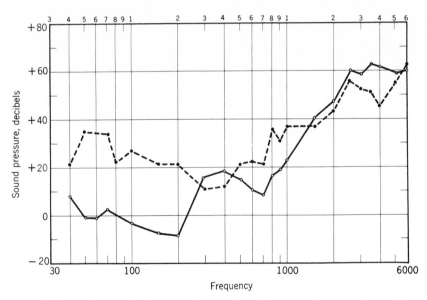

Fig. 15-10. Aerial sensitivity in two specimens of *Amphiuma m. tridactylum*. Shown is the sound pressure, in db relative to 1 dyne per sq cm, required for an inner ear potential of 0.1 μv.

16. THE AMBYSTOMATIDAE

The family Ambystomatidae contains three genera, *Ambystoma*, *Dicamptodon*, and *Rhyacotriton*, the first with 17 species and subspecies and the others with a single species each (Bishop, 1974). All occur in North America. In this group is found the most complete form of sound-receptive mechanism along the salamanders, with important developments in the course of transformation from larval to adult stages.

In the larva a columella is present, bearing an extended process, the stylus, that makes a connection with the squamosal, or sometimes with the palatoquadrate, or with both these elements. Also the base of the columella (the fenestral plate) connects anteriorly with the otic capsule. In the process of transformation the columella becomes more completely fused with the otic capsule, and posteriorly a portion of this capsule separates to form a second ossicular element, the operculum, that is free in the oval window but is connected to a muscle (the opercular muscle) running posteriorly to the suprascapula.

Ambystoma maculatum.   A common form is *Ambystoma maculatum*, the spotted salmander, portrayed in Fig. 16-1. This is a medium large species distributed over most of the eastern half of the United States and extending a short distance into Canada. It is terrestrial in habit except in the breeding season, living well hidden under logs and litter in deciduous forest areas that contain pools and sluggish streams suitable as breeding places. In the spring the mature animals migrate to the breeding sites, often in considerable numbers, enter the water, and engage in a ritual of interaction in which sperm packets are released by the males and picked up by the females, after which the fertilized eggs are deposited in small masses attached to underwater vegetation or other fixed objects. The larvae hatch out and live on a variety of animal food to be found in the water, including worms of various kinds and the larvae of insects and crustaceans. By autumn these animals have developed to the stage at which they transform and come out on land; and about this time they acquire the rows of yellow or orange spots along the sides of the body that give them their specific name. The body length is about 17.5 cm for adult males and a little less for females.

This species has been extensively studied, especially by Kingsbury and Reed (1909) under the older name of *A. punctatum* and was dealt with ear-

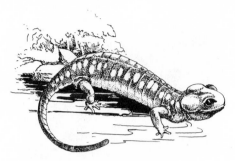

Fig. 16-1. The spotted salamander, *Ambystoma maculatum*, about one-half natural size. Drawing by Anne Cox.

lier in Chapter 11 as a representative of the whole urodele group. The form of the skull was shown in Figs. 11-2 and 11-3, and the general structure of the auditory mechanism was presented in a series of transverse sections (Figs. 11-6 to 11-10). This structure is here shown further in this same plane in Figs. 16-2 to 16-5 and in a frontal section in Fig. 16-6.

Figs. 16-2 to 16-5 present for a particular specimen a series of four sections progressing posteriorly through the ear region and showing the relations to the broad expanse of squamosal and palatoquadrate plates and the overlying muscle and skin tissues. Figures 16-2 and 16-3 indicate the linkage formed by the stylus between the columella and these lateral bony and cartilaginous plates, and Figs. 16-5 and 16-6 represent the appearance of the operculum farther posteriorly.

The sensitivity of the ear in this species as measured in terms of the inner ear potentials in response to aerial sounds is indicated for an adult specimen in Fig. 16-7, where a broad maximum appears in the middle frequencies, with the best region between 400 and 600 Hz.

Results for a larva of this species are presented in Fig. 16-8 for aerial stimulation (solid line) and also for vibratory stimulation (broken line), produced by means of a blunt probe applied to the left side of the face, after exploring the region to determine the most effective spot for this application. Both these forms of stimulation show a fair degree of sensitivity in the middle range, with only moderate peaking at 400 Hz for the aerial form.

The further use of vibratory stimulation in an adult specimen, in which the vibrating prod was applied to the operculum, is indicated in Fig. 16-9. Here the solid line represents the sensitivity function obtained under normal conditions, in which a sharp maximum appears at 400 Hz. The sensitivity falls off as the frequency rises until a secondary peak of moderate size appears at 1500 Hz, after which the response falls with great rapidity. When the contralateral operculum was immobilized by molding over it a lump of

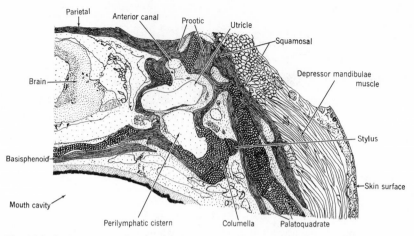

Fig. 16-2. A transverse section through the right side of the head in a specimen of *Ambystoma maculatum* at an anterior position, passing through the utricle. Scale 15X.

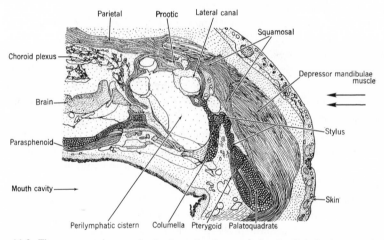

Fig. 16-3. The same specimen as in the foregoing figure at a level a little farther posteriorly. Scale 15X.

modeling clay, the broken curve of this figure was obtained, in which responses to the middle frequencies were greatly impaired. This observation reflects the high degree of interaction between the two ears of salamanders in the response to acoustic stimuli.

The results of hair-cell counts made in 4 specimens of *Ambystoma maculatum*, in 3 of them on both ears, are shown in Table 16-I. The fourth

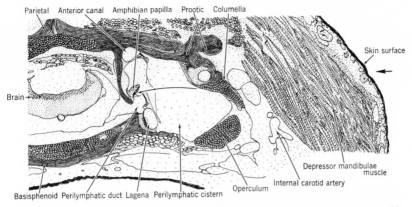

Fig. 16-4. The above specimen sectioned still farther posteriorly, passing through the amphibian papilla. Scale 15X.

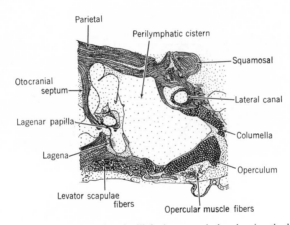

Fig. 16-5. The above specimen sectioned still farther posteriorly, showing the lagenar papilla. Scale 15X.

specimen was a larval form, and it is evident that the hair-cell population reaches a high level early in development.

Ambystoma tigrinum. A second species that has been extensively studied is the tiger salamander, which occurs in four forms considered as subspecies distributed over most of the United States except the west coast areas. The form examined was the eastern subspecies, *A. tigrinum tigrinum*, which is found over most of the area east of the Great Plains except the Appalachian and northeastern regions.

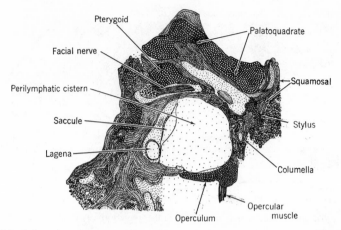

Fig. 16-6. A frontal section in another specimen of *Ambystoma maculatum*, showing the relations between columella and operculum, and the attachment of the opercular muscle. Scale 15X.

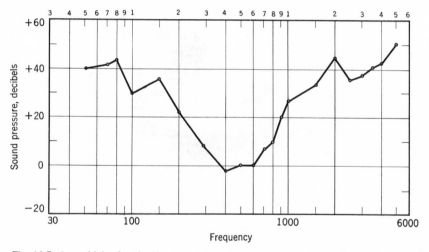

Fig. 16-7. A sensitivity function for a specimen of *Ambystoma maculatum*, for aerial stimulation. Shown is the sound pressure, in decibels relative to a zero level of 1 dyne per sq cm, required for an inner ear potential of 0.1 μv.

The auditory structures for this species are shown in Fig. 16-10 in a dorsolateral view, drawn from a dissected specimen. The skin and lateral muscle mass were removed, and now exposed to view are the prootic bone above, in which the inner ear structures are mainly contained, with the squamosal extending ventrally and mostly covering the palatoquadrate. The quadrate

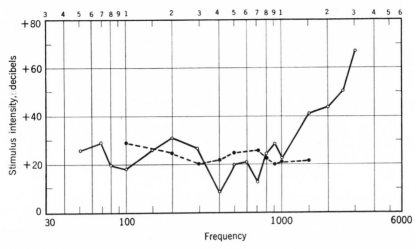

Fig. 16-8. Sensitivity curves for a larva of the species *Ambystoma maculatum* for aerial (solid line) and vibratory (broken line) stimulation. For aerial stimulation the reference is 1 dyne per sq cm, and for vibratory stimulation it is an amplitude of 1 mμ.

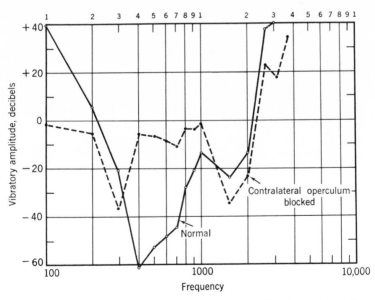

Fig. 16-9. Sensitivity for a specimen of *Ambystoma maculatum* with vibratory stimulation under normal conditions (solid line) and with the contralateral operculum immobilized (broken line). Shown is the amplitude in db relative to 1 mμ required for a response of 0.1 μv.

Table 16-I

Hair-Cell Populations in the Ambystomatidae

In the Amphibian Papilla		In the Lagenar Papilla	
Left	Right	Left	Right
#1 *Ambystoma maculatum*			
168	140	49	42
#2 *Ambystoma maculatum*			
202	164	58	65
#3 *Ambystoma maculatum*			
135	172	66	73
#4 *Ambystoma maculatum* (larva)			
152	†	62	†
#1 *Ambystoma tigrinum*			
249	262	60	47
#2 *Ambystoma tigrinum*			
228	253	†	†
#3 *Ambystoma tigrinum*			
284	375	96	82
#4 *Ambystoma tigrinum*			
275	†	80	†
#1 *Ambystoma texanum*			
218	215	34	37
#2 *Ambystoma texanum*			
187	196	32	32
#3 *Ambystoma texanum*			
167	157	30	25
#1 *Dicamptodon ensatus*			
236	231	70	85
#2 *Dicamptodon ensatus*			
180	211	63	89

† Not determined.

portion of this bone forms the articulation with the lower jaw, which has been separated here by displacing the jaw mechanism ventralward (as indicated by the heavy arrow). The operculum is shown at the posterior end of the ear region, to which the opercular muscle is attached. The posterior end

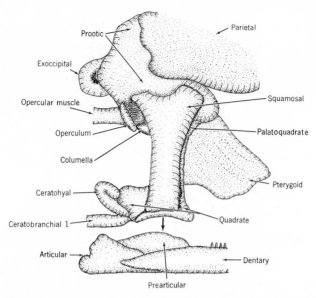

Fig. 16-10. The auditory structures viewed dorsolaterally in a dissected specimen of *Ambystoma tigrinum*. Scale 6X.

of the columella appears from behind the palatoquadrate and makes contact with the edge of the operculum.

A frontal section through this region that cuts across the otic capsule is shown in Fig. 16-11. The columella is fused anteriorly with the otic capsule, but is separated from the operculum, which forms the posterior boundary of this capsule and is also fused to it medially. The palatoquadrate is closely attached to the squamosal at its posterolateral end and at its other end lies in close relation to the anterior wall of the capsule.

As shown in the frontal section of Fig. 16-12 from the same specimen, the opercular muscle extends from an attachment to the lateral surface of the suprascapula and runs to the operculum at its anterior end. Also seen is the quadrate muscle, which runs beneath the skin (from the attachment posterior and ventral to the level of this picture). This muscle at another level makes an attachment to the columella.

In a specimen of this salamander an aerial stimulus applied to the left side of the head produced the solid-lined curve of Fig. 16-13 when responses were recorded from an electrode in the left lateral semicircular canal, and gave the broken curve with the recording from the right canal. Both functions show best sensitivity in the region from 200 to 700 Hz, but for the most part with small or moderate differences in favor of the ipsilateral recording site. Similar results for another specimen are given in Fig. 16-14; here again the stim-

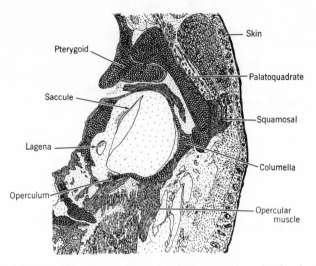

Pterygoid

Saccule

Lagena

Operculum

Skin

Palatoquadrate

Squamosal

Columella

Opercular
muscle

Fig. 16-11. A frontal section of a specimen of *Ambystoma tigrinum*, showing the ear region. Scale 10X.

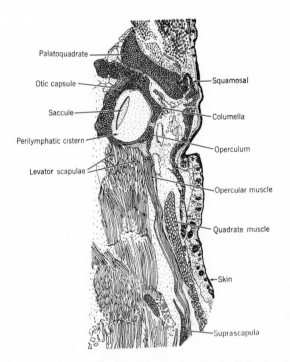

Palatoquadrate

Otic capsule

Saccule

Perilymphatic cistern

Levator scapulae

Squamosal

Columella

Operculum

Opercular muscle

Quadrate muscle

Skin

Suprascapula

Fig. 16-12. The same specimen as the foregoing in a broader view, showing the extent of the opercular muscle. Scale 8X.

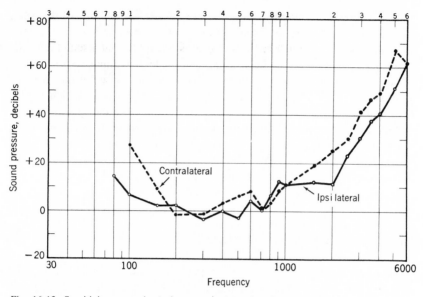

Fig. 16-13. Sensitivity curves in *Ambystoma tigrinum* for stimulation with aerial sounds presented to one ear, with both ipsilateral and contralateral recording.

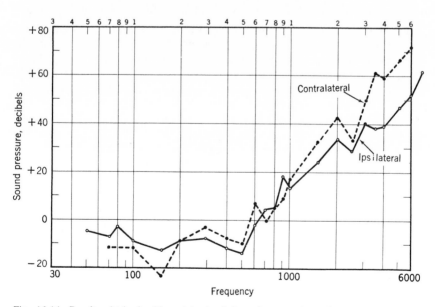

Fig. 16-14. Results obtained with aerial stimulation of a second specimen of *Ambystoma tigrinum* with ipsilateral and contralateral recording.

ulus was an aerial sound applied through a tube to the left side of the head, and the responses were recorded from both ears. These results show clearly what the anatomical relations would lead us to expect: that a sound reaching one ear passes through the head to the other side and stimulates the contralateral ear as well. The differences shown in the responses of the two ears indicate that there is a loss in transmission across the head, so that for the most part the ipsilateral ear is favored, though only to a moderate extent.

Similar results were obtained in another specimen of *A. tigrinum* by the use of vibratory stimulation applied to the right columella, with recording from electrodes on both sides, one in the right saccule and the other in the left saccule. These response curves are given in Fig. 16-15 and again show similar results from the two ears, with the ipsilateral responses generally favored.

Hair-cell populations were counted for one or both papillae in four specimens, with results shown in Table 16-I.

Ambystoma texanum.   The small-mouthed salamander *Ambystoma texanum* has a range over the mideastern and midsouthern states, defined by a rough quadrilateral running from Ohio westward to the edge of Kansas, then south to eastern Texas, next east through Louisiana, and finally northeast to Ohio once more. It is of medium size, around 12 cm in length, of generally a dusky color but variable from brown to black. Figure 16-16 gives a lateral view of the ear showing the enclosing structures consisting of the squamosal along the anterior side, the quadrate below and the prootic above, and the operculum forming the posterior boundary. The opercular muscle is large, extending posteriorly to attach to the suprascapula. The columella is seen wedged between the opercular plate and the squamosal. Figure 16-17 gives

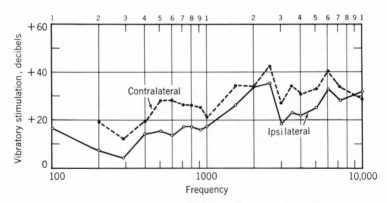

Fig. 16-15. Ipsilateral and contralateral functions for a third specimen of *Ambystoma tigrinum* in which the stimulation was with a mechanical vibrator applied to the right operculum.

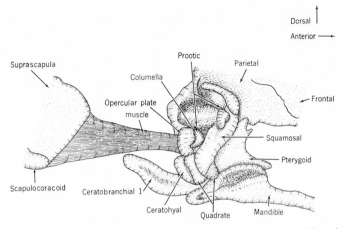

Fig. 16-16. A lateral view of the ear structures of *Ambystoma texanum*, from a dissection. Scale 8X.

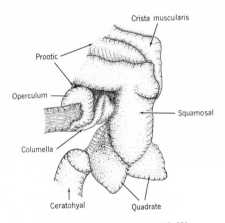

Fig. 16-17. A further view of the above specimen. Scale 12.5X.

an oblique view from the dorsolateral aspect showing more clearly the operculum and the anterior end of the opercular muscle, with the columella lying between operculum and squamosal.

The sensitivity in response to aerial sounds is shown for three specimens of *Ambystoma texanum* in Figs. 16-18 and 16-19. These curves indicate good sensitivity in the low tones between 100 and 800 Hz, with the maximum varying from 200 to 400–500 Hz in the three ears. The decline in sensitivity is rapid for the high tones, with a rate approaching 25 db per octave.

In one specimen the number of hair cells in the amphibian papilla was 218

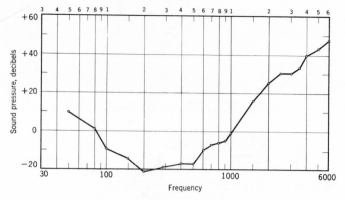

Fig. 16-18. Sensitivity to aerial sounds in a specimen of *Ambystoma texanum*. Shown is the sound pressure, in db relative to 1 dyne per sq cm, required for an inner ear potential of 0.1 μv.

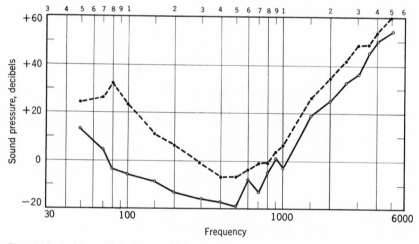

Fig. 16-19. Aerial sensitivity in two additional specimens of *Ambystoma texanum*, expressed as in the foregoing.

on the left side and 215 on the right, and the corresponding numbers for the lagenar papilla were 34 and 37 respectively. These results and similar observations on two other specimens are shown in Table 16-I.

Ambystoma mexicanum. This species occurring in mountainous areas of Mexico was given the common name of axolotl (which comes from an Aztec word meaning a water monster) and at first was assigned to a distinct genus of *Siredon*, but is now known to be an ambystomid exhibiting an unusual degree of neoteny. Its development and the amount of transformation from larval to adult stages are dependent on environmental conditions and the food supply, especially the amount of iodine present. It has been found possible to control the development in some degree by altering the chemical intake.

The specimens used in the present experiments show a moderate degree of neoteny: many larval features are present but there are adult characters also that appear in the usual transformation process. The specimen whose ear structures are represented in Fig. 16-20 in a frontal section exhibits a relatively free columella, attached medially to the otic capsule by short ligamentary strands and attached anteriorly to the squamosal by a heavy ligament (the squamoso-columellar ligament); this ligamentary attachment appears to be a larval character. Posteriorly, attached loosely to the columella at one end and to the rather tenuous prootic bone at the other end, is a cartilage that probably is to be identified as the operculum. This is an element that is absent in larvae and ordinarily appears only in the transformation process; its presence indicates that this process has set in but is incomplete, for there is no sign of an opercular muscle. Another specimen, also with a well-developed columella, is shown in a transverse section in Fig. 16-21.

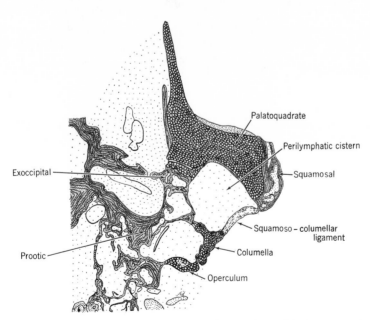

Fig. 16-20. A frontal section through the ear region in a specimen of *Ambystoma mexicanum*. Scale 9X.

Sensitivity curves are given for two specimens of *A. mexicanum* in Fig. 16-22. These functions indicate only relatively poor sensitivity for the low tones and progressively poorer sensitivity as the frequency is increased.

OTHER SPECIES OF *Ambystoma*

The species *Ambystoma cingulatum, gracile, mavortium,* and *opacum* were examined more briefly. In general their auditory structures resemble the ones

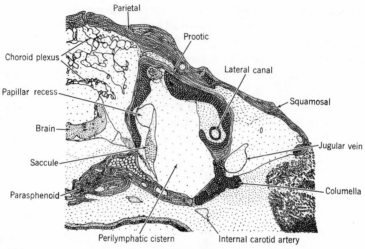

Fig. 16-21. A transverse section through the right side of the head in a specimen of *Ambystoma mexicanum*. Scale 9X.

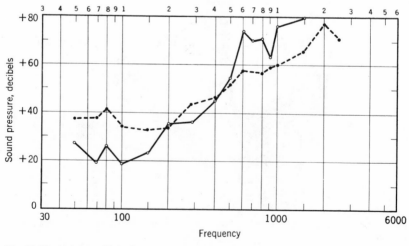

Fig. 16-22. Aerial sensitivity in two specimens of *Ambystoma mexicanum*, expressed as above.

already described, but there are variations in their courses of bodily development and in the degrees of neoteny shown. Figure 16-23 presents sensitivity functions for two specimens of *A. cingulatum*, found mainly in coastal areas from southeast Alabama to the southern edge of Mississippi, in cypress woods near ponds. Figure 16-24 shows similar results on *Ambystoma opacum*, which inhabits a wide range in the eastern United States from Maine to Florida and west to the edges of Oklahoma and Texas. Both species are most sensitive in the medium low frequencies, with *A. cingulatum* showing

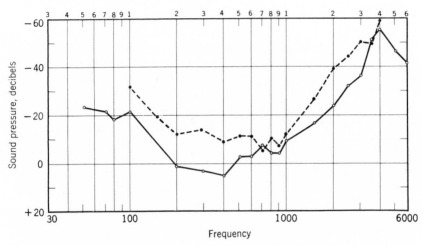

Fig. 16-23. Aerial sensitivity functions for two specimens of *Ambystoma cingulatum*.

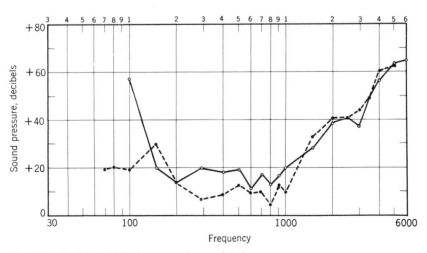

Fig. 16-24. Aerial sensitivity in two specimens of *Ambystoma opacum*.

a curve centered around 400 Hz and *A. opacum* with one extending more broadly around 300-1000 Hz.

Dicamptodon ensatus. The ambystomatid genus *Dicamptodon* contains the single species *ensatus*, a large form attaining lengths up to 30 cm, occurring along the Pacific coastline from the edge of British Columbia to about the middle of California and in a few isolated spots in the mountains of California, Idaho, and Montana.

A lateral view of the ear region, drawn from a dissected specimen, is given in Fig. 16-25. As will be seen, the bone structures have a massive character. A cross section showing the relations of the peripheral tissues, especially the palatoquadrate, to the otic regions and the amphibian papilla is presented in Fig. 16-26. The palatoquadrate is particularly thick and connects through a stylus with the columella, which occupies the dorsolateral portion of the oval window. Further occupying this window ventrally is the operculum, recognizable as a separate element though fused with the columella to form a continuous wall from prootic to parasphenoid. Both amphibian and lagenar papillae are present at this level, and from these two receptor organs a foramen leads into the brain cavity.

The size of the hair-cell population was determined in two specimens, with results shown in Table 16-I. These numbers for the amphibian papilla fall within the range of other ambystomids, but for the lagenar papilla they are

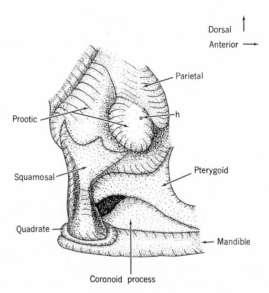

Fig. 16-25. The ear structures of *Dicamptodon ensatus*, viewed laterally in a dissected specimen. Scale 5X.

relatively large, exceeded by only a few among the various species of *Ambystoma* examined.

The reception of aerial sounds is poor, as shown for two specimens in Fig. 16-27. The low tones, over a range from 100 to 600 Hz, seem somewhat better received than the others, but at best these ears require strong stimuli to produce the standard response.

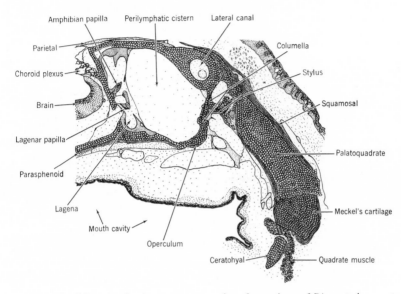

Fig. 16-26. The right ear region in a transverse section of a specimen of *Dicamptodon ensatus*. Scale 12.5X.

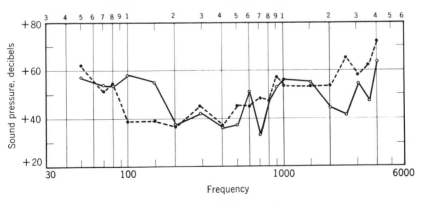

Fig. 16-27. Aerial sensitivity in two specimens of *Dicamptodon ensatus*.

Tests made by bone conduction, with a vibratory stimulus applied to the surface of the prootic bone (broken line) or to the squamosal (solid line), gave the results shown in Fig. 16-28. It is evident that stimulation of the prootic is the more favorable mode in the low-tone region of 290 to 500 Hz.

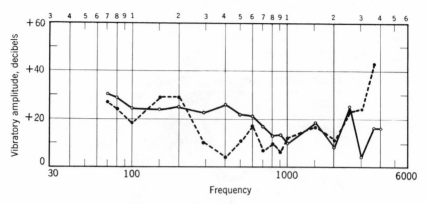

Fig. 16-28. Sensitivity to vibratory stimulation in *Dicamptodon ensatus*, with stimulation applied to the squamosal bone (solid line) or to the prootic (broken line). Shown is the vibratory amplitude, expressed in db relative to 1 millimicron, required for a response of 0.1 μv.

17. THE PLETHODONTIDAE:

THE LUNGLESS SALAMANDERS

The Family Plethodontidae is not only the largest among the caudates, containing 23 genera with about 175 species, but also exhibits the greatest diversity, its members occupying a wide variety of habitats ranging from brooks and streams to forest areas and rocky regions containing water sources, and including even holes in trees and basins of bromeliads. Members of this family have gone farthest of all the salamanders in reducing the dependence on an aquatic environment, and a few have achieved complete terrestriality, laying their eggs on land in protected places such as chinks in logs or in caves where they hatch out as well-formed individuals without having gone through an aquatic larval stage. Others take an intermediate course and lay their eggs on land in the vicinity of water sources, into which the hatchlings find their way and continue their development as aquatic larvae.

The seven species in the genus *Desmognathus* have sometimes been considered as making up a separate family, but the close relationships with other members of the group favors the arrangement used by Wake (1964) in which two subfamilies are recognized, Desmognathinae (including the *Desmognathus* species and two other genera, *Laurognathus* and *Phaeognathus*, along with a genus now extinct) and the subfamily Plethodontinae (including all the others).

The most striking characteristic of members of this family is the absence of lungs, though lungs are lacking occasionally in species of other families as well.

A second characteristic pointed out early by Kingsbury and Reed (1909) is the form of the ear structures: the presence in the oval window of a bony plate from which a rod extends anterolaterally to the suspensorium (usually to the squamosal, and often to quadrate or palatoquadrate as well), and an attachment to the ceratohyal by means of a ligament.

The bony plate has been regarded variously as a modified columellar footplate or as an operculum; Kingsbury and Reed considered this plate as combining the functions of both these structures, and most others have followed this lead. The extended rod is regarded by Kingsbury and Reed as representing the columella, or specifically its stylus, and this view has been generally accepted.

Whereas in salamanders in general an opercular muscle going from the operculum to an attachment on the suprascapula is considered to be derived as a slip of the levator scapulae muscle, it was pointed out by Dunn (1941) that in plethodontids the muscle in this position has a different origin: it is a derivative of the cucullaris muscle. Moreover, Monath (1965) found considerable variability in the nature and form of this muscle in plethodontids. In most of these a slip of the cucullaris muscle runs to the operculum as just indicated, but in some species (as in *Ensatina*) the cucullaris muscle as a whole is reduced or even vestigial (as it is in *Batrachoseps*). A still more aberrant condition is found in the species *Pseudotriton*, with variations among particular specimens. Sometimes the levator scapulae sends fibers to the fenestral plate just as in most other salamander species, and there is in addition a slip from the cucullaris major going to an adjoining area of the fenestral plate (Monath, 1965).

Attention will now be given to the auditory structure and its performance in response to sounds in those twelve species of plethodontids available for study. This treatment will involve four groups, beginning with the Desmognathinae that are considered the more primitive, and continuing through the tribe Hemidactyliini (3 species) that are a little further advanced, then the tribe Plethodontini (6 species) that are still more advanced, and finally to the tribe Bolitoglossini (1 species) considered the most specialized of all.

Desmognathus species. *Desmognathus f. fuscus*, the Northern Dusky salamander, is a medium-sized species of regular body form with the hind legs noticeably larger than the forelegs and a sturdy, tapering tail. This species occurs over most of the region from Maine to Alabama, except for the southeastern seaboard area where the southern subspecies holds sway, and extends through the Appalachian region as far as Ohio, Kentucky, and Tennessee. It inhabits the margins of streams and other moist areas and enters the water only rarely. The eggs are laid in clusters under stones or in vegetable litter in damp places, usually beside streams or springs.

The ear region as viewed from the right and somewhat posteriorly is represented in Fig. 17-1. The opercular muscle is shown running to the columella that is partially hidden by the squamosal. The opercular region is more fully revealed in Fig. 17-2 which indicates the footplate in its window and also shows the stylus going to an attachment on the squamosal. Figure 17-3 presents a cross section through the ear region showing the otic capsule containing the perilymphatic cistern, with the saccule forward and the amphibian papilla located on the otocranial septum.

The frontal section of Fig. 17-4 shows the fenestral plate occupying the posterolateral wall of the otic capsule and running anteriorly to the immediate vicinity of the squamosal, which in turn attaches to the palatoquadrate. No direct contact with the squamosal was seen, but ligamentary fibers made a connection with this bone.

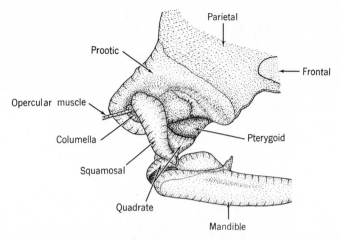

Fig. 17-1. The ear structures in *Desmognathus f. fuscus*. Scale 12.5X.

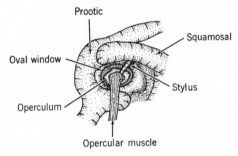

Fig. 17-2. Detail of the opercular region in *Desmognathus f. fuscus*. Scale 25X.

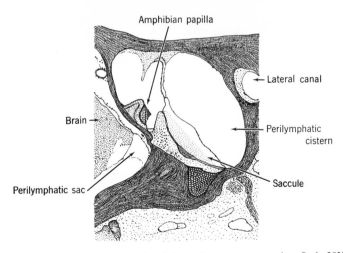

Fig. 17-3. The ear region of *Desmognathus f. fuscus* in a transverse section. Scale 30X.

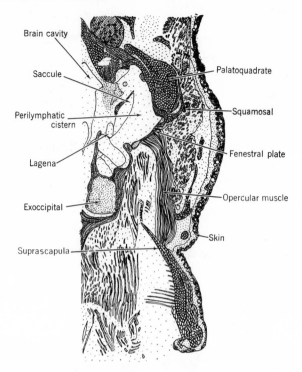

Fig. 17-4. Auditory structures of *Desmognathus fuscus* in a frontal section. Scale 20X.

The number of hair cells in the amphibian papilla was determined in three specimens of *Desmognathus f. fuscus*, with the results shown in Table 17-I. The mean of these six determinations is 106.7, a moderate population size. This ear, like that of all plethodontids, lacks a lagenar papilla.

The sensitivity curve of Fig. 17-5 indicates best responses to sounds in the region of 700 to 2000 Hz, with a peak at 800 Hz.

Desmognathus quadramaculatus, the Black-bellied Salamander. This is a relatively large species occurring in a limited region running through the mountains from West Virginia to Georgia, touching on three other neighboring states. It lives along rocky streams at moderately high elevations.

The ear structure of this species resembles that already described for the preceding species, with a fenestral plate mainly composed of bone occupying the broad posterolateral wall of the otic capsule and sending a short stylus laterally to make a ligamentary connection with both squamosal and palatoquadrate.

The numbers of hair cells in the amphibian papilla were determined in

Table 17-I

Hair-Cell Populations in the Plethodontidae *

In the Amphibian Papilla		In the Amphibian Papilla	
Left	*Right*	*Left*	*Right*
#1 *Desmognathus f. fuscus*		#3 *Plethodon c. cinereus*	
108	130	138	147
#2 *Desmognathus f. fuscus*		#4 *Plethodon c. cinereus*	
106	103	114	100
#3 *Desmognathus f. fuscus*		#1 *Ensatina e. eschscholtzii*	
97	96	159	151
#1 *Plethodon c. cinereus*		#2 *Ensatina e. eschscholtzii*	
127	114	139	115
#2 *Plethodon c. cinereus*		#3 *Ensatina e. eschscholtzii*	
119	134	177	157

* Family lacks the lagenar papilla.

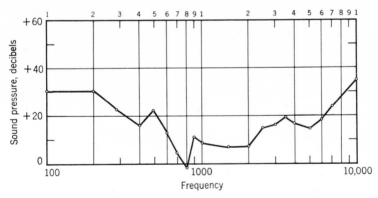

Fig. 17-5. An aerial sensitivity function in *Desmognathus fuscus.* Shown is the sound pressure, in db relative to 1 dyne per sq cm, required for a response of 0.1 μv.

two specimens; these were found to be 151 and 142 in right and left ears in one specimen and 145 and 139 correspondingly in the other, giving an average number of 144.2. These numbers are considerably larger than those observed for *Desmognathus f. fuscus.*

Auditory sensitivity for aerial sounds was measured for four specimens, shown in Figs. 17-6 and 17-7. The first of these figures indicates a fairly

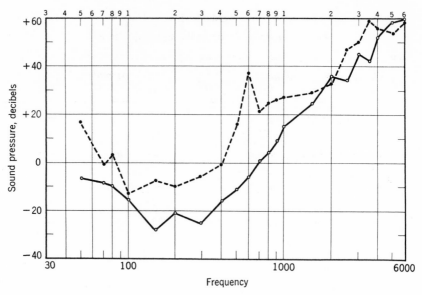

Fig. 17-6. Aerial sensitivity curves for two specimens of *Desmognathus quadramaculatus*, represented as in the foregoing figure.

good level of sensitivity in the low frequencies from 100 to 400 Hz, notably for one of these animals, and then a rapid decline for lower and especially for higher tones. The second animal in this same graph is less sensitive, but still shows a creditable level in the 100–400 Hz region. The two animals shown in Fig. 17-7 present much the same form of functions, but are in general somewhat less sensitive.

Tribe Hemidactyliini

Pseudotriton r. ruber, the Northern Red Salamander. This moderately large salamander occupies various regions from cold springs to streams and swamps in a wide area from Pennsylvania southwestward to the northern portions of Alabama and Georgia, avoiding the coastal plains except at the northern end of this range. Other subspecies occur south of this area and also displace this one in portions of the Blue Ridge mountain region. The eggs are laid in the water of springs and streams, usually attached to submerged stones, and in these waters the larvae develop and finally transform to the adult stage.

The ear structure is shown at one level in the transverse section of Fig. 17-8 and then at a more posterior level in Fig. 17-9. The first of these views shows the broad fenestral plate and the stylus going laterally to its connection to the squamosal. The next view shows the ligamentary connection be-

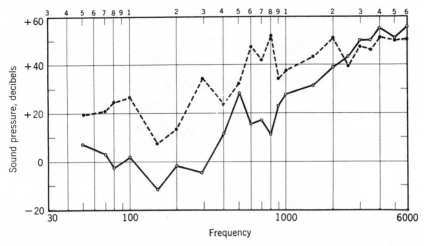

Fig. 17-7. Aerial sensitivity curves for two additional specimens of *Desmognathus quadrama-culatus* represented as in the foregoing figures.

tween this plate and the ceratohyal, and the quadrate muscle coming from the throat region to its attachment to the ceratohyal. It is clear that contractions of this muscle will produce tensions that are transmitted through the ceratohyal and continue through the connecting ligament to the fenestral plate, thus serving to immobilize this element and protect the ear against excessive sounds.

There is also an opercular muscle, consisting of an otic division of the cucullaris major, as represented in Fig. 17-10. A further view of this muscle and its attachment to the operculum, and also the extension of the operculum as the columella, appear in the next figure (Fig. 17-11).

Counts of hair cells in the amphibian papilla were made in two specimens giving numbers of 135 and 134 in left and right ears in one of these, and 128 and 115 in the two ears of the other specimen.

Sensitivity curves for two specimens are presented in Fig. 17-12. These functions agree in showing fairly good sensitivity in the low frequencies, with the maximum around 200 to 400 Hz.

Eurycea b. bislineata, the Northern Two-lined Salamander. This small, slender salamander occurs over a large area of the United States and extends northward to some parts of Canada. Its westward range is along the edge of Illinois and its southern extent runs roughly along the borders of North Carolina and Tennessee. It lives mainly along the banks of small streams, often entering the water.

A frontal section through the ear region is presented in Fig. 17-13 show-

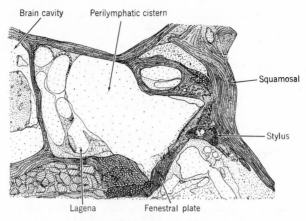

Fig. 17-8. A transverse section through the ear region of *Pseudotriton ruber*. Scale 25X.

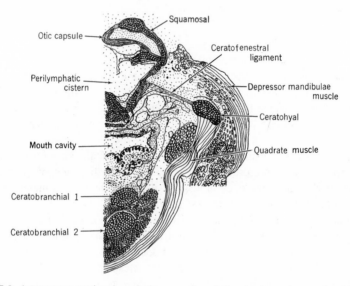

Fig. 17-9. A transverse section through the ear region of *Pseudotriton ruber* at a level farther posteriorly than the preceding. Scale 15X.

ing the otic capsule with the relatively thin fenestral plate filling the oval window, the rather heavy stylus extending anterolaterally, and the large opercular muscle running posteriorly to its attachment on the suprascapula. The stylus makes a firm connection with the squamosal.

The size of the hair-cell population was determined in two specimens, one

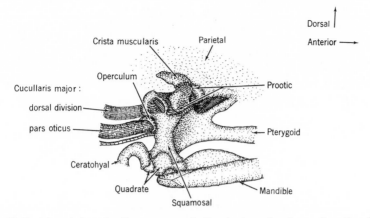

Fig. 17-10. A lateral view of auditory structures in *Pseudotriton ruber*, from a dissection. Scale 8X.

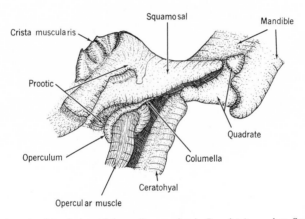

Fig. 17-11. A posterolateral view of the auditory region in *Pseudotriton ruber*. Scale 12X.

of which showed 124 and 115 hair cells in left and right ears, and the other corresponding numbers of 138 and 133 hair cells.

Eurycea longicauda guttolineata, the Three-lined Salamander. This salamander occurs in an arc extending from Virginia along the coast south and westward as far as Mississippi and a small portion of Louisiana, passing through a part of the Florida panhandle but missing the remainder of this state.

Only limited study was made of this species, as it closely resembles *Eu-*

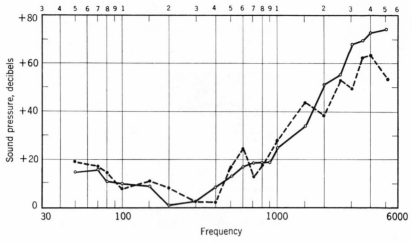

Fig. 17-12. Sensitivity functions in two specimens of *Pseudotriton ruber*. Shown is the sound pressure, in db relative to 1 dyne per sq cm, required for an inner ear potential of 0.1 μv.

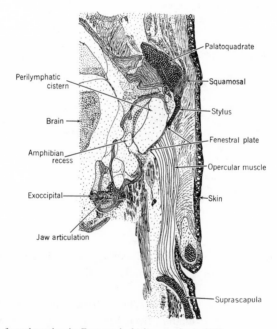

Fig. 17-13. A frontal section in *Eurycea b. bislineata*. Scale 25X.

rycea bislineata in general structure. Measurements of the hair-cell population were made in two specimens and gave the numbers of 163 and 146 hair cells in left and right ears of one, and the numbers 133 and 121 in the other.

Sensitivity curves for these two specimens are given in Fig. 17-14. These curves show a fair level of sensitivity, with the maximum in the low tones around 400 to 700 Hz.

TRIBE PLETHODONTINI

Plethodon c. cinereus, the Red-backed Salamander. This salamander occurs in the northeastern part of the United States and extends into Canada as far north as South Labrador; in the west its range reaches Wisconsin and in the south the borders of South Carolina and Georgia. This species is largely terrestrial and often is found in woods and forests remote from water sources. The eggs are deposited in chinks in logs or cavities formed by stones, a few in each of several places, and it is said that the female guards them.

As Fig. 17-15 shows, the ear structure follows the usual plethodontid form, with an osseous plate covering the posterolateral portion of the otic capsule and extended anterolaterally as a relatively thick stylus consisting of cartilage with a thin osseous sheath and with its end fused to the squamosal. The opercular muscle covers the posterolateral face of the bony plate and can be traced to an attachment on the suprascapula.

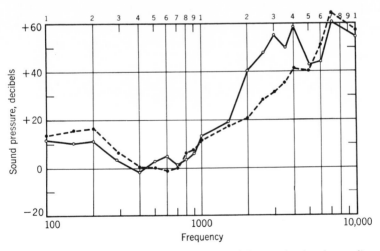

Fig. 17-14. Aerial sensitivity curves in two specimens of *Eurycea longicauda guttolineatus.* Shown is the sound pressure, in db relative to 1 dyne per sq cm, required for an inner ear potential of 0.1 μv.

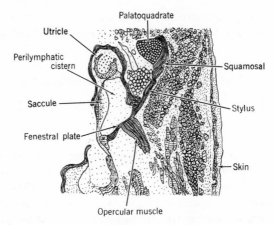

Fig. 17-15. A frontal sextion through the ear region of *Plethodon c. cinereus*. Scale 20X.

Counts of hair cells in the amphibian papilla were made in four specimens as indicated in Table 17-I; the mean number in this papilla was 124.

Sensitivity curves in response to aerial sounds are given for two of these specimens in Fig. 17-16. The sensitivity is somewhat poor for these ears, with the better responses in the low tones up to 2000 Hz, and in one specimen noticeably improved responses around 290 and 2000 Hz.

Plethodon glutinosus, the Slimy Salamander. The species *Plethodon glutinosus* occurs over a broad area of the eastern United States from parts of New York south to half of Florida, then west to the edge of Texas and north to parts of Missouri, Indiana, and Ohio. It lives in woodlands and ravines where there is at least a moderate amount of moisture.

Counts of hair cells in the amphibian papilla were made in two specimens and showed 175 and 168 hair cells in right and left ears for one, and 131 and 136 hair cells for the other.

A sensitivity curve for one specimen is shown in Fig. 17-17. This function shows a sharp maximum at 800 Hz, where a fair degree of sensitivity is indicated.

Plethodon jordani shermani. A third *Plethodon* species is found in woodland areas in the Nantahala mountains of North Carolina. Hair-cell counts in one specimen showed 229 hair cells in the left papilla and 195 in the right one.

Sensitivity curves for two specimens of this subspecies are given in Fig. 17-18. These functions show somewhat poor sensitivity, with the best region around 400–1000 Hz, where one curve has a peak at 700 Hz.

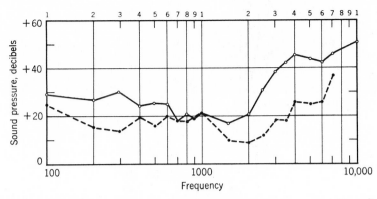

Fig. 17-16. Aerial sensitivity curves in two specimens of *Plethodon c. cinereus*. Shown is the sound pressure, in db relative to 1 dyne per sq cm, required for an inner ear potential of 0.1 μv.

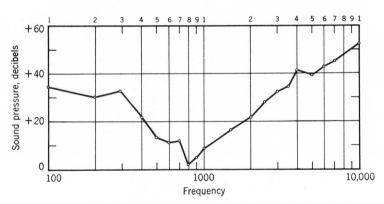

Fig. 17-17. Auditory sensitivity in *Plethodon glutinosus*, represented as in the preceding figure.

Ensatina e. eschscholtzii, the Red Salamander. This species occurs in forest areas along the Pacific coastline from Washington to California usually in moist areas concealed under forest debris. The eggs are laid in cavities under logs or in other protected places, and the female is often reported to be present in their vicinity.

Figure 17-19 shows the ear region in a frontal section, where the stylus appears as an extension of the fenestral plate and makes a close attachment to the palatoquadrate. The squamosal lies lateral to the palatoquadrate, but does not have a direct connection with the stylus. Numbers of hair cells in the amphibian papilla of three specimens are shown in Table 17-I.

A sensitivity curve for one specimen of *Ensatina e. eschscholtzii* is shown

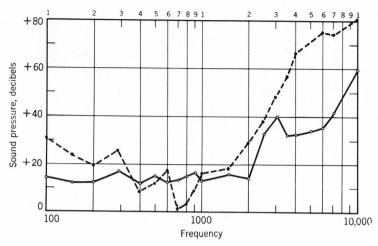

Fig. 17-18. Aerial sensitivity curves in two specimens of *Plethodon jordani shermani*, represented as in the preceding figure.

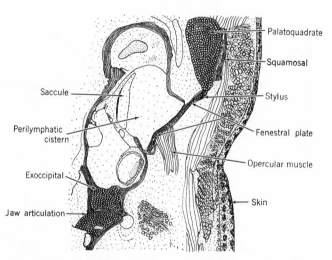

Fig. 17-19. A frontal section through the ear region in *Ensatina e. eschscholtzii*. Shown is the sound pressure, in db relative to 1 dyne per sq cm, required for an inner ear potential of 0.1 μv.

in Fig. 17-20. The sensitivity is fairly good in a range from 100 to 800 Hz and falls off rapidly for lower and higher tones.

Aneides lugubris, the Arboreal Salamander. These salamanders occur along most of the coastal region of California, often occupying cavities in trees,

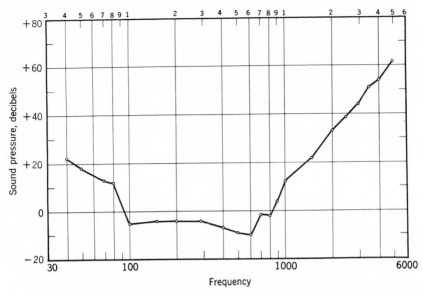

Fig. 17-20. A sensitivity function in *Ensatina e. eschscholtzii*. Shown is the sound pressure, in db relative to 1 dyne per sq cm, required for an inner ear potential of 0.1 μv.

sometimes at considerable heights. They are sometimes found on the ground also, hidden under logs, bark, and other debris. A striking feature is the large, protruding eyes, as shown in the sketch of Fig. 17-21. The eggs are usually laid in crevices in trees, though ground cavities also are employed.

A view of the ear structures seen from the right side is given in Fig. 17-22. The posterodorsal view in Fig. 17-23 shows the operculum with its muscle, and the connection of the columella (or stylus) on one side. The relations to the fenestral plate are not indicated here, but are similar to those for the second species of *Aneides* next to be represented.

The numbers of hair cells were determined in one specimen as 174 for the left ear and 172 for the right ear.

Sensitivity for two specimens as represented in Fig. 17-24 appears to be poor all along the low-frequency range up to 900 Hz, and is further reduced for higher tones.

Aneides flavipunctatus, the Shasta Salamander. This species occurs along the coastal region of the northern half of California, living mainly in burnt-over areas of redwood forests, usually adjacent to a spring or stream. It is listed among the climbing salamanders, but spends most of its time on the ground, often hidden under rocks or forest debris.

A frontal section through the ear region is pictured in Fig. 17-25 which shows the fenestral plate and its attachment through a heavy stylus to the

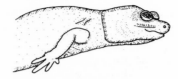

Fig. 17-21. A lateral view of the fore part of the body of *Aneides lugubris*. Natural size.

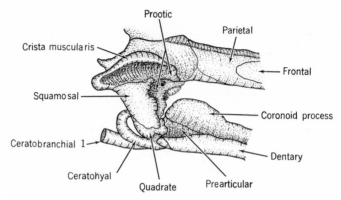

Fig. 17-22. The ear region of *Aneides lugubris*. Scale 4X.

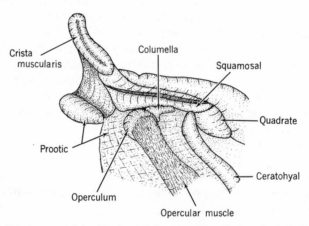

Fig. 17-23. Ear structures of *Aneides lugubris* in a posterolateral view. Scale 7.5X.

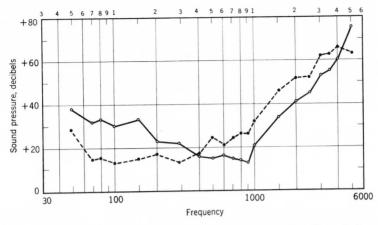

Fig. 17-24. Aerial sensitivity curves in two specimens of *Aneides lugubris*. Shown is the sound pressure, in db relative to 1 dyne per sq cm, required for an inner ear potential of 0.1 μv.

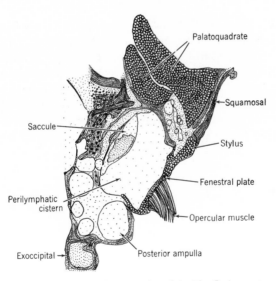

Fig. 17-25. A frontal section through the ear region of *Aneides flavipunctatus*. Scale 25X.

squamosal. The upper end of the opercular muscle is seen also, extending from the fenestral plate. The transverse section of Fig. 17-26 passes through the anterior part of the ear region and shows the amphibian papilla. This view also includes the fenestral plate and its connections with both squamosal and palatoquadrate, and cuts across the stylus.

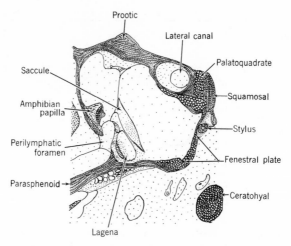

Fig. 17-26. A transverse section through the ear region of *Aneides flavipunctatus*. Scale 25X.

The numbers of hair cells in the amphibian papilla were determined for one specimen as 223 on the left side and 205 on the right.

Sensitivity to aerial sounds was determined for one specimen, and as indicated in Fig. 17-27 it appears to be very poor. Another specimen was tested with vibratory stimuli, applied with a needle on the otic capsule immediately over the saccule. The sensitivity function is indicated in Fig. 17-28. These results indicate a fair degree of sensitivity in the middle tones, those around 700–1000 Hz.

Batrachoseps a. attenuatus, the Worm Salamander. Six species and subspecies of *Batrachoseps* are recognized (Bishop, 1943), with *Batrachoseps a. attenuatus* by far the most widespread; it is said to be the most abundant salamander in California. It occurs in Oregon in the coastal region, continues southward through California as far as Los Angeles County, and also has been reported on Santa Cruz Island. Other subspecies occur to the south and southwest of this range, one extending into Mexico; a separate species is reported in Alaska. These salamanders are distinguished by an extremely slender body, a very long tail that tapers sharply near the end, and the presence of four toes on each of the four feet.

The ear structures are shown in a frontal section in Fig. 17-29, where the relatively thick fenestral plate is extended laterally by the cartilaginous stylus, which is fused to both the squamosal and the palatoquadrate. Seen a little farther ventrally in this same series, in Fig. 17-30, the opercular muscle is attached to this plate, and the stylus is fused to the outer wall of the otic capsule.

The number of hair cells in the amphibian papilla was determined in two

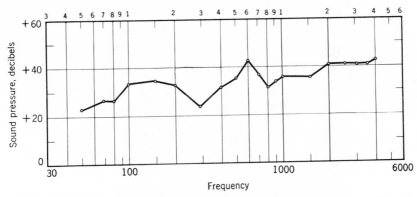

Fig. 17-27. An aerial sensitivity function for a specimen of *Aneides flavipunctatus*. Shown is the sound pressure, in db relative to 1 dyne per sq cm, required for an inner ear potential of 0.1 μv.

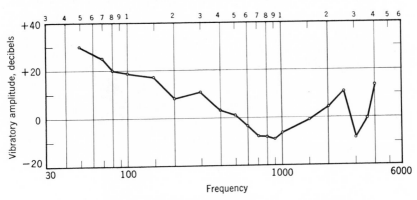

Fig. 17-28. Vibratory sensitivity in a specimen of *Aneides flavipunctatus*. Shown is the vibratory amplitude, in db relative to 1 millimicron, required for an inner ear potential of 0.1 μv.

specimens; this number was 81 in the left ear and 91 in the right ear of one of these and was considerably greater, 142 and 148, in the two ears of the other specimen.

A sensitivity function for a specimen of *Batrachoseps attenuatus* is shown in Fig. 17-31 where a low degree of sensitivity to tones in the middle range around 290 to 900 Hz is indicated.

THE PLETHODONTIDAE AND THE HEARING PROCESS

An important development traceable through the plethodontid family as emphasized by Wake (1964) and others is a reduction in the dependence on water sources and thus an increase in terrestriality. The order of treatment

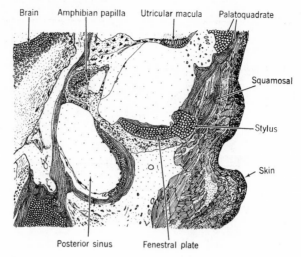

Brain Amphibian papilla Utricular macula Palatoquadrate

Squamosal

Stylus

Skin

Posterior sinus Fenestral plate

Fig. 17-29. A frontal section through the ear region of *Batrachoseps a. attenuatus*. Scale 50X.

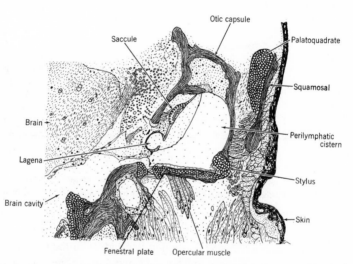

Otic capsule

Saccule Palatoquadrate

Squamosal

Brain Perilymphatic cistern

Lagena

Stylus

Brain cavity

Skin

Fenestral plate Opercular muscle

Fig. 17-30. A frontal section through the ear region of *Batrachoseps a. attenuatus* at a level farther ventrally than in the preceding figure. Scale 50X.

of the several species here examined has been arranged to test whether this terrestrial trend is in any respect reflected in the form and functioning of the ear. Thus this order has been from species that are relatively aquatic in habit and can be considered more primitive in this respect, like the Desmognath-

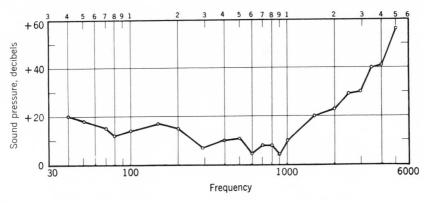

Fig. 17-31. Aerial sensitivity in a specimen of *Batrachoseps a. attenuatus.* Shown is the sound pressure, in db relative to 1 dyne per sq cm, required for an inner ear potential of 0.1 μv.

idae and members of the tribe Hemidactyliini, to those that are chiefly terrestrial and use water sources in feeding, breeding, and other essential activities to only a limited extent or not at all, such as the Plethodontini and the Bolitoglossiini.

It can be said at once that no special trend in the ear structure is evident in the series of species here considered. This structure shows no significant variations; the form of the fenestral plate and its attachments are much the same throughout, and everywhere present is a mechanism that is suited to the reception of aquatic vibrations: a broad skin surface underlaid by large muscle masses that are coupled to bony and cartilaginous plates—the squamosal and palatoquadrate structures—to which are attached a columella (usually simply a stylus) from the fenestral plate. This mechanism altogether provides an avenue for the entrance of sounds and their conveyance to the auditory papilla, with the usual contralateral escape path so that the sounds can flow over the auditory sense cells and stimulate them. The structural arrangement in the bolitoglossines does not differ significantly from the one in the more primitive desmognathines, and the effectiveness of the system seems much the same.

On further consideration this condition is altogether to be expected: an increase in terrestriality does not carry with it any improvement in the mode of reception of aerial sounds; the middle ear mechanism that is presumed to have served this purpose in the ancestral forms (as it still does in most anurans) was lost long ago in the early differentiation of the caudates and cannot now be regained. These animals when in the water—and many of them do enter bodies of water on occasion, some with great frequency—can make effective use of their aquatic ears. When out of water and in the aerial medium these animals must be practically deaf: presumably they perceive only the most intense aerial sounds.

Such social activity that these animals engage in, which no doubt is chiefly the assembly for breeding purposes and choice of a mate, must be carried out by use of other sensory channels, such as smell and contact.

PART IV. THE CAECILIANS

18. THE CAECILIAN EAR

INTRODUCTION

Of the three kinds of amphibians now existing, by far the most poorly known are the caecilians, of the order *Gymnophiona*. They occur in only a few remote places around the world, mainly in tropical and subtropical regions and are exceedingly difficult to collect and study. This difficulty comes from the habit of most species of living in burrows underground, often far below the surface; the remaining few that are not fossorial but aquatic dwell almost as well concealed in bogs, pools, or streams.

Caecilians occur in Asia and especially its adjoining islands, in the Seychelles in the Indian Ocean, in many scattered places in Africa, and in widely distributed ranges in Central and South America. A general requirement is an abundance of moisture, though a few species are adapted to a moderate degree of dryness. These more terrestrial species are ovoviviparous, producing living young that have passed the larval stage and are capable of taking up an existence on land at once. For most species, however, the eggs must be laid close to water, into which the hatchlings make their way and feed upon insect and crustacean larvae and the like until they reach the stage of transformation into adults.

As mentioned in Chapter 1 (p. 15) the caecilians were for many years considered to constitute a single family, but later study and the perception of systematic differences among them has led to a splitting of the group into five families, with two of these further separated into subfamilies.

The caecilians are distinguished from the other amphibians by the presence of a peculiar sense organ, the tentacle, the absence of legs, and the presence of a series of rings or folds in the skin that give them a worm-like appearance; other features vary greatly with species. Thus the eye is present in many species, but in others is either absent or covered over by skin and other soft tissues, or even by bone. The teeth lie in three or four semicircular or U-shaped rows, two on each side of the upper jaw and one or two on the lower jaw. A tail is usually present but in some species is very short or obscure. Scales are commonly present, but are inconspicuous; these are located in the body folds beneath the skin, and can be discovered only by dissection or the use of X-rays.

Caecilians live mainly in wet, marshy areas near bodies of water or streams, often burrowing in the muddy banks; the adults feed on worms of various sorts and the larvae of insects and crustaceans. Those of the family Typhlonectidae are aquatic, living in ponds and streams. A drawing of the species *Dermophis mexicanus* is shown in Fig. 18-1, and displays the typical body form.

Available for study were the following five species, representing three out of the seven basic groups (families and subfamilies) now recognized: 3 specimens of *Ichthyophis glutinosus* and 1 specimen of *Ichthyophis orthoplicatus* of the family Ichthyophiidae; 2 specimens of *Geotrypetes seraphini* and 3 specimens of *Dermophis m. mexicanus* of the family Caeciliidae; and 3 specimens of *Typhlonectes anguillaformis* of the family Typhlonectidae. All were tested for hearing capability in terms of the electrical potentials of the ear and then prepared by serial sectioning for a microscopic study of the ear structures.

GENERAL STRUCTURE: *TYPHLONECTES*

The auditory structures of the caecilians will be examined first in the species *Typhlonectes anguillaformis* as a representative of this group, much as has already been done for anurans and urodeles. Because particularly good series of sections were obtained for this species, it is chosen for this special treatment, even though being of aquatic habit it is not typical of the group in its mode of life. The ear, however, does not display any aberrant features, but appears to be well advanced in its development. As before, the ear structures will be followed in a specimen sectioned transversely, from anterior to posterior, with drawings that show particular stages along this course.

The first figure of this series, Fig. 18-2, which will serve for general orientation, passes through the head at a level just anterior to the otic capsule and shows the cranial cavity with the posterior portion of the brain, the mouth cavity below with the quadrate bones on either side, and the hyoid cartilages that serve to support the tongue and its muscles.

Fig. 18-1. A drawing of *Dermophis mexicanus.*

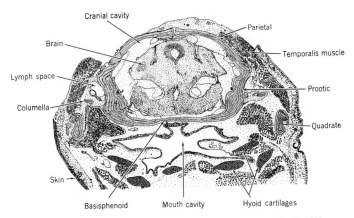

Fig. 18-2. A section through the head of *Typhlonectes anguillaformis*. Scale 10X.

As in most specimens used in these experiments, the skin was removed from the dorsal surface of the head along with portions of the temporalis muscle to permit the drilling of a small hole in the otic capsule over the anterior semicircular canal for the insertion of a recording electrode.

The anterior portions of the otic labyrinth first appear in Fig. 18-3, enclosed in the greatly thickened wall of the prootic. These are the utricle above and the saccule below, separated from one another here only by loose limbic tissue.

Two foramina between prootic and basisphenoid bones appear at this level: a lateral one filled with ligamentary tissue into which the inner end of the columella is protruding and a medial one that transmits fibers of the anterior branch of the eighth nerve that enter the auditory endorgans farther posteriorly.

In Fig. 18-4 the maculae of both utricle and saccule come into view. The utricular organ consists of a plate of epithelial cells resting on a moderately thick web of limbic tissue and supporting a layer of hair cells, which here appears as a row of about 20 cells, over whose free ends lies a mass of tectorial material in which are embedded an irregular jumble of elongated crystals, the statoliths.

The cavity below contains the saccule, along with a perilymph space that takes up more than half the room. The saccule lies on the lower medial side of this cavity, and its macula like the other consists of an array of hair cells sustained on an epithelial plate with a loading mass above. This mass differs somewhat from the utricular one, consisting of a large globular body whose lower portion is in contact with the hair-cell layer and whose main portion is a reticulum of delicate tissue in which are embedded scattered groups of crystals.

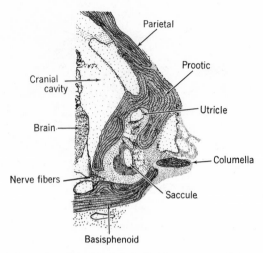

Fig. 18-3. The anterior region of the labyrinth in *Typhlonectes anguillaformis*, showing utricle and saccule. Scale 20X.

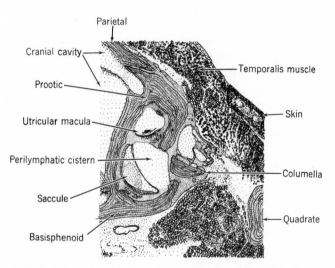

Fig. 18-4. The labyrinth of *Typhlonectes anguillaformis* showing utricular and saccular maculae. Scale 20X.

The columella appears much broadened at this level and presents a well-developed footplate in the oval window, held there by dense ligamentary tissue. Its shaft portion is short and blunt and is well buried beneath the skin and muscle layers.

In Fig. 18-5 the anterior semicircular canal comes into view, and at its anteromedial edge is the bed of epithelial cells from which the crista organ extends as seen a little farther posteriorly. The maculae of both utricle and saccule continue at their middle levels of development.

In Fig. 18-6 the anterior crista is presented longitudinally, bridging across the cavity from medial to lateral sides. This cavity is still incompletely separated from the utricular region. This latter region shows the beginning of a division into two parts, the utricle proper on the medial side and its recess that contains the endorgan disposed laterally. Only a thin floor separates the utricular cavity from the space below, which contains the saccule along with the large perilymphatic cistern.

In Fig. 18-7 the ampullated part of the anterior canal continues, with the crista still showing on its lateral wall, and the utricular recess gives way to the lateral ampulla with its crista. The utricle is now a separate cavity on the medial side. At this level also appears the endolymphatic duct, which comes off from the dorsal end of the saccule a little anterior to this point, then passes in a rather wayward course alongside the otocranial septum, and finally at its upper end bends sharply and enters a foramen leading into the brain cavity. The columellar footplate is much reduced at this level.

In the next figure, 18-8, the anterior and lateral ampullae have ended, and the section cuts across the two semicircular canals. A foramen has opened between utricle and saccule and is partially obstructed by a small mass of tectorial tissue as here indicated; the function of this obstructing material will be considered presently. Also a thickened part of the utricular floor is seen, which will be the site of a sensory structure encountered farther posteriorly.

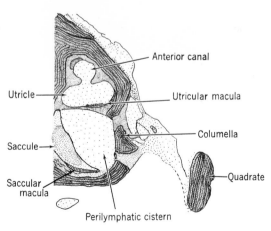

Fig. 18-5. The appearance of the anterior semicircular canal in *Typhlonectes anguillaformis* and the macular organs farther posteriorly. Scale 20X.

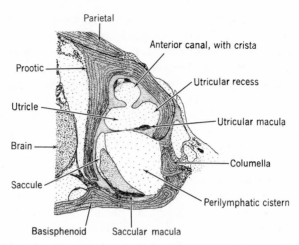

Fig. 18-6. The crista of the anterior semicircular canal in *Typhlonectes anguillaformis*. Scale 20X.

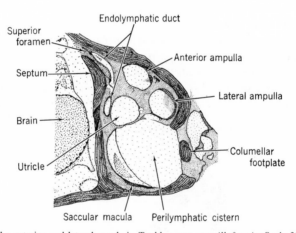

Fig. 18-7. The anterior and lateral canals in *Typhlonectes anguillaformis*. Scale 20X.

The columella has come to an end, and the oval window contains only the dense ligamentory tissue that holds this ossicle in place.

Figure 18-9 presents a number of important features. The utriculosaccular foramen has ended, and on the floor of the utricle a sense organ is present that on careful consideration appears to be an auditory receptor; it will hereafter be referred to as the utricular papilla. The tectorial mass represented in the preceding figure continues also and occupies the lower corner of the utri-

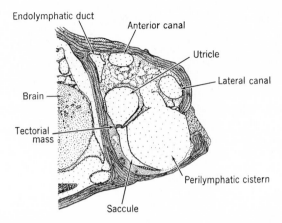

Fig. 18-8. The anterior and lateral semicircular canals in *Typhlonectes anguillaformis*, and a passage between utricle and saccule. Scale 20X.

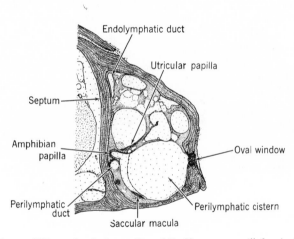

Fig. 18-9. The amphibian and utricular papillae of *Typhlonectes anguillaformis*. Scale 20X.

cular cavity. Below, at the upper end of the saccule, is the beginning of another sense organ, the amphibian papilla; the opening seen here is its recess. The oval window is further reduced, and in the section following this one it disappears altogether. The perilymphatic duct is seen just below the recess for the amphibian papilla; it is the initial part of the path of outflow from this region as will presently be indicated.

In Fig. 18-10 the two papillae are further represented, with an enlarged drawing of these organs in Fig. 18-11. The utricular papilla appears as a broad plate of hair cells surmounted by a tectorial body of some complexity.

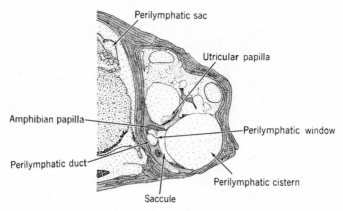

Fig. 18-10. A further view of the two papillae in *Typhlonectes anguillaformis* at a slightly more posterior position. Scale 20X.

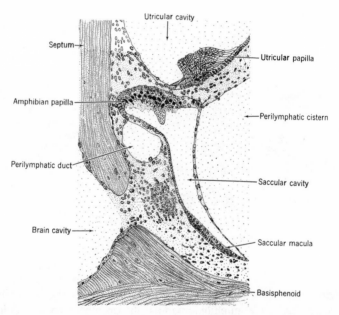

Fig. 18-11. A more detailed view of the two papillae in *Typhlonectes anguillaformis*. Scale 100X.

The amphibian papilla is a broad cup-like organ with a tectorial body suspended from epithelial cells on the limbic roof, and its cavity connects through a thin membranous window with the perilymphatic duct below.

Figure 18-12 shows the utricular papilla as somewhat broadened, with its tectorial body in increased prominence. The amphibian papilla continues also, and the perilymphatic duct below now opens through a broad foramen into the brain cavity. Loose limbic tissue is seen here as partially obstructing the passage from duct to brain cavity, but a little farther posteriorly this passage is completely clear. Thus there is a continuous channel from the perilymphatic cistern alongside the amphibian papilla to the cranial cavity except for the presence of two thin membranes: the lateral wall of the saccule, which is formed by a single layer of squamous cells, and the perilymphatic window, which likewise appears as a single layer of greatly flattened cells. At this level the anterior semicircular canal has made its connection with the superior sinus.

A little farther posteriorly, as Fig. 18-13 shows, the utricular papilla presents its region of maximum expansion. The amphibian papilla is a little smaller than before, but it still communicates through its window with the perilymphatic duct, now widely expanded into the brain cavity.

At this level the lagena is seen; it is only vestigial in this species. As shown it is a ventromedial diverticulum of the middle wall of the saccule; it extends as a little pocket that runs posteriorly and ventrally for a short distance (about 100 μ) and then ends; it is devoid of sensory cells and has no nerve supply.

The last figure in this series is 18-14, showing the extreme posterior end of the labyrinth. Here the posterior ampulla appears with its crista, contained in the terminal portion of the exoccipital bone. The jugular foramen between exoccipital and basisphenoid transmits the glossopharyngeal and vagal groups of nerves.

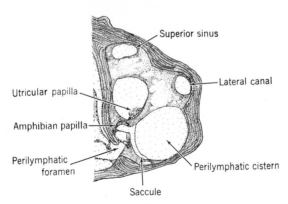

Fig. 18-12. The perilymphatic foramen in *Typhlonectes anguillaformis*. Scale 20X.

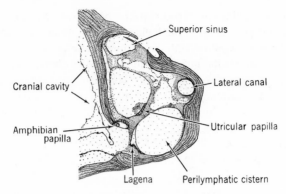

Fig. 18-13. The lagena of *Typhlonectes anguillaformis*. Scale 20X.

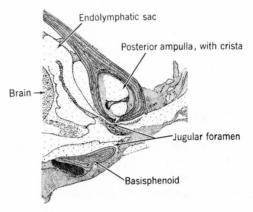

Fig. 18-14. The posterior end of the labyrinth in *Typhlonectes anguillaformis*. Scale 20X.

DETAILED ANATOMY OF THE CAECILIAN EAR

THE SENSORY PAPILLAE

Seven sense organs have been encountered in this labyrinth: the anterior, lateral, and posterior cristae, the utricular and saccular maculae, and the utricular and amphibian papillae. The first five of these organs require no special notice as they follow the form generally seen in vertebrates and are closely similar to the ones already described in anurans and urodeles. These organs serve as equilibrial receptors, operating as usual in the sensing and regulation of head position and motion. The two papillae, however, are deserving of further consideration since they appear to be specific organs for the reception of sounds and exhibit some special characteristics.

The presence of one of these auditory receptors, the amphibian papilla, has long been known. It is closely adjacent to the saccule and appears to be a derivative of that organ, as evidently is the case also in the other two orders of Amphibia. Further, this organ has certain structural resemblances to the ones in the other amphibians that make it seem entirely appropriate to consider it homologous.

The other papilla, however, appears to be a unique structure; it is different in place of origin in the labyrinth and in structural design from the second papilla in anurans and urodeles and thus seems to deserve a distinctive designation; it will be referred to from its location in the labyrinth as the utricular papilla. This organ is not be confused with the utricular macula, which has a different location, arising in the utricular recess rather than in the utricle proper, and presenting a wholly different structure.

In earlier treatments of this group of amphibians this second auditory receptor in the caecilians had been missed (Wever, 1975, 1978), yet its identification is actually a rediscovery rather than an original one. The presence of two papillae in the caecilian labyrinth was reported by the Sarasins in 1887, and to them belongs the priority. It happened, however, that following the Sarasins' report the identity of their second sense organ was questioned by Retzius, who asserted (quite improperly as we now know) that no other amphibian is equipped with more than one auditory receptor and that the caecilians could hardly be distinctive in this respect. Retzius had failed to find these organs in two specimens that he examined and indeed had found no sense organ at all and thus concluded that the ear in this amphibian was undeveloped. The influence of this widely recognized authority in the field of acoustic anatomy was such that the clear evidence presented by the Sarasins was ignored or lost sight of, and it has been generally accepted ever since that the caecilians possess a single auditory receptor. The reason for this historical event is now clear: the specimens examined by Retzius were in so poor a state of preservation as to be worthless for his purpose.

That the two sense organs observed by the Sarasins and represented here are indeed auditory in function is evidenced by their particular structure and their location in vibratory fluid pathways, as will now be brought out. Detailed consideration will first be given to the amphibian papilla.

The Amphibian Papilla.   Arising as a medial diverticulum of the upper end of the saccule, as Fig. 18-10 has shown, is a cavity that in transverse section as in this figure appears cup-shaped, but is oval in lateral view and about three times longer than it is wide. This papilla is formed of epithelial cells that support a layer of hair cells whose ciliated ends extend out into the open end of the structure; covering the tips of these cells is a tectorial body. This structure, seen in further detail in Fig. 18-11, extends downward as a sort of flap that nearly closes the passage from saccule to papillar cavity and covers the entrance to the thin membranous window leading into the perilymphatic duct below. This duct opens into the brain cavity a little farther posteriorly (as shown in Fig. 18-12) and is the initial part of the relief route for the fluid vibrations. Fluid movements beginning at the columellar footplate and passing first across the perilymphatic cistern and then from the saccule to the perilymphatic duct must first wash over the dependent edges of the tectorial flap and thus will transmit a portion of their energy to the hair cells above.

This pathway is represented by the arrows in Fig. 18-15; the identity of the parts will be clear from preceding figures, especially 18-9 and 18-13. Shown is the course of vibratory fluid flow that extends first across the perilymphatic cistern to enter the saccule through its thin lateral wall and then across the upper end of the saccule and through a thin perilymphatic window into the perilymphatic duct, which leads into the brain cavity. From the brain cavity the vibrations have two possible outward paths: they can cross over to the opposite side of the head and traverse the same portions of the otic structures in reverse order, finally reaching the other columella, or they can pass along the brain cavity a short distance to the contrary face of the columella on the same side. The ipsilateral escape path is difficult to follow and requires a careful reconstruction by the use of both transverse and frontal sections. Its course was most thoroughly followed in the species *Geotrypetes seraphini* and will be treated more fully in the description of that form. The other species were examined in sufficient detail to determine that this path is functionally similar in all. The effectiveness of this path, as well as the contralateral one, will be considered below in the general discussion of the sound stimulation processes.

It will now be apparent that in structure and location the amphibian pa-

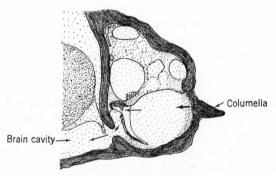

Fig. 18-15. The sound path through the amphibian papilla of *Typhlonectes anguillaformis*. Scale 20X.

pilla in caecilians corresponds closely to the one known by this name in anurans and urodeles, so that it seems proper to use a common designation.

The Utricular Papilla.   Close by, but located in the utricular cavity rather
than in the saccule, is the second auditory organ, already seen in Figs. 18-9
to 18-11 and now shown in further detail in Fig. 18-16 as a bed of hair cells
surmounted by a complex tectorial structure that presents a dense, finely re-
ticulated central core from which radiates an extensive array of canals spreading
out over the hair cells. This structure is best seen in a view at right angles
to this one, as in the sagittal section of Fig. 18-17. This figure shows the
relations to neighboring structures and indicates the presence of a sensing
membrane that extends over the utriculosaccular foramen. Figure 18-18 gives
a more detailed picture of the papilla from this aspect.

As will be noted, the body of the papilla lies on the outer floor of the
utricle, with its medial side adjacent to the utriculosaccular foramen and a
membranous process, indicated as the sensor, extending over the foramen
and attached to its medial wall. This membrane thus forms a barrier across
the path from utricle to saccule and should readily respond to motions of
fluid particles in this path. There is also a thin reticular mass in this area that
probably enhances the sensitivity. The action, consisting of a direct response
of the tectorial body to the vibratory stream, reinforced by movements transmitted by the sensing membrane to at least a portion of the hair cells, should
provide a highly effective means of sound detection.

A schematic drawing to show the vibratory path through the utricular papilla is presented in Fig. 18-19. This schema is based upon several of the
labyrinthine sections already presented, especially 18-6, 18-8, and 18-13. The
route through this utricular papilla can best be comprehended by examining
these figures in the order 18-6, 18-13, 18-8, and then 18-13 again. The route
as shown in 18-19 begins with the vibratory displacements of the columellar
footplate, passes through the perilymphatic cistern and its dorsal extension
into a number of irregular perilymph spaces that run alongside the lateral

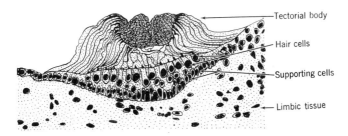

Fig. 18-16. A detailed view of the utricular papilla in *Typhlonectes anguillaformis*. Scale 20X.

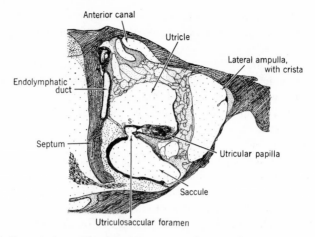

Fig. 18-17. The utricular papilla of *Typhlonectes anguillaformis* shown with its sensing membrane (*s*). Scale 25X.

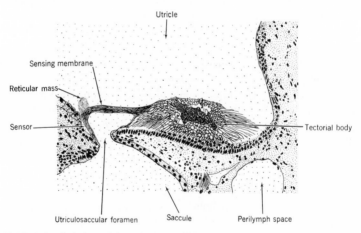

Fig. 18-18. A detailed view of the utricular papilla in *Typhlonectes anguillaformis*. Scale 100X.

wall of the utricle, and then passes through this wall in regions where it becomes particularly thin. After the vibratory motion reaches the utricular cavity its further course is through the utriculosaccular foramen into the upper end of the saccule and then out through the perilymphatic duct to the brain cavity. The latter part of the course, from the saccule on, is thus the same as the outlet for the amphibian papilla.

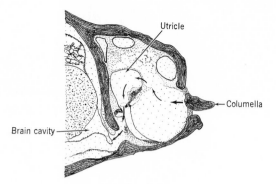

Fig. 18-19. The sound path through the utricular papilla of *Typhlonectes anguillaformis.*

Usually when an animal possesses two sense organs serving the same modality it is expected that the actions of the two will be in some way complementary, perhaps with one of them being the more delicate, responding to the faintest stimuli, and the other being more rugged, coming into action when the intensity is great and the first organ has reached its limit of response. In the present instance we would expect the amphibian papilla to be the more sensitive because of its direct location in the sound path from the columella and because the path through the utricle appears somewhat less favorable, not so much because it is more circuitous but because of the presence of numerous strands of tissue along the path that must reduce the transmission through friction. Yet the presence of the sensor mechanism in this papilla may compensate for any frictional deficiencies, so that it is possible that this papilla performs on something like the same level of effectiveness as the other. At any rate, this caecilian appears to be well provided with receptor organs for sound reception.

SENSITIVITY

The sensitivity of this ear was determined in the usual way, in terms of the electrical potentials in response to sounds. These determinations used airborne sounds, presented through a tube sealed over the ear region at the side of the head so as to enclose the area containing the columella just beneath the skin surface. Tones were presented over a range from 50 to 4000 Hz at the level required to produce a potential of 0.1 μv as recorded from an electrode in the form of a fine silver wire, insulated except at the tip, inserted through a hole the location of which varied in the two specimens illustrated: for the uppermost curve of Fig. 18-20 the hole was drilled through the wall of the posterior semicircular canal and for the other curve it was in the superior sinus. This second location gave the better results both in terms of

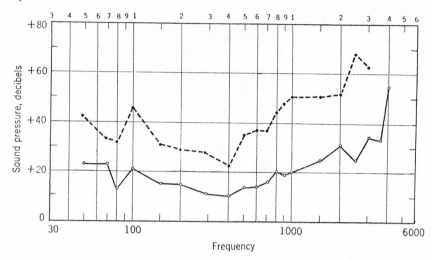

Fig. 18-20. Aerial sensitivity for two specimens of *Typhlonectes anguillaformis*. Shown is the sound pressure, in db relative to 1 dyne per sq cm, required for an inner ear potential of 0.1 μv.

regularity of response over the frequency range and in the degree of sensitivity indicated. For both curves the point of greatest sensitivity was 400 Hz, and for the more sensitive of these two ears the response is fairly good for about two octaves in the low-frequency range and rather gradually becomes poorer for both lower and higher tones, until in the high-frequency range the sensitivity fails rapidly.

19. THE CAECILIAN EAR

CONTINUED

THE CAECILIIDAE

Dermophis mexicanus. The second caecilian to be examined is *Dermophis mexicanus*, a relatively large, heavy-bodied species occurring in southern Mexico and neighboring regions. Figure 18-1 above shows the body form, and Fig. 19-1 gives a view of the head after the removal of superficial tissues to expose the bone structures in the posterior region. The columella presents a broad footplate in the oval window and gives off an anterior process that makes a ligamentary attachment to the quadrate.

In general form and arrangement the labyrinthine structures are much the same as described for *Typhlonectes*. Two auditory organs are present—the amphibian and utricular papillae—as in that species. As represented in Fig. 19-2, a transverse section through the right side of the head, these two papillae are closely adjacent, separated at this level by a septum of limbic tissue that extends horizontally across the otic cavity and divides it into dorsal and ventral halves, with the utricle above and the perilymphatic cistern and saccule below. Opening into the upper end of the saccule and extending deep into the thick layer of limbic tissue along the lower portion of the otocranial

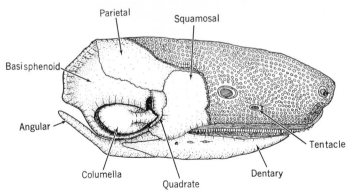

Fig. 19-1. The head of *Dermophis mexicanus*, dissected to expose the skull surface in posterior and ventral regions. Scale 6X.

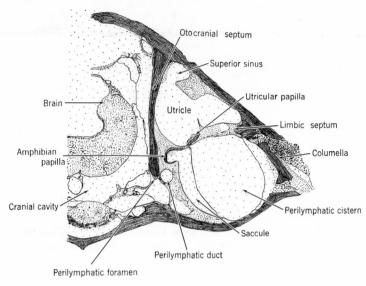

Fig. 19-2. A transverse section through the ear region of *Dermophis mexicanus* on the right side. Scale 20X.

septum is the amphibian papilla. In the floor of the utricle, and separated from the perilymphatic cistern by a thin portion of the limbic septum, is the utricular papilla. An enlarged view of this region is presented in Fig. 19-3.

The Amphibian Papilla. — Further details of the amphibian papilla are presented in Fig. 19-4, where about 25 rows of hair cells are seen, sustained by processes extending from the many supporting cells around the periphery. On close examination of selected specimens, as in Fig. 19-5, the ciliary tufts of the hair cells are seen to have an orderly arrangement, with the longer cilia in each tuft oriented toward the middle of the semicircular row. In one specimen 368 hair cells were counted in the organ on one side, and 384 in the one on the other side.

As usual in such structures, the supporting cells send up thin columns that expand to embrace the outer ends of the hair cells, leaving the ciliary tuft to protrude as through a collar. These tufts then extend into little pockets or canals in the mass of tectorial material at the center of the organ, or some-times—and especially for those cells at the outer fringe of the array—the ciliary tufts are more remotely connected to the central mass by long, thin filaments.

The Sound Circuit for the Amphibian Papilla. — As may be observed in Fig. 19-6, sound vibrations transmitted by the columella reach the amphibian papilla directly on passing across the perilymphatic cistern and then through

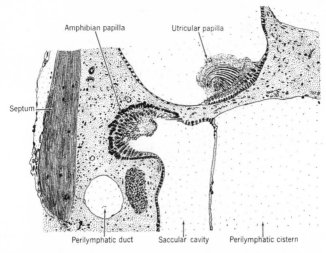

Fig. 19-3. An enlarged view of the two auditory papillae in *Dermophis mexicanus*. Scale 90X.

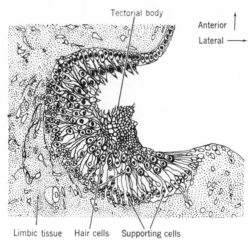

Fig. 19-4. The amphibian papilla in *Dermophis mexicanus* in a frontal section. Scale 200X.

the lateral wall of the saccule; they then continue to the interior of the papillar cavity where the motions are transmitted through the tectorial body to
the hair cells. This oscillatory movement of fluid and tissue in response to
the imposed sound pressures is made possible, as always, by the presence
of a further path to the outside of the capsule that serves as a relief route. In
this labyrinth two avenues of relief appear possible, and it is likely that both

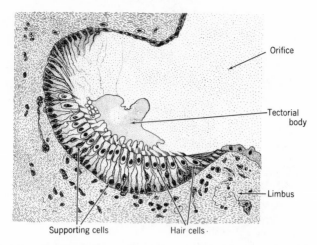

Fig. 19-5. A frontal section in *Dermophis mexicanus* showing the orientation of the ciliary tufts of the hair cells. Scale 200X.

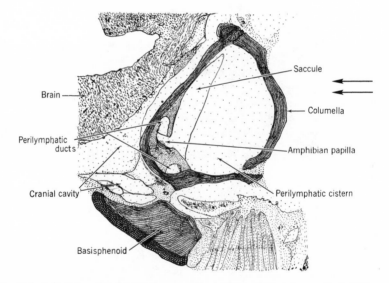

Fig. 19-6. A frontal section through the ear region in *Dermophis mexicanus*. Scale 20X.

are utilized. One of these paths is through the dorsal branch of the perilymphatic duct (shown as the upper one in Fig. 19-7) that leads far posteriorly into the cranial cavity, then passes across this cavity to the opposite side and back through a pathway corresponding to this one but in reverse so as finally to reach the contralateral columella and the exterior.

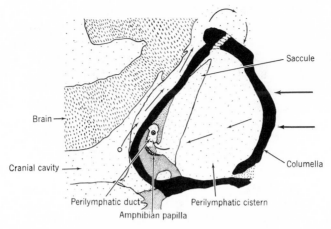

Fig. 19-7. A frontal section in a specimen of *Dermophis mexicanus* showing the path of access of acoustic vibrations to the amphibian papilla. Also indicated is one of the relief paths. Scale 20X.

For this stimulating system to be effective it is necessary that the contra-lateral columella either be shielded from the actuating sound waves, so that the pressures exerted upon it are less than for the other incident path, or that these waves are in a different phase relation. This same problem is encountered in the salamander ear, in which it has been shown experimentally that the relief pathway for sounds impressed upon one ear passes through the ear on the opposite side. With such a structural arrangement the ideal situation for hearing is that in which the sound waves reaching the two ears are in opposite phase; but this condition can be expected to obtain but rarely: it occurs only for particular sound frequencies and certain favorable orientations of the head.

A second relief pathway for the amphibian papilla of *Dermophis* is along the brain cavity in a circuitous course to the outside surface of the columella on the same side, as indicated in Fig. 19-7. As this figure shows, the sound waves incident on the columella pass across the amphibian papilla at the posterior end of this cavity and finally go through the perilymphatic window into the perilymphatic duct. The symbol (cross in circle) in this duct represents a retreating arrow in a passageway that runs into the brain cavity to a region indicated by the plain circle. The sound thereafter follows a dorsolateral path between brain and otic capsule that leads to the outer face of the columellar footplate, thus completing the circuit.

This is the type of relief that occurs commonly in many reptiles such as snakes, turtles, and a few lizards in which the round window is lacking. This reentrant circuit is probably the more useful of the two relief paths in *Dermophis*.

The Utricular Papilla. — In *Dermophis*, as in *Typhlonectes*, a utricular papilla is present, and here as in that species it is almost certain that this organ is an auditory receptor, though the path of access of sounds is less obvious and the performance is probably less efficient. This second papilla in *Dermophis* has already been seen in Figs. 19-2 and 19-3 in its relation to other structures, and its location is further indicated in the sagittal section of Fig. 19-8. It lies on the anteroventral floor of a recess that may be seen (at a more lateral level) to come off from the posterior ampulla. More medially, as Fig. 19-9 shows, its cavity is confluent with that of the utricle, so that it can be regarded as an extension of that division of the labyrinth. This sensory structure consists of a plate of hair cells and supporting cells borne on a limbic web that expands to a pillar on the floor of the otic capsule and is also anchored to the dorsal and posterior walls of this capsule.

Details of structure of the utricular papilla are shown in Fig. 19-10, which represents a frontal section, and in Fig. 19-11 from a transverse one. This papilla contains as many as 20 rows of hair cells in the transverse direction, and the total number of cells as determined in one specimen was 186, which is nearly half the number in the amphibian papilla of the same ear.

This array of hair cells is surmounted by a tectorial body of some complexity, consisting of a canaliculate structure whose lower openings lie over the hair cells, apparently with every hair cell in the main array having an opening into which its ciliary tuft is inserted. These canals run upward, taking a curving course, but maintaining their essentially parallel relations, though often with turns and twists that are difficult to follow. For the most part the outer ends of the canals seem simply to end at the surface, though occasionally threads of tissue may be seen that run for short distances over the elements. This binding probably serves to strengthen the structure sufficiently to enable it to resist fraying from mechanical forces.

The Sound Circuit for the Utricular Papilla. — The vibratory pathway for the utricular papilla is represented in Fig. 19-12. Fluid motions arising from the inner face of the columellar footplate pass through the perilymphatic cistern into the papillar cavity and sweep over the tectorial body of the utricular papilla in their path outward through the superior branch of the perilymphatic duct. This duct then leads into the brain cavity as already seen in Fig. 19-7, and the circuit is complete. This is the same outward path as for the amphibian papilla, having two terminal routes, one through a channel alongside the otic capsule on the same side, and the other by way of the brain cavity to a symmetrical outlet on the opposite side.

Further consideration will be given to the utricular papilla in its relations to the utriculosaccular foramen. No special sensing membrane is found in this region comparable to that in *Typhlonectes*, but there are scattered strands of tectorial material in this vicinity as shown in Fig. 19-9. These strands retain at least a tenuous connection with the main body of the organ and probably convey any vibratory fluid motions through and around the utricu-

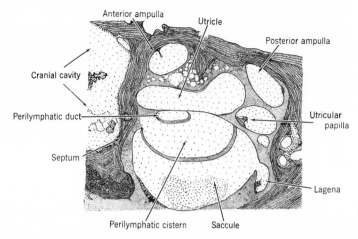

Fig. 19-8. The ear region of *Dermophis mexicanus* in a sagittal section. Scale 20X.

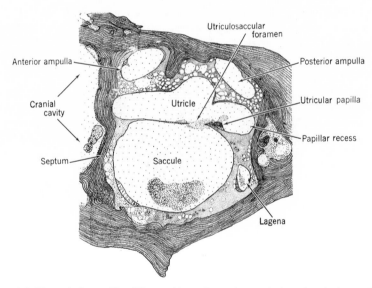

Fig. 19-9. The utricular papilla of *Dermophis mexicanus* in a sagittal section, farther medially. Scale 20X.

losaccular foramen to the papillar hair cells. There is also the possibility that eddies are produced in this region that spread widely enough to involve this second papilla, much as was inferred from an examination of the more elaborate structure in *Typhlonectes*.

A lagena is present in *Dermophis* as a medial and posterior expansion of

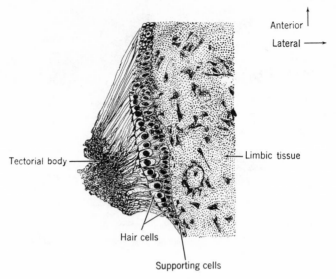

Fig. 19-10. The utricular papilla of *Dermophis mexicanus* in a frontal section. Scale 200X.

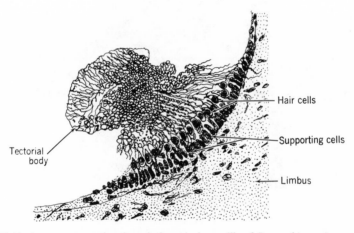

Fig. 19-11. A transverse section through the utricular papilla of *Dermophis mexicanus*. Scale 200X.

the wall of the saccule, represented in Figs. 19-8, 19-9, 19-13, and 19-14. The enlarged view in Fig. 19-14 shows on the floor of the lagenar cavity a moderate number of hair cells and above them a mass of tectorial tissue that includes a considerable number of statolithic crystals. This lagenar organ is relatively small, but evidently is a functional receptor in this species.

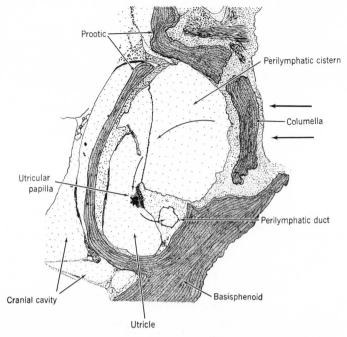

Fig. 19-12. A frontal section showing the ear region in *Dermophis mexicanus*, with an indication of the sound pathway through the utricular papilla. Scale 30X.

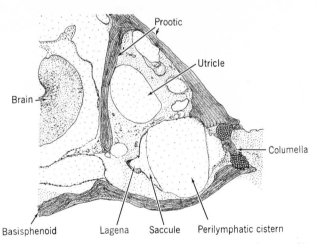

Fig. 19-13. A transverse section through the right side of the head of *Dermophis mexicanus*, showing the lagena medial to the saccule. Scale 20X.

Sensitivity. — The sensitivity of this ear was measured in terms of inner ear potentials in the usual way, with the sounds applied through a tube at the side of the head, and gave results as shown for one specimen in Fig. 19-15. The response is fairly uniform over the low-frequency range up to 1500 Hz, varying between 0 and +10 db, and thereafter falls off rapidly as the

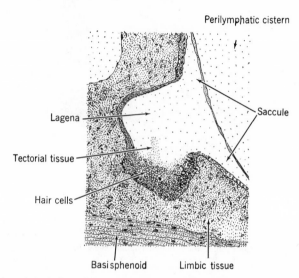

Fig. 19-14. An enlarged view of the lagena shown in the preceding figure. Scale 80X.

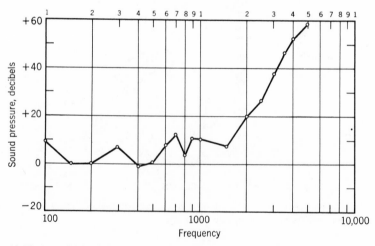

Fig. 19-15. A sensitivity function for aerial sounds in a specimen of *Dermophis mexicanus*.

frequency is raised, reaching a level of +60 db at 5000 Hz. The papillae in *Dermophis* clearly are low-frequency receptors.

Geotrypetes seraphini. Available for study were two specimens of a second species belonging to the family Caeciliidae, *Geotrypetes seraphini*, a small, slender form from southwest Africa. One specimen was 123 mm long with a body diameter of 4.0 mm, and the other was 147 mm long with a diameter of 4.2 mm. These were evidently subadults of this species. Both were first tested for inner ear potential sensitivity and then prepared for sectioning in the frontal plane.

The Amphibian Papilla. — A frontal section through the head for one of the specimens of *Geotrypetes seraphini* is shown in part *a* of Fig. 19-16. The columellar footplate is seen lying close beneath the skin and a thin muscle layer at the side of the head, with its widely flaring footplate filling the lateral opening of the otic capsule. Deep in this capsule, on its posteromedial wall, is the amphibian papilla, located in a cup-like recess facing anterolaterally. It consists of an assembly of hair cells with their ciliary tufts extending outward into a dense network of tectorial tissue. This papillar structure is shown somewhat enlarged in part *b* of this figure, and in further detail in Fig. 19-17. In one specimen of *Geotrypetes* the number of hair cells in the amphibian papilla was found to be 219 for one ear and 241 for the other.

In part *a* of Fig. 19-16 the heavy arrows on the right represent sound waves

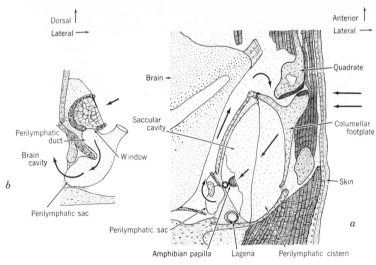

Fig. 19-16. The ear region of *Geotrypetes seraphini*. At *a*, the paths of vibratory fluid flow from columellar footplate to amphibian papilla; *b*, the return path from papillar cavity to the brain cavity. Scale for part *a* 37.5X; for part *b* 125X.

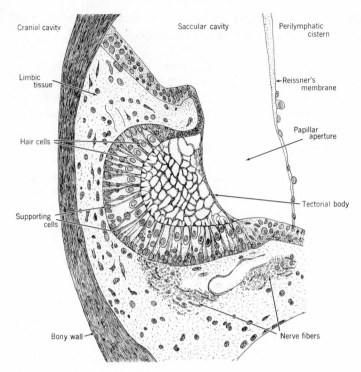

Fig. 19-17. The amphibian papilla of *Geotrypetes seraphini*. Scale 300X.

whose pressures act through the superficial tissue layer over the columellar footplate and are then transmitted across the perilymphatic cistern through the saccular cavity to the amphibian papilla as indicated. Here the pressure waves enter the lateral opening of the papilla and act on the tectorial tissues, and through these on the hair cells.

The course of this action is followed further in part *b* of this figure, which is oriented at right angles to the main part to show the outlet for the pressure pulses. On the ventral side of the papilla is a thin membranous window that opens into a perilymphatic sac that in turn leads through a thin membrane into the brain cavity. A passage alongside the brain (as part *a* shows) then leads anterolaterally to the outside surface of the columellar footplate.

Thus as this footplate is pushed inward by a positive pressure of the sound wave it displaces a quantity of fluid first in the perilymphatic cistern, then in the saccule and papillar cavity. By a deflection of the perilymphatic window this displacement is transferred to the perilymphatic duct and sac and then through the cranial fluid to the front face of the columellar footplate, completing the circuit.

By means of this continuous fluid pathway the sound pulsations pass through the papilla, vibrating its tectorial body and through this body stimulating the hair cells whose cilia are enmeshed in the tectorial strands.

The Utricular Papilla. — A second auditory papilla is located on the posterolateral wall of the utricle as shown in Fig. 19-18, at a level much farther dorsally than shown in the two preceding figures. Sound vibrations transmitted to the columella pass through the perilymphatic cistern to the saccule and through this cavity to the utricle, as this figure together with Fig. 19-19 will show.

As seen in Fig. 19-18, the utricular papilla rests on a limbic pillar between perilymphatic cistern and utricular cavity, and this pillar contains a perilymphatic duct with a medial wall that is particularly thin. Pressure waves passing from the utricle into this perilymphatic duct sweep over the utricular papilla, transmitting their movements to the tectorial body that extends out into the passage, thus stimulating these papillar hair cells.

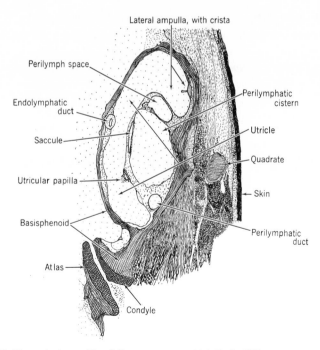

Fig. 19-18. The utricular papilla of *Geotrypetes seraphini*. Scale 40X.

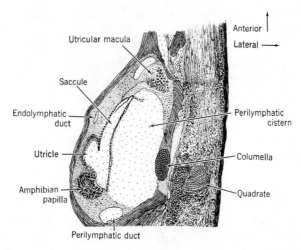

Fig. 19-19. The ear region of *Geotrypetes seraphini* in a frontal section. Scale 40X.

THE ICHTHYOPHIIDAE

The family Ichthyophiidae is of relatively wide distribution, with representatives in both hemispheres: in South and Central America, in Asia and the Philippines, and in the Indo-Australian archipelago. Taylor recognized four genera, of which *Ichthyophis* has by far the greatest number of species. Nussbaum, however, has split off the South American species, placing them in a separate family.

The two species of Ichthyophiidae studied were *Ichthyophis glutinosus* (3 specimens) and *Ichthyophis orthoplicatus* (1 specimen), all collected in Sri Lanka (Ceylon). These are relatively large forms, with body lengths up to 40 cm in *I. glutinosus* and up to 22.5 cm in *I. orthoplicatus*; both species, like all members of this family, possess well-defined tails.

Ichthyophis glutinosus. For general orientation a skull of *Ichthyophis glutinosus* is shown in Fig. 19-20 in a lateral view, drawn from a prepared specimen, with the skin retained over the frontal and nasal surfaces, but removed posteriorly to show the relations between columella and quadrate; these elements are closely adjacent and united by ligamentary fibers. The next figure (Fig. 19-21), taken from one presented by the Sarasins, gives a dorsal view of the skull structure. The labyrinths are contained in the prootic bone at the posterior end of the skull and exhibit a degree of surface molding. This skull is compact, with its elements closely joined, forming a firm structure well suited to its use as a burrowing instrument.

The form of the labyrinth is shown in two drawings taken from the Sara-

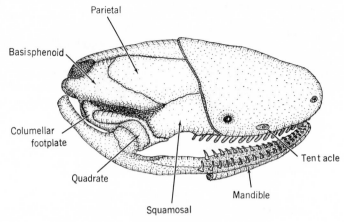

Fig. 19-20. The head of *Ichthyophis glutinosus*. Scale 5X.

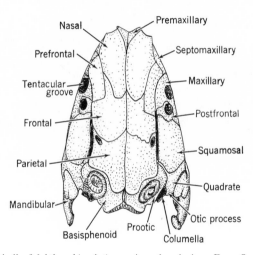

Fig. 19-21. The skull of *Ichthyophis glutinosus* in a dorsal view. From Sarasin and Sarasin, 1890.

sins' report, one (Fig. 19-22) a lateral view of the right side, and the other (Fig. 19-23) a medial view in which the various innervation areas for the saccule are particularly extensive; four divisions of these saccular areas are indicated. Figure 19-24 presents a frontal section through the right ear region showing the columella in the oval window and both amphibian and utricular papillae lying medial to the large perilymphatic cistern into which the fluid pulsations caused by vibratory movements of the columella are ra-

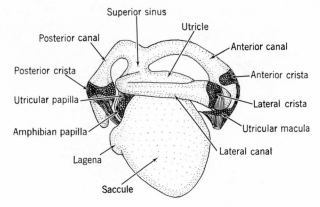

Fig. 19-22. The labyrinth of *Ichthyophis glutinosus*, in a lateral view of the right side. Scale 17X.

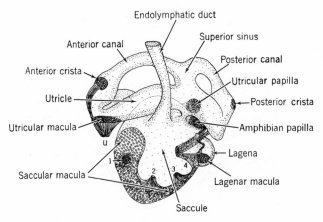

Fig. 19-23. The labyrinth of *Ichthyophis glutinosus* in a medial view. Scale 17X.

diated. The amphibian papilla is reached directly by these pulsations, with an outlet path through the perilymphatic duct, whereas the course to the utricular papilla is more circuitous (as described earlier in the treatment of *Typhlonectes*) and is somewhat hindered by the presence along its path of a large amount of tissue material. On the other hand, perhaps compensating in some degree for this frictional restraint, is the presence in the utricular papilla of numerous fine filaments or fringes that extend out into the fluid surrounding this receptor organ and perhaps aid in transmitting the fluid motions to the hair cells. The next figure (Fig. 19-25) represents a transverse

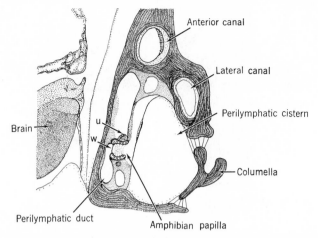

Fig. 19-24. The ear region of *Ichthyophis glutinosus* in a frontal section: *u*, the utricular papilla; *w*, the perilymphatic window. Scale 25X.

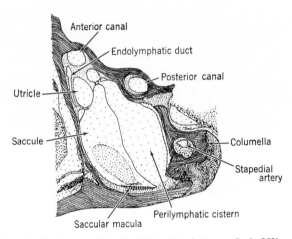

Fig. 19-25. A section farther posteriorly in *Ichthyophis glutinosus*. Scale 25X.

section farther posteriorly where the columella is passed through near its midportion and the stapedial artery is seen.

A section still farther posteriorly is represented in Fig. 19-26, where the amphibian papilla is adjacent to a window leading into the perilymphatic duct, and the utricular papilla lies dorsally to it, separated by limbic tissue. In the next section (Fig. 19-27), taken from this same specimen at a level still farther posteriorly, the utricular papilla has ended and the perilymphatic duct

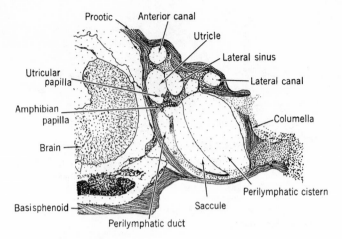

Fig. 19-26. The utricular papilla in *Ichthyophis glutinosus*. Scale 25X.

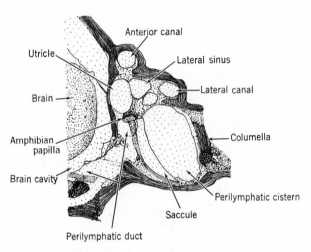

Fig. 19-27. A section in *Ichthyophis glutinosus* farther posteriorly showing the connection of the perilymphatic duct to the brain cavity. Scale 25X.

that serves as the outlet for this pathway opens freely into the brain cavity.

The detailed structure of the utricular papilla is shown in Fig. 19-28 in a frontal section where the fringed structure of the tectorial body is clearly evident.

A sensitivity curve for this species is presented in Fig. 19-29, where it is seen that the low-frequency tones are greatly favored, with the maximum sensitivity at 200 Hz, and there is a progressive decline for higher tones that

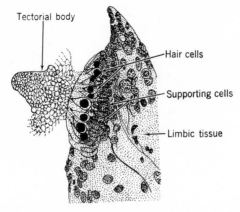

Fig. 19-28. Detail of the utricular papilla in *Ichthyophis glutinosus*. Scale 250X.

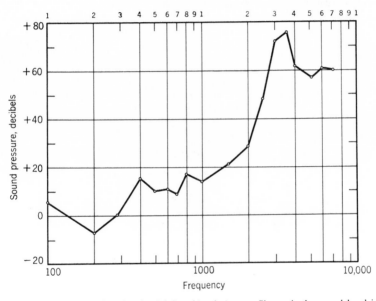

Fig. 19-29. A sensitivity function for *Ichthyophis glutinosus*. Shown is the sound level in db relative to 1 dyne per sq cm required for an inner ear potential of 0.1 μv. From Wever and Gans, 1976.

becomes especially rapid above 2000 Hz. Figure 19-30 shows sensitivity curves for this specimen obtained at two different body temperatures, at 26° and 31°. An increase in sensitivity is shown for the higher temperature except for the uppermost tones.

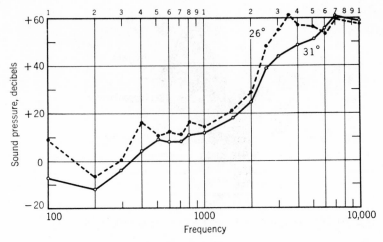

Fig. 19-30. Sensitivity curves for two additional specimens of *Ichthyophis glutinosus*, expressed as in the preceding figure. From Wever and Gans, 1976.

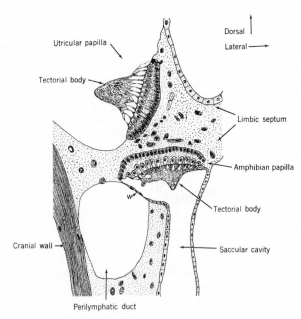

Fig. 19-31. Detailed representation of the utricular and amphibian papillae of *Ichthyophis orthoplicatus*. Scale 75X. From Wever and Gans, 1976.

Ichthyophis orthoplicatus. The auditory structures in this second species of *Ichthyophis* are closely similar to those just described for *I. glutinosus.* Figure 19-31 gives a detailed view of the papillar region in which both the utricular and amphibian organs are seen to be well developed. Also in this species the lagena is well formed; it is shown for this species in Fig. 19-32.

Auditory sensitivity is represented for this second species in Fig. 19-33,

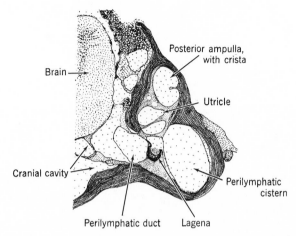

Fig. 19-32. A section through the right side of the head of *Ichthyophis orthoplicatus* showing the lagena. Scale 25X.

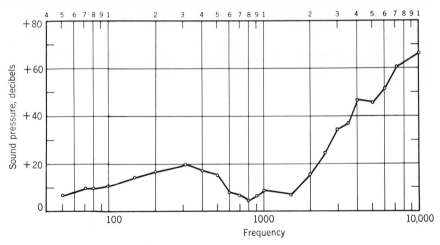

Fig. 19-33. A sensitivity function for a specimen of *Ichthyophis orthoplicatus.* From Wever and Gans, 1976.

which shows a region of fair response between 600 and 1500 Hz, with a rapid decline in the higher frequencies, and first a decline and then a degree of recovery at the lower end of the frequency scale. The sensitivity displayed for aerial sounds is only moderate, and it seems likely that these ears are best adapted to the reception of vibrations in the water or in semisolid substances such as mud or soft earth.

PART V. THE EVOLUTION
OF THE EAR

20. THE AMPHIBIAN EAR
IN EVOLUTION

The sense of hearing at the advanced stage in which it occurs in ourselves and the other higher vertebrates carries such a wealth of information about the outside world (and its loss occasions so severe a burden) that its appearance in the evolution of animal life seems almost an inevitable event. Even among the very lowly forms of life this sense appears early and plays a highly significant role in such fundamental matters as the finding of food, the choice of mates, the care of the young, and escape from dangers. An earlier treatment (Wever, 1978b) was concerned with this sense in the reptiles, where it takes a form closely similar to the one that we ourselves possess; the present concern is with the next lower group in the vertebrate series, the amphibians, in which this sense takes a very different and largely unfamiliar form and yet serves a closely similar set of functions. Again a consideration of the roles taken by this sense in the lives of these animals makes us aware of the powerful evolutionary forces that have been involved in bringing this capability of sound reception into being, and in extending its range and sensitivity.

Traditional accounts of the origin of the vertebrate ear present a series of progressive developments conforming in the main to the course of evolution as conceived for the vertebrates as a whole, beginning with the fishes and extending through the amphibians to the reptiles, and then proceeding along separate lines to the birds and mammals (see Fig. 20-1).

This traditional view will now be challenged: the consideration given in the foregoing chapters to both the structure of the amphibian ear and its manner of functioning in response to sounds casts doubt on this conception of a single, continuous development and points to the amphibian ear as set apart— as in many respects unique in derivation and with its own peculiar form and manner of operation. This view is in agreement with that of Lombard (1980) who has suggested that the ears of amphibians and amniotes developed along two separate lines out of the labyrinths of fishes.

To be sure, the separation between amphibian and reptilian lines of evolutionary development is not complete: these two types of ears have a number of features in common: the hair cells, supporting cells, and tectorial tissues are similar and readily recognized as linked in their past history. This

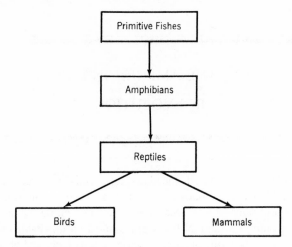

Fig. 20-1. The conventional chart of vertebrate evolution.

linkage, however, most likely is indirect; the similarities only reflect a common derivation of these structural elements from the labyrinthine organs of fishes and do not indicate a continuous line of development along these auditory organs themselves. The labyrinthine structures arose in the primitive fishes as essential organs of equilibrium to enable these animals to maintain bodily orientation and to execute orderly movements in an aquatic medium where often other sensory modes such as touch and vision were unavailable. Apart from their early derivation by the use of tissue elements from the equilibrium system, the amphibian and reptilian types of ears appear to have remained completely separate; their lines of development are distinctive both in structure and function.

The evolution of the amphibian ear can be conceived of in at least three distinct ways, the conventional way represented in the scheme of Fig. 20-1 and two other ways that may be followed in Fig. 20-2. (The usually accepted course of development may also be traced in Fig. 20-2 by following the dashed line marked "1," which makes explicit the derivation of the modern amphibians from the ancient ones.)

A second conception is represented in Fig. 20-2 by following the broken line marked "2," in which it is supposed that the primitive fishes gave rise to a group of early amphibians distinct from the ones that produced the reptiles, birds, and mammals, and that this other group of ancient amphibians gave rise to the existing ones.

A third conception further separates the amphibian line from the other vertebrates as represented by the column marked "3"; here it is supposed that primitive fishes distinct from the ones indicated on the left gave rise to

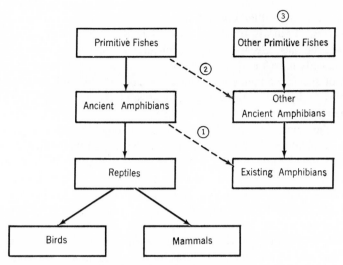

Fig. 20-2. Alternative conception of vertebrate development.

a group of ancient amphibians other than the ones indicated in the main line of evolution of higher vertebrates, and produced the existing amphibians.

All three of these conceptions are possible; the fossil evidence now available is not sufficient to permit a clear choice among them. Much more study is needed on fossils now at hand but still resting unprepared in museum cases, and further collections are needed of material in pertinent strata around the world that might yield valuable information on this problem.

A COMPARISON OF AMPHIBIAN AND REPTILIAN EARS

THE AMPHIBIAN TYPE OF EAR

In the ears of amphibians the sensory cells lie on an immobile base: on a wall of the papillar cavity or on a shelf of limbic tissue extending outward from this wall. These cells then send their ciliary tufts into the interior of the fluid cavity through which vibratory waves pass under the influence of alternating pressures caused by a sound. These ciliary tufts are embedded in a body of tectorial tissue so that the whole mass is washed over by the fluid currents resulting from sound, and thus the mass is set in motion. The deflections of the cilia relative to the bodies of these cells are stimulative, whereby the sound is perceived.

THE REPTILIAN TYPE OF EAR

In the reptilian form of ear, on the contrary, the hair cells are borne on the surface of a special membrane—the basilar membrane—that is a flexible barrier stretched across the fluid conduit along which the sound waves are transmitted; this membrane is forced to move with the vibratory currents and the bodies of the hair cells follow the membrane vibrations. But because the cilia of these cells have their tips attached to a stationary side wall, usually by means of special elements—a tectorial membrane or strands of tectorial tissue—these cilia are restrained, and a relative motion is set up between them and their cell bodies that is stimulative.

Thus the amphibian and reptilian modes of sound reception are similar in that in each instance a relative displacement between the bodies of the hair cells and their ciliary tufts is produced that sets these cells in action, yet physically these two processes are almost the reverse of one another. More significantly, there is a difference in the effectiveness of these two modes of stimulation: hair cells borne on the surface of a membrane lying athwart the path of vibratory fluid flow are forced to move relative to their embedded cilia, whereas a fluid wave that merely washes over a group of hair cells along one wall of a conduit will transmit only a fraction of its vibratory energy to the ciliary tips of the cells along its path.

Thus from the standpoint of sensitivity alone the reptilian design for an ear presents an outstanding advantage. Yet there are other positive attributes also. A membrane in the path of sound vibrations can be differentiated in its physical characters such as length, mass, and tension so as to be selective in its response in relation to stimulus frequency. Thus this membrane can operate segmentally, with some portions responding to certain periodicities more readily than others, thereby achieving at least the preliminary stage of a frequency analysis of a complex sound. The membrane thus will respond differently over its extent, at greater amplitudes in some regions than in others, further facilitating the analysis of a sound into its components. Such a tuning that extends over a range of frequencies and also involves differential amplitude of action provides a substantial basis not only for improved sensitivity but also for frequency and amplitude analysis. These valuable characteristics are possessed by the reptilian ear that we are so fortunate to have inherited and render this form of auditory receptor vastly superior to the amphibian type.

SOME ELABORATIONS OF AMPHIBIAN EAR STRUCTURES

In view of the inherent deficiencies just indicated in the basic character of the amphibian type of ear it is not surprising that over the long history of this group of animals there have been a number of additions and modifications of structure that have served to improve this ear's performance in the reception and discrimination of sounds. Indeed, such improvements are the rule, and only a few species of salamanders seem to be lacking in them.

Two forms of improvement are found: the supplementation of existing structures and the creation of special devices for enhancing the sensitivity of the receptive process.

THE ADDITION OF RECEPTOR ORGANS

The most general improvement of the amphibian sound receptor system is the addition of a second auditory papilla. The primary sound receptive organ almost certainly is the amphibian papilla, occurring in all species in the same general location alongside the otocranial septum. Because of this common location and a generally similar form this appears to be a homologous structure throughout the amphibian series, most likely derived from the equilibrial system as an outgrowth from the superior division of the saccule.

A second auditory papilla appears in each of the three amphibian orders, but these do not seem to be of common origin: the three additional organs found in anurans, urodeles, and caecilians differ somewhat in form and most definitely in their locations in the head. It is likely that in the anurans this second papilla is an outgrowth from the inferior saccule, in the urodeles is derived from the lagena, and in the caecilians is developed from the utricle.

In some of the urodeles this second papilla is absent: it has not been found in *Necturus*, in the Sirenidae, in *Notophthalmus* among the newts, or in any of the Plethodontidae so far examined; but it is present in all the other amphibian species studied.

Traditionally the second auditory receptor of anurans has been referred to as the "basilar papilla"; it seems best to retain this ancient name even though it is an inappropriate one; it was originally applied in the erroneous belief that this organ is associated with a basilar membrane corresponding to the one that had long been known in the higher animals. The term is now so thoroughly entrenched in the amphibian literature that it would be confusing to attempt to replace it.

For the second papillar organ in salamanders the term "lagenar papilla" is appropriate, and for the one in the caecilians the term "utricular papilla" will be applied; both indicate the general locations and apparent derivations of these two structures.

In the present study the utricular papilla was examined most thoroughly in *Typhlonectes* where it takes an elaborate form, with numerous canals and filamentary processess as shown in Figs. 18-10 to 18-13. Further study involving a variety of caecilian species is needed to determine the range of appearance of this structure.

Sensing Membranes. — A membrane suspended across the path of vibratory fluid flow through the ear region with connections to a group of hair cells can greatly augment the ear's sensitivity to sounds. This detection method is utilized most effectively in one region of the amphibian papilla of many anurans (in all but the more primitive species) in which a web of tectorial material is stretched about halfway across the lumen of an acoustic passage-way so as partly to occlude it, and thus the vane is made to vibrate in synchronism with the transmitted wave. Along one wall of the passage where the web is attached are rows of hair cells that are stimulated by the membrane's movements. As has been noted, the sensing membrane principle is also utilized in the basilar papilla of anurans. A sensing membrane is clearly shown in Figs. 18-17 and 18-18 for *Typhlonectes anguillaformis*.

Sensing Masses. — Many species show also a mass of tenuous tectorial material that further enhances the transmission of fluid vibrations to the papillar hair cells. Such masses are prominent in the species *Desmognathus quadramaculatus, Pseudotriton ruber, Plethodon cinereus,* and *Typhlonectes anguillaformis;* examples are given in Figs. 18-8, 18-9, 18-10, 18-17, and 18-18.

A FINAL COMMENT

This consideration of the forms of the auditory receptors of existing amphibians gives insight into the extensive developments that must have come about in this group of vertebrates over their long course of differentiation and testifies to the great evolutionary pressures represented in the value to living creatures of effective means for the reception and utilization of auditory information. The presence of an amphibian papilla of much the same type in all three amphibian orders points to a common ancestor at some early date.

Then the appearance of three forms of supplementary organs arising in these different orders emphasizes the evolutionary pressures still further and at the same time shows that these orders diverged early and have pursued independent lines of development for a very long time.

Especially striking is the level of the ear's performance among the anurans in comparison with the others, the salamanders and apodans. Through the development of the anuran middle ear the scope of operation of the auditory sense is significantly extended and enriched, with substantial aid both to the survival of individuals and to the assisting of mating activities and thus the propagation of the species.

By the production of aerial sounds by males the female frog is informed about the presence, the degree of receptivity, and perhaps the state of vigor of males in the vicinity, and no doubt she makes use of this information in her choice of mates, for it seems likely that sexual qualities are reflected in the mating signals.

In this respect the anurans seem definitely superior to the other two amphibian groups; the frogs make extensive use of vocal signals and have a large repertoire of vocalizations. These features have not been covered as extensively as is desired in observations thus far and need to be investigated further in field studies.

Some fundamental problems of the early origin and evolution of the vertebrate ear remain unsolved. It seems certain from the evidence now at hand that the vertebrate sense of hearing developed out of the organs of equilibrium and posture that emerged among the fishes and that this development occurred separately along two lines, one leading to the ears of the amphibians as here treated in some detail and another of a different character that appeared in the higher animals, the reptiles, birds, and mammals. Whether these two lines of evolution can be further identified and characterized is still to be discovered. The fossil record presents faint hope of specific clues to such developments, and chief attention must be turned to the animals now living. The existing fishes present an enormous range of diversification, with many directions of specialization, and it remains to be determined whether among them the early separation into reptilian and amphibian types of ear can be worked out.

GLOSSARY

Amphicoelous. A type of vertebra in which the anterior and posterior surfaces of the centrum are concave.

Anthracosaurs. A genus of labyrinthodonts from the coal measures of England, considered to be early ancestors of the amphibians.

Aponeurosis. A flat sheet of connective tissue forming the termination of a muscle.

Arcifera. Frogs in which the epicoracoids overlap on the two sides and are fused along their anterior edges.

Ascending process (of the columella). A cartilaginous process running from the outer end of the columella anteromedially to the parotic process.

Auditory sensitivity. For aerial sounds, the sensitivity is herein expressed as the sound pressure, in decibels relative to a standard (zero level) of 1 dyne per sq cm, required to produce an inner ear response with a root-mean-square amplitude of 0.1 microvolts. For vibrations, the sensitivity is expressed as the amplitude, in decibels relative to a standard of 1 millimicron (mμ) required to produce an inner ear potential with a root-mean-square amplitude of 0.1 μv.

Bromeliad. One of a group of tropical plants that retain water in their calyxes and serve as breeding sites for some frogs.

Clavicle. The anterior of two bones of the shoulder girdle connecting the scapula to the sternum. See Fig. G-1.

Coracoid. The posterior of two bones of the shoulder girdle connecting the scapula to the sternum. See Fig. G-1.

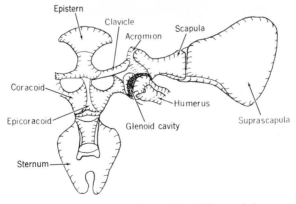

Fig. G-1. The shoulder girdle of *Rana catesbeiana*. About 0.3 natural size.

Cornu. A horn or projection.

Cricoid cartilage. A cartilage of the larynx that articulates with the thyroid and connects also with the arytenoid cartilages.

Crossopterygians. Fishes belonging to the Teleostomi considered to be ancestors of the terrestrial vertebrates.

Depressor mandibulae muscle. A muscle serving to open the mouth.

Enchondral. Related to or derived from cartilage.

Enchondrosis. An outgrowth of cartilage.

Epicoracoid cartilage. An element of the sternal end of the coracoid in the shoulder girdle of some vertebrates. See Fig. G-1.

Epicoracoid horns. Extensions of the epicoracoid cartilages posteriorly beyond their point of fusion, where they run in lateral grooves on either side of the sternum.

Epicoraco-sternal complex. The posterior portion of the shoulder girdle; see Fig. G-1.

Ethmoid. A bone forming part of the walls of the nasal cavity.

Firmisternia. A division of the amphibian suborder Linguata in which (in adults) the epicoracoids on the two sides meet and fuse in the median ventral line.

Holochordal centra. A form of vertebrae in anurans in which the original notochord is replaced by bone, forming a solid centrum.

Holochordate. Having a notochord that extends the full length of the body.

Intercalary phalanx. An extra finger or toe joint.

Labyrinthodonts. A primitive group of amphibians, or their ancestral stock, present in the late Paleozoic and early Mesozoic periods.

Lepospondyls. Primitive amphibians with spool-shaped ventral segments of the vertebral column, regarded as the earliest land vertebrates. They are considered as having arisen from the labyrinthodonts and to have produced the aistopods, nectrideans, and microsaurians as well as the modern urodeles and apodans.

Microsaurs. A group of extinct stegocephalians containing salamander-like forms: the lepospondyls.

Nectrideans. Two types of early amphibians, one of eel-like form, the other with horned skulls.

Neoteny. The persistence of larval or juvenile characters in the adult.

Opisthocoelous. Designating animals with vertebral segments whose anterior end is flat or convex and whose posterior end is concave.

Pectoral girdle. The bony or cartilaginous arch supporting the forelimbs of a vertebrate. Primitively it is a single cartilage on each side. In higher forms it is ossified and divided into the scapula above, the precoracoid and coracoid below, and complicated by the addition of membrane bones.

Pedicel. A short stalk or stem. In the vertebrae it forms a basal part on each side, connecting the laminae with the centrum.

Pedicellate teeth. Teeth consisting of two portions, a pedicle and a crown, connected by a flexible joint. See p. 36 and Fig. 3-9.

Precoracoid. An anterior ventral bone of the pectoral girdle of the higher vertebrates.

Presacral vertebrae. The more anterior vertebral segments, including those of the cervical and trunk regions.

Procoelous. Having vertebrae with the anterior end of the centrum concave, and the posterior end usually convex.

Pseudobasal process. The outer end of the basitrabecular process to which is added the postpalatine commissure to form a part of the jaw suspension. This process replaces the commissura quadrato-cranialis present in early development and then later is destroyed. See Pusey, 1938.

Raniformia. One of three infraorders of frogs as suggested by Cope (1864) and modified by Boulenger (1882). It included only the Ranidae.

Rhachitomous. Applied to a primitive form of arch vertebrae in which the centrum is composed of parts that remain separate. This condition is seen in the oldest amphibia and in the main line of the labyrinthodonts.

Sternal apparatus. The sternum and parts immediately connected with it.

Sternum. The breast bone or cartilage. See Fig. G-1.

Temnospondyls. Primitive amphibians in which some of the vertebrae are rhachitomous. The intercentrum is ventral and the pleurocentra remain distinct, above and behind the intercentrum.

REFERENCES

Adrian, E. D., Craik, K.J.W., and Sturdy, R. S. 1938. The electrical response of the auditory mechanism in cold-blooded vertebrates. *Proc. Royal Soc. London*, Ser. B., *125*, 435–455.

Alexander, I. E., and Githler, F. J. 1951. Histological examination of cochlear structure following exposure to jet engine noise. *J. Comp. Physiol. Psychol.*, *44*, 513–524.

————. 1952. Chronic effects of jet engine noise on the structure and function of the cochlear apparatus. *J. Comp. Physiol. Psychol.*, *45*, 381–391.

Axelrod, F. S. (as reported by J. Y. Lettvin and H. R. Maturana). 1960. Hearing senses in the frog. Mass. Inst. Technol., Res. Lab. Electron., *Quart. Prog. Report*, *57*, 167–168.

Bajandurov, B. I., and Pegel, W. A. 1932–33. Der bedingte Reflex bei Fröschen. *Zeits. wiss. Biol.*, Part C, *18*, 284–297.

Baker, C. L. 1945. The natural history and morphology of Amphiumidae. *Tennessee Acad. Sci.*, *Journal*, *20*, 55–91.

Bast, T. H. 1928. The utriculo-endolymphatic valve. *Anat. Rec.*, *40*, 61–64.

————. 1934. Function of the utriculo-endolymphatic valve. *Arch. Otolaryngol.*, *19*, 537–550.

Birkmann, K. 1940. Morphologisch-anatomische Untersuchungen zur Entwicklung des häutigen Labyrinthes der Amphibien. *Zsch. Anat. Entwicklungsges.*, *110*, 443–488.

Bishop, S. C. 1943. *Handbook of salamanders*. Comstock Publ. Assoc., Ithaca and London, 555 pp.

Blair, W. F. 1955. Differentiation of mating call in spadefoots, genus *Scaphiopus*. *Texas J. Sci.*, *7*, 183–188.

————. 1956. Mating call and possible stage of speciation of the Great Basin spadefoot. *Texas J. Sci.*, *8*, 236–238.

————. 1958. Mating call in the speciation of anuran amphibians. *Amer. Nat.*, *92*, 27–51.

————. 1962. Non-morphological data in anuran classification. *Syst. Zool.*, *11*, 72–84.

————. 1963. Acoustic behavior of Amphibia. In R. G. Busnel, ed., *Acoustic Behavior of Animals*, pp. 694–708, Elsevier Publ. Co., New York, N.Y.

Blair, W. F., and Pettus, D. 1954. The mating call and its significance in the Colorado River toad (*Bufo alvarius* Girard). *Texas J. Sci.*, 72–77.

Blankenagel, F. 1931. Untersuchungen über die Grosshirnfunktionen von *Rana temporaria*. *Zool. Jahrb.*, *Abt. Allg. Zool. Physiol.*, *49*, 271–322.

Bogert, C. M. 1960. The influence of sound on the behavior of amphibians and reptiles. In W. E. Lanyon and W. N. Tavolga, eds., *Animal Sounds and Communication*, No. 7, pp. 137–320, Amer. Inst. Biol. Sci., Arlington, Va.

Bolt, J. R. 1969. Lissamphibian origins; possible protolissamphibian from the Lower Permian of Oklahoma. *Science, 166*, 888–891.

———. 1977. Dissorophoid relationships and ontogeny, and the origin of the Lissamphibia. *J. Paleontol., 51*, 235–249.

Buytendijk, F.J.J., and Eerelman, J. 1930. La réaction galvanique de la peau. *Arch. neérl. Physiol., 15*, 358–380.

Capranica, R. R. 1965. *The evoked vocal response of the bullfrog*. Research Monog. No. 33. MIT Press, Cambridge, Mass., 110 pp.

———. 1966. Vocal response of the bullfrog to natural and synthetic mating calls. *J. Acoust. Soc. Amer., 40*, 1131–1139.

———. 1976. Morphology and physiology of the auditory system. In R. L. Llinas and W. Precht, eds., *Handbook of Frog Neurobiology*, pp. 561–575. Springer Verlag, New York.

Capranica, R. R. , and Moffat, A.J.M. 1975. Selectivity of the peripheral auditory system of spadefoot toads (*Scaphiopus couchi*) for sounds of biological significance. *J. Comp. Physiol., 100*, 231–249.

Capranica, R. R., Sachs, M. B., and Murray, M. J. , 1963. Auditory discrimination in the bullfrog. Mass Inst. Technol., Res. Lab. Electron., *Quart. Progr. Report, 74*, 245–249.

Conant, R. 1958. *A field guide to reptiles and amphibians*. Houghton Mifflin Co., Boston, 366 pp.

Courtis, S. A. 1907. Response of toads to sound stimuli. *Amer. Naturalist, 41*, 677–682.

DeBurlet, H. M. 1929. Zur vergleichenden Anatomie und Physiologie des perilymphatischen Raumes. *Acta otolaryngol., 8*, 153–187.

DeVilliers, C. G. S. 1934. Studies of the cranial anatomy of *Ascaphus truei* Stejneger, the American leiopelmid. *Bull. Mus. Comp. Zool.* (Harvard), *77*, 1–38.

———. 1938. A comparison of some cranial features of the East African Gymnophiones *Boulengerula boulengeri* Tornier and *Scolecomorphus ulugurensis* Boulenger. *Anat. Anz., 86*, 1–26.

Diebschlag, E. 1935. Zur Kenntnis der Grosshirnfunktionen einiger Urodelen und Anuren. *Zeits. f. vergl. Physiol., 21*, 343–394.

Dowling, H. G. 1975. *1974 Yearbook of herpetology*. HISS Publ., Amer. Museum Nat. Hist. New York, N. Y.

Dowling, H. G., and Duellman, W. E. 1974–1978. *Systematic herpetology, a synopsis of families and higher categories*. HISS Publ., New York.

Druner, L. 1902. Studien zur Anatomie der Zungenbein-Kiemenbogen und

Kehlkopfmuskeln der Urodelen, I. *Zool. Jahrb. Anat. Abt.*, *15*, 435–622 (see pp. 521–522); II, 1904, *19*, 361–690.

Duellman, W. E. 1970. *The hylid frogs of middle America*, vols. 1,2. Monog. Museum Nat. Hist., Univ. of Kansas, Lawrence, Kans.

———. 1975. On the classification of frogs. *Occas. Papers Mus. Nat. Hist.* (Univ. of Kansas), No. 42, 1–14.

Dunn, E. R. 1941. The "opercularis" muscle of salamanders. *J. Morphol.*, *69*, 207–215.

Eaton, T. H. 1959. The ancestry of modern Amphibia: a review of the evidence. *Univ. Kansas Publ. Museum Nat. Hist.*, *12*, 155–180.

Eiselt, J. 1941. Der Musculus opercularis und die mittlere Ohrsphäre der anuren Amphibien. *Arch. f. Naturges.*, *10*, 179–230.

Estes, R. 1965. Fossil salamanders and salamander origins. *Amer. Zoologist*, *5*, 319–334.

Estes, R., and Wake, M. H. 1972. The first fossil record of caecilian amphibians. *Nature* (London), *239*, 228–231.

Fejérváry, G. J., de. 1923. Ascaphidae, a new family of the tailless Batrachians. *Annales Musei Nationales Hungarici*, *20*, 178–181.

Feng, A. S., Narins, P. M., and Capranica, R. R. 1975. Three populations of primary auditory fibers in the bullfrog (*Rana catesbeiana*.) *J. Comp. Physiol.*, *100*, 221–229.

Ferhat-Akat, S. 1938. Untersuchung über den Gehörsinn der Amphibien. *Zeits. f. vergl. Physiol.*, *26*, 253–281.

Frishkopf, L. S., and Geisler, C. D. 1966. Peripheral origin of auditory responses recorded from the eighth nerve of the bullfrog, *J. Acoust. Soc. Amer.*, *40*, 469–472.

———. and Goldstein, M. H. Jr. 1963. Responses to acoustic stimuli from single units in the eighth nerve of the bullfrog. *J. Acoust. Soc. Amer.*, *35*, 1219–1228.

Gadow, H. 1901. Amphibia and reptiles. In S. F. Harmer and A. E. Shipley, eds., *Cambridge Natural History*, MacMillan, London.

Gans, C. 1973. Sound production in the Salientia: mechanism and evolution of the emitter. *Amer. Zool.*, *13*, 1179–1194.

———. 1974. *Biomechanics, an approach to vertebrate biology*, pp. 192–245. J. B. Lippencott Co., Philadelphia.

Gans, C., and Wever, E. G. 1975. The amphisbaenian ear: *Blanus cinereus* and *Diplometopon zarudni*. *Proc. Nat. Acad. Sci.*, *72*, 1487–1490.

Gaupp, E. 1896. In A. Ecker and R. Wiedersheim, *Anatomie des Frosches*, pt. 1, 3rd ed., Vieweg und Sohn, Braunschweig.

Glekin, G. V., and Erdman, G. M. 1960. Discrimination of a useful signal by the auditory analyser; I. Potential is from the elements of the frog auditory nerve. *Biophysics*, *5*, 412–419.

Griffiths, I. 1963. The phylogeny of the Salientia. *Biol. Rev.*, *38*, 241–292.

Herre, W. 1935. Die Schwanzlurch der Mitteleocänen. *Zoologica*, *33*, 1–85.

Holmgren, N. 1949. Contributions to the question of the origin of tetrapods. *Acta Zoologica*, *30*, 459–484.

Hubl, L., and Schneider, H. 1979. Temperature and auditory thresholds: bioacoustic studies of the frogs *Rana r. ridibunda*, *Hyla a. arborea*, and *Hyla a. savignyi* (Anura, Amphibia). *J. Comp. Physiol.*, *130*, 17–27.

Jarvik, E. 1942. On the structure of the snout of crossopterygians and lower gnathostomes in general. *Zool. Bidr.*, *21*, 235.

———. 1955. The oldest tetrapods and their forerunners. *Sci. Monthly*, *80*, 141–154.

———. 1960. *Théories de l'evolution des vertebrés reconsidérées à la récentes déconvertes sur les vertébrés inférieurs.* Masson, Paris, 104 pp.

———. 1968. Aspects of vertebrate phylogeny. In T. Orvig, ed., *Current problems of lower vertebrate phylogeny*, pp. 497–527, Almquist and Wiksell, Stockholm.

Kingsbury, B. F., and Reed, H. D. 1908. The columella auris in amphibia. *Anat. Rec.*, *2*, 81–91.

———. 1909. The columella auris in amphibia; second contribution. *J. Morphol.*, *20*, 549–628.

Kleerekoper, H., and Sibabin, K. 1959. A study on hearing in frogs (*Rana pipiens* and *Rana clamitans*). *Zeits. f. vergl. Physiol.*, *41*, 490–499.

Kohlrausch, A., and Schilf, E. 1922. Der galvanische Hautreflex beim Frosch auf Sinnesreizung. *Pflüg. Arch. ges. Physiol.*, *194*, 326–329.

LeConte, J. E. 1825. Remarks on the American species of the genera *Hyla* and *Rana*. *Ann. Lyceum Nat. Hist.*, *N.Y.*, *1*, 278–282.

Liem, S. S. 1970. The morphology, systematics, and evolution of the Old World treefrogs (Rhacophoridae and Hyperoliidae). *Fieldiana, Zool.*, *57*, vii–145.

Littlejohn, M. J. 1961. Mating call discrimination by females of the spotted chorus frog (*Pseudacris clarki*). *Texas J. Sci.*, *13*, 49–50.

Littlejohn, M. J., Fouquette, M. J., and Johnson, C. 1960. Call discrimination by female frogs of the *Hyla versicolor* complex. *Copeia*, 47–49.

Littlejohn, M. J., and Michaud, T. C. 1959. Mating call discrimination by females of Strecker's chorus frog (*Pseudacris streckeri*). *Texas J. Sci.*, *11*, 86–92.

Litzelmann, E. 1923. Entwicklungsgeschichtliche und vergleichend-anatomische Untersuchungen über den Visceralapparat der Amphibien. *Zeits. f. Anat. Entwicklungsges.*, *67*, 457–493.

Lombard, R. E. 1971. A comparative morphological analysis of the salamander inner ear. Thesis, Univ. of Chicago.

————. 1980. The structure of the amphibian auditory periphery: a unique experiment in terrestrial hearing. In A. N. Popper and R. R. Fay, eds., *Comparative studies of hearing in vertebrates*, pp. 121–138, Springer Verlag, New York.

Lombard, R. E., and Bolt, J. R. 1979. Evolution of the tetrapod ear: an analysis and reinterpretation. *Biol. J. Linnean* Soc., *11*, 19–76.

Lynch, J. D. 1971. Evolutionary relationships, osteology, and zoogeography of leptodactyloid frogs. *Misc. Publ. Mus. Nat. Hist.* (Univ. of Kansas), *53*, 1–238.

McAlister, W. H. 1959. The vocal structures and method of call production in the genus *Scaphiopus* Holbrook. *Texas J. Sci.*, *11*, 60–77.

Maslin, T. P. 1950. The production of sound in caudate amphibia. *Univ. of Colo. Studies*, Biol. Sci. Series, *1*, 29–45.

Monath, T. 1965. The opercular apparatus of salamanders. *J. Morphol.*, *116*, 149–170.

Nicholls, G. C. 1916. The structure of the vertebral column in the anura phaneroglossa and its importance as a basis of classification. *Proc. Linn. Soc. London, Z.*, *128*, 80–92.

Noble, G. K. 1922. The phylogeny of the Salientia, I. *Bull. Amer. Mus. Nat. Hist.*, *46*, 1–87.

————. 1931. *The biology of the amphibia*. McGraw-Hill, New York, 577 pp. (Dover reprint, 1954).

Noble, G. K., and Putnam, P. G. 1931. Observations on the life history of *Ascaphus truei* Stejneger. *Copeia*, No. 3, 97–101.

Nussbaum, R. A. 1977. Rhinatrematidae: a new family of Caecilians (Amphibia: Gymnophiona). *Occas. Papers Mus. Zool.* (Univ. of Mich.), No. 682, 1–30.

————. 1979. The taxonomic status of the caecilian genus *Uraeotyphlus* Peters. *Occas. Papers Mus. Zool.* (Univ. of Mich.) No. 687, 1–20.

Olson, E. C. 1965. Evolution and relationships of the Amphibia (Symposium). *Amer. Zoologist*, *5*, 263–318.

Özeti, N., and Wake, D. B. 1969. Morphology and evolution of the tongue and associated structures in salamanders and newts (family Salmandridae). *Copeia*, 91–123.

Pace, A. E. 1972. Systematic and biological studies of the leopard frogs (*Rana pipiens* complex) of the United States. PhD. dissertation, Univ. of Michigan.

Parker, H. W. 1934. *A monograph of the frogs of the family Microhylidae*. British Museum, (Nat. Hist.) London, 208 pp.

Parsons, T. A., and Williams, E. E. 1963. The relationships of the modern Amphibia: a re-examination. *Quart. Rev. Biol.*, *38*, 26–53.

Paterson, N. F. 1960. The inner ear of some members of the Pipidae (Amphibia). *Proc. Zool. Soc. London*, *134*, 509–546.

Patterson, W. C. 1966. Hearing in the turtle. *J. Auditory Res.*, *6*, 453–464.

Peter, K. 1898. Die Entwicklung und funktionelle Gestaltung des Schädels von *Ichthyophis glutinosus*. *Morphol. Jahrb.*, *25*, 555–627.

Pusey, H. K. 1938. Structural changes in the anuran mandibular arch during metamorphosis, with reference to *Rana temporaria*. *Quart. J. Micros. Sci.*, *80*, 479–552.

Reed, H. D. 1909. Systematic relations of the Urodela as interpreted by a study of the sound-transmitting organs. *Science*, *29*, 715.

———. 1914. Further observations on the sound-transmitting apparatus in urodeles. *Anat. Rec.*, *8*, 112.

———. 1915. The sound transmitting mechanism in Necturus. *Anat. Rec.*, *9*, 581.

———. 1920. The morphology of the sound-transmitting apparatus in caudate amphibia and its phylogenetic significance. *J. Morphol.*, *33*, 325–387.

Reig, O. A. 1958. Proposiciones para una neuve macrosistematica de los anures. *Physis*, *21*, 109–118.

Romer, A. S. 1962. *The vertebrate body*. W. B. Saunders Co., Philadelphia, 635 pp.

———. 1966. *Vertebrate paleontology*, 3rd ed. (earlier editions 1933, 1945). Univ. of Chicago Press, Chicago.

Sarasin, P., and Sarasin, F. 1890. *Ergebnisse naturwissenschaftlicher Forschungen auf Ceylon*, vol. II, pt. 4: *Zur Entwicklungsgeschichte und Anatomie der ceylonischen Blindwühle, Ichthyophis glutinosus*, pp. 153–263. Wiesbaden.

———. 1892. Über das Gehörorgan der Caeciliiden. *Anat. Anzeiger*, *7*, 812–815.

Savage, J. M. 1973. In J. L. Vial, ed., *Evolutionary biology of the anurans*, pp. 351–445, Univ. of Missouri Press, Columbia, 1973.

Schmalhausen, I. I. 1959. The origin of the amphibia. *Proc. 15th Internat. Congress of Zoology*, *15*, 455–458.

———. 1968. *The origin of terrestrial vertebrates*. Translated by L. Kelso. Academic Press, New York, 314 pp.

Stebbins, R. C. 1966. *A field guide to western reptiles and amphibians*. Houghton Mifflin Co., Boston, 295 pp.

Stejneger, L. 1899. Description of a new genus and species of Discoglossoid toad from North America. *Proc. U. S. Nat. Museum*, *21*, 899–901.

Stephenson, E. M. 1960. The skeletal characters of *Leiopelma hamiltoni* (McCulloch) with particular reference to the effects of heterochrony on the genus. *Trans. Roy. Soc. New Zealand*, *88*, 473–488.

Strother, W. F. 1959. The electrical response of the auditory mechanism in the bullfrog (*Rana catesbeiana*). *J. comp. Physiol. Psychol.*, *52*, 157–162.

———. 1962. Hearing in frogs. *J. Auditory Res.*, *2*, 279–286.

Taylor, Edward H., 1968. *The caecilians of the world.* Univ. of Kansas Press, Lawrence, 848 pp.

———. 1969. A new family of African Gymnophiona. *Univ. Kansas Sci. Bull.*, *48*, 297–305.

———. 1969. Skulls of Gymnophiona and their significance in the taxonomy of the group. *Univ. Kansas Sci. Bull.*, *48*, 585–687.

Theron, J. G. 1952. On the cranial morphology of *Ambystoma maculatum* (Shaw). *So. Afr. J. Sci.*, *48*, 343–365.

Thorn, R. 1968. *Les salamandres.* Paris, 376 pp.

Vial, J. L., ed. 1973. *Evolutionary biology of the anurans.* Univ. of Missouri Press, Columbus, 477 pp.

Wagner, D. S. 1934. On the cranial characters of *Liopelma hochstetteri. Anat. Anz.*, *79*, 65–77.

———. 1935. The structure of the inner ear in relation to the reduction of the middle ear in the Liopelmidae (Noble). *Anat. Anz.*, *79*, 20–36.

Wake, D. B. 1964. Comparative osteology and evolution of the lungless salamanders, family Plethodontidae. Ph.D. dissertation, Univ. of Southern Calif.

Walker, C. F. 1938. The structure and systematic relationships of the genus Rhinophrynus. *Occas. Papers Mus. Zool.* (Univ. of Mich.), No. 372, 1–11.

Weiss, B. A., and Strother, W. F. 1965. Hearing in the green treefrog (*Hyla cinerea cinerea*). *J. Auditory Res.*, *5*, 297–305.

Wever, E. G. 1965. The degenerative processes in the ear of the Shaker mouse. *Ann. Otol. Rhinol. Laryngol.*, *74*, 5–21.

———. 1973a. Tectorial reticulum of the labyrinthine endings of vertebrates. *Ann. Otol. Rhinol. Laryngol.*, *82*, 277–280.

———. 1937b. The labyrinthine sense organs of the frog. *Proc. Natl. Acad. Sci.*, *70*, 498–502.

———. 1973c. The ear and hearing in the frog. *J. Morphol.*, *141*, 461–477.

———. 1975. The caecilian ear. *J. Exper. Zool.*, *191*, 63–72.

———. 1978a. Sound transmission in the salamander ear. *Proc. Nat. Acad. Sci. USA*, *75*, 529–530.

———. 1978b. *The reptile ear.* Princeton Univ. Press, Princeton, N.J., 1036 pp.

———. 1979. Middle ear muscles of the frog. *Proc. Nat. Acad. Sci.*, *USA*, *76*, 3031–3033.

———. 1981. The role of the amphibians in the evolution of the vertebrate ear. *Amer. J. Otolaryngol.*, *2*, 145–152.

Wever, E. G., and Gans, C. 1976. The caecilian ear: further observations. *Proc. Nat. Acad. Sci. USA*, *73*, 3744–3746.

Wiedersheim, R. 1877. Das Kopfskelet der Urodelen. *Morphol. Jahrb.*, *3*, part I, 352–448; part II, 459–548.

――――. 1890. Beiträge zur Entwicklungsgeschichte von Proteus anguineus. *Arch. f. mikr. Anat.*, *35*, 121–140.

Wright, A. H., and Wright, A. A. 1949. *Handbook of frogs and toads of the United States and Canada*, 3rd ed. Comstock Publ. Assoc., Ithaca, 640 pp.

Yerkes, R. M. 1904. Inhibition and reinforcement of reaction in the frog *Rana clamitans. J. Comp. Neurol. Psychol.*, *14*, 124–137.

――――. 1905. The sense of hearing in frogs. *J. Comp. Neurol. Psychol.*, *15*, 279–304.

INDEX

LIBRARY OF CONGRESS CATALOGING IN PUBLICATION DATA

Wever, Ernest Glen, 1902-
The amphibian ear.

Bibliography; p.   Includes index.

1. Amphibians—Physiology. 2. Amphibians—Evolution.
3. Ear. 4. Ear—Evolution. I. Title.
QL669.2.W47   1985   597.6′041825   84-17767
ISBN 0-691-08365-7

ERNEST GLEN WEVER is Professor Emeritus of Psychology and Director of the Auditory Research Laboratories at Princeton University. He is the author of *The Reptile Ear* (Princeton, 1978) and has written more than two hundred articles on the ear and hearing in professional journals.